Lego® Mindstorms™ Mechatronics

Don Wilcher

McGraw-Hill
New York Chicago San Francisco Lisbon
London Madrid Mexico City Milan New Delhi
San Juan Seoul Singapore Sydney Toronto

The McGraw-Hill Companies

Cataloging-in-Publication Data is on file with the Library of Congress.

1 2 3 4 5 6 7 8 9 0 DOC/DOC 0 9 8 7 6 5 4 3

P/N 141744-3
Part of
ISBN 0-07-141745-1

The sponsoring editor for this book was Judy Bass and the production supervisor was Pamela Pelton. It was set in Century Schoolbook by MacAllister Publishing Services, LLC.

Printed and bound by RR Donnelley.

This book is printed on recycled, acid-free paper containing a minimum of 50 percent recycled de-inked fiber.

This book is dedicated to my supportive family who allowed me to work on this project without any distractions.

CONTENTS

PREFACE

In my first book, *LEGO Mindstorms Interfacing*, I used lab projects and experiments to explain how basic electronic circuits are used to control tabletop robots. Software controls were introduced as well, to illustrate the importance of a multidisciplinary technology intertwined to create intelligent machines.

This concept brings us to the subject of mechatronics. Mechatronics is a multidisciplinary technical field encompassing electronics, mechanics, and software technologies for the creation of smart devices and intelligent machines. The LEGO Mindstorms *Robotic Invention System* (RIS) is an inexpensive development toolbox that enables the hobbyist and experimentalist to explore the world of mechatronics on the desktop. With gears, beams, motors, sensors, and software, the LEGO RIS is a powerful mechatronics development system. It is well equipped to enable the use of electronics and software to improve a mechanical system's operation.

This book will investigate, by means of lab projects and experiments, how control systems are designed and built. Programming languages like Java, *Not Quite C* (NQC), *Interactive C* (IC), Python, *Robot Command Explorer* (RCX code), and *Visual BASIC for Applications* (VBA) assist in the monitoring and management of the electronics *input/output* (I/O) circuits used in creating mechatronics-based LEGO robots and machines. In addition, Robolab, a graphical programming language, will be investigated in terms of creating a *Hardware in the Loop* (HIL) tool for developing robots and intelligent machines. The format of the book follows the first book, but with more tutorials to explain complex controls concepts and theories.

Software Tutorials

Mechanical *Compuer Aided Design* (CAD) tutorials will be part of the mechatronics lab project experience using *Michael Lachmann Computer Aided Design* (MLCAD). The reason for adding this topic to the discussion is that LEGO Mindstorms robot and machine designs need a fundamental beginning. The origin of a robot or machine should start with building a 3-D virtual model, using a CAD software package to capture mechanical design concepts, and MLCAD will serve this purpose.

LEGO Mindstorms Interfacing introduced Electronics Workbench and Circuit Maker simulation tools in order to experiment with virtual circuits before building them on a breadboard. By using simulation tools to build and experiment with electronic circuits, the hobbyist and experimentalist will immediately know what to expect from the actual prototype circuit, based on the results obtained from the virtual design. Therefore, troubleshooting should be kept to a minimum, thereby speeding up the actual robot-building activity. Following the same format explained in the mechanical CAD description, this book provides detailed tutorials on using the Circuit Maker simulation tool. The projects and lab experiments discussed in this book will be explained in the following "Chapter Descriptions" section.

Chapter Descriptions

Now that the software tutorials format has been discussed, here is the format the book will use in discussing LEGO Mindstorms mechatronics.

Section 1, "Mindstorms Embedded Systems Tools"

The following chapters explain the various programming language tools and software available, and how to use them in developing Mindstorms Mechatronic robots and intelligent machines.

Chapter 1, "Software Tools for LEGO Mindstorms Embedded Systems Development"

This chapter discusses Integrated Development Environment (IDE) tools for *Interactive C Version 4* (IC4), Java (RCX Tools), NQC (BricxCC), Scout *Programmable Brick* (P-Brick) IDE, Python, VBA *Graphical User Interfaces* (GUIs), Robolab, and the Basic Stamp Windows programming languages. Hands-on lab projects illustrate programming conventions and direct controls of the RCX, Basic Stamp, and Scout-based robots and intelligent machines.

Chapter 2, "Electronics Hardware Add-Ons"

The chapter discusses how the Basic Stamp, Palm Computer, and Texas Instruments (TI) Graphics Calculator (with Calculator-Based Laboratory [CBL] and *Digital Control Unit* [DCU]) can enhance RCX and Scout P-Brick robots and intelligent machines (mechatronics). Electronic interfaces are designed and built for power supply conversion (linear switching and buck and boost circuits), sensing, and direct control techniques. Advanced data acquisition techniques, Microsoft Excel VBA, Java, and IC4 programming languages with Mindstorms P-Bricks and the Basic Stamp are explained using hands-on laboratory projects, experiments, and test procedures.

Chapter 3, "Instrumentation and Robots"

The chapter discusses the importance of using *Digital Multimeters/Volt-Ohm Meters* (DMMs/VOMs), oscilloscopes, function generators, and low-cost data acquisition systems to test mechatronic designs of Mindstorms robots and intelligent machines. Machine-vision is introduced using the Vision Command LEGO Camera, which is followed by several hands-on robotics laboratory projects and experiments.

Chapter 4, "Electronic Circuit Simulators and MLCAD"

The chapter discusses *Electronics Design Automation* (EDA), with an explanation of OrCad, Electronics Workbench, and Circuit Maker circuit-schematic simulator software packages. Electronic interfacing circuits for robotic control is demonstrated via laboratory design projects and experiments. MLCAD software packages for designing LEGO mechanical parts and robots are explained using lab projects and experiments as well.

Section 2, "Robot and Intelligent Machine Projects"

The following chapters provide detailed instructions for developing Mindstorms-embedded systems used in Mechatronic robots and intelligent machines.

Chapter 5, "Telerobotics"

The chapter discusses how to build and program remote-control vision-based robots. The LEGO Vision Command camera, along with Java and VBA programming languages, will be used to develop a basic telerobotics system. Step-by-step instructions for building the telerobotics system are also provided.

Chapter 6, "Smart Hand Controllers"

The chapter discusses how to build mechatronics-based remote controls for Mindstorms robots. The fundamentals of hand controller development are provided for building a steering controller for a Mindstorms robot as well as step-by-step instructions for building the telerobotics system.

Chapter 7, "Hybrid Robots"

The chapter discusses other motorized construction sets, such as Capsela and the Mecanno Erector Set, and how they can be used with a Basic Stamp controlling Scout/Droid Development P-Bricks to build hybrid robots. A *Society of Automotive Engineers* (SAE) educational construction kit is used as the mobile base for one of the robotics projects discussed in the chapter. A Systems Block diagram explains the fundamental theory behind building a hybrid robot, and instructions for building the robotic systems are also provided.

Chapter 8, "Mechatronic Bots"

This chapter explores the fundamentals of a Mechatronic robot using the hybrid technique discussed in Chapter 7. Mechanical fabrication is explained via the hands-on robot construction project, and Java programming language will test the function of the robot for correct I/O functions. Step-by-step instructions for building the robotic systems are also provided.

Chapter 9, "Walking Bots"

This chapter is a culmination of Chapters 5, 6, and 8 because it deals with the design techniques used in building a walking Mindstorms robot. Vision and telerobotics control techniques will be used in manipulating the robot's walking motion, and instructions for building the robotic systems are provided.

Chapter 10, "Audio Bots"

This chapter discusses sound-generation techniques using the Basic Stamp as an intelligent audible-sound indicator interface for a Mindstorms robotic warning system. Discrete circuit designs are explored using electronics hardware breadboard projects, and the chapter also includes instructions for building the robotic systems.

Appendices: LEGO Mindstorms Embedded Systems Resources

This section lists web addresses, books, articles, design equation derivations, and freeware/shareware software.

Overview

The objective of this book is to teach mechatronics techniques using inexpensive development software tools, electronic components, and LEGO bricks with detailed laboratory procedures for advanced Mindstorms robotics applications. This book will make references to *LEGO® Mindstorms™ Interfacing* for historical and introductory information, thereby setting the stage for the advanced concepts that will be be discussed. Purchasing the first book will allow the reader to transfer prior knowledge more effectively when reading this second Mindstorms volume. This book will provide advanced topics with detailed technical information along with the tutorials. It will take the amateur roboticist, hobbyist, and experimentalist to a new level in the Mindstorms robotics interfacing experience by building LEGO mechatronics systems.

Don Wilcher

ACKNOWLEDGMENTS

I would like to thank Judy Bass, Sr. Acquisition Editor of McGraw-Hill, for her enthusiasm as well as gentle nudge during the writing phase of this book project. Also, a warm Hello and thank you to Beth Brown of MacAllister Publishing for a wonderful job in layout and editing of this book. Last, I would also like to thank Dr. Chan-Jin Chung of Lawrence Technological University for allowing me to judge at the 2003 Robofest competition. His vision and joy of robotics building to teach kids about technology has inspire me to continue with my educational mission in engineering education as well.

Don Wilcher
August 7, 2003

ABOUT THE AUTHOR

Don Wilcher is an Electrical Engineer and the author of the bestseller *LEGO® Mindstorms™ Interfacing*, also from McGraw-Hill.

CHAPTER 1

Software Tools for LEGO Mindstorms Mechatronics Embedded Systems Development

Developing software for LEGO Mindstorms robots requires an *Integrated Development Environment* (IDE) that enables you to read analog values from sensory devices using simple text commands or a *Graphical User Interface* (GUI). These sensory devices are connected to the inputs of the *Robot Command Explorer* (RCX) *Programmable Brick* (P-Brick), as well as the control switching and actuator components wired to their output bricks.

Upon entering programming instructions using a command prompt, the action or required data should be executed or viewed immediately. Several IDEs are available free of charge from several Web sites that enable these basic read/write instructions to be executed on the RCX and Scout P-Bricks. Chapter 6 of *LEGO Mindstorms Interfacing*[1] examines several examples of IDE tools for Mindstorms' robotics embedded control. This chapter will continue Chapter 6's discussion by exploring the tools' remote control capabilities from a GUI and text command approach. This chapter discusses the following topics:

- *Interactive C Version 4* (IC4) IDE
- Robolab
- Java (RCX Tools)
- *Not Quite C* (NQC) (BricxCC)

The objective of the chapter is to illustrate how to remotely control Mindstorms robotics software using an IDE tool. Detailed hands-on lab projects and experiments ensure this objective is carried out. After completion of the lab activities, you should appreciate the varied embedded systems development tools available for remote robot-based control software.

The IC4 IDE

In this section Interactive C version 4 software development history will be discussed along with several hands-on laboratory experiments.

Introduction of IC4

Interactive C (IC) is the dynamic compiler that Randy Sargent and Fred Martin developed specifically for students participating in the *Mass-*

---

[1]Wilcher, Don, *LEGO Mindstorms Interfacing*, McGraw-Hill Publishing, New York, 2003.

achusetts Institute of Technology (MIT) 6.270 robot builder's course. The intercession course allowed students to design, build, and debug a 68HC11-based LEGO robot using IC for ease of software design and development. Fred and Randy developed IC to allow students to program LEGO robots using this interactive C programming language. They called the language Interactive C to highlight its interactivity.[2] IC was created to run on small self-contained robot controllers like the Handy Board or the LEGO RCX P-Brick.

IC enables you to interact with the language using a command prompt text box. The interactive command prompt text box enables the developer to control the RCX P-Brick's *input/output* (I/O) features. IC is based on *procedural programming,* whereby procedure calls interface with the motors, *liquid crystal display* (LCD), sound generator, and sensors using the RCX P-Brick. Procedural programming is a structured top-down method of writing code for embedded microcontrollers and processors. No *jump*, *branch* or *goto* statements are used for logic flow or the conditional control of data.

The unique feature of IC is the ability of the amateur roboticist to test an applet's code by typing instructions into the command prompt text box and the seeing the results immediately on the PC or notebook screen. An *applet* is a small application that can be programmed *procedurally*, whereby each line of code is sequenced from the previous command or instruction statement. The ability to test code while writing it helps reduce debug time and enables experimentation in optimization to be done concurrently with the development of the software control application. IC can serve as a prototyping language because sections of code can be tested immediately before the final code is compiled. Four revisions have been made to IC, with the fourth one allowing Mindstorms robotics programming using the RCX P-Brick. IC4 is a free programming language that can be obtained from `www.kipr.org/ic/download/`.

Controlling an RCX P-Brick Using IC4

As discussed, a RCX P-Brick can be controlled immediately by typing function commands into the command prompt text box. The IC4 function not only manages small blocks of code, but it also provides direct control of the RCX P-Bricks' physical layer or the I/O structure. The RCX P-Brick will perform the action immediately upon receiving the function command from

---

[2]Circuits to Control: Learning Engineering by Designing LEGO Robots, PhD Dissertation, Fred G.Martin, MIT Media Lab, June 1994.

the infrared tower. Although commands are typed using the prompt, the IC4 IDE is still an effective software tool for testing logic and code directly when using a target RCX P-Brick. To explore the IC4 programming language, the following lab project will illustrate how to use the IC4 IDE to control the RCX P-Brick's internal and I/O devices.

Driving a Seven-Segment *Light-Emitting Diode* (LED) Display and LCD Control Lab Project This project explores driving a seven-segment LED display and displaying text on the RCX P-Brick's LCD screen using the IC4 programming language. Figure 1-1 shows the building blocks for this project. The software development station is a typical setup for programming LEGO Mindstorms robots. The RCX P-Brick's input will use a touch sensor for turning the output of the programmable brick on and off. The output will be wired to a seven-segment LED display using a 7805 linear voltage regulator circuit. Upon completing the project, you will be able to control an external electronic display circuit and display the status of the RCX P-Brick's output using the *printf(string)* function command.

The *Bill of Materials* (BOM) for this project is as follows:

One 7805 linear voltage regulator *integrated circuit* (IC)

Two 0.1 µF (microfarad) capacitors

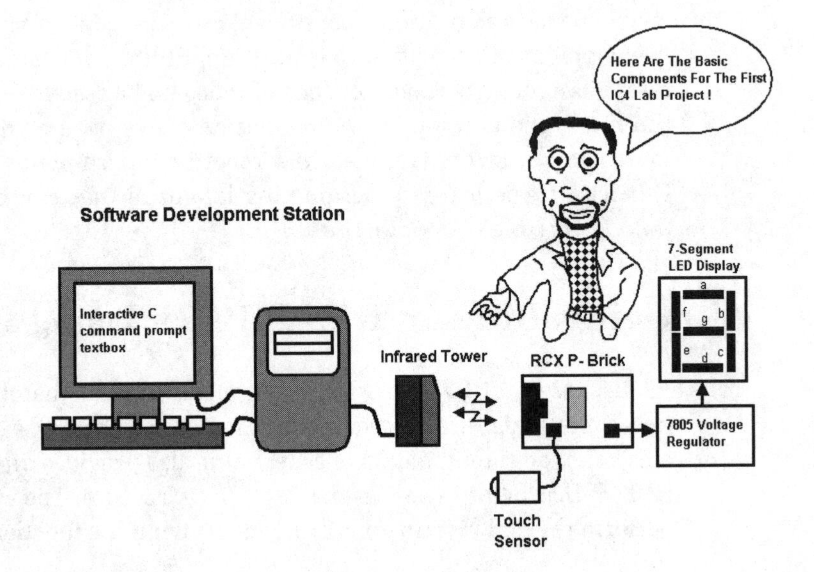

Figure 1-1 Mr. Don showing the components for the first IC4 lab project

One 470 Ω resistor, one-quarter watt

One seven-segment LED display

One solderless breadboard

One modified LEGO electric wire

One touch sensor

One standard LEGO electric wire

One RCX P-Brick

One infrared tower

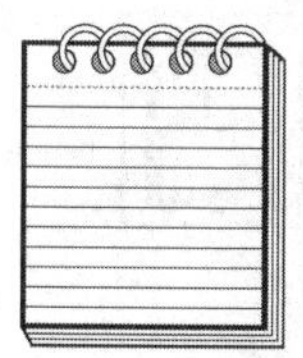

TECH NOTE: *Bill of Materials, or BOM, is an engineering term for a parts list. In order to manage the millions of parts used in electronic modules and mechanical assemblies of consumer products, computerized systems and special software are used to maintain the massive inventory of parts for these components.*

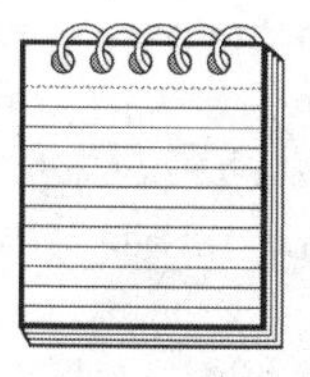

TECH NOTE: *In* LEGO Mindstorms Interfacing, *the electronic circuits were built on the Radio Shack Electronic Learning Lab breadboarding system (Cat. No. 28-280) for ease of RCX interfacing and rapid prototyping. This breadboarding unit not only makes it convenient for building circuits, but it has an ample supply of electronic components as well.*

Circuit Description As shown in Figure 1-2, the electronic driver circuit for switching a red LED is basically a 7805 linear voltage regulator. The input voltage is provided by the RCX P-Brick output interface that has a *Direct Current* (DC) signal level of +7V. This output voltage is stepped down to a constant +5V DC that provides a *supply bus rail* for driving the series 470 Ω resistor and red LED indicator circuit. A supply bus rail is a common electrical node used to provide a negative or positive source of DC voltage for an electronic circuit.

The two 0.1 μF capacitors attached to the input and output pins of the voltage regulator IC provide noise suppression from the +7V internal DC source of the RCX P-Brick and output switching of the seven-Segment LED display circuit.

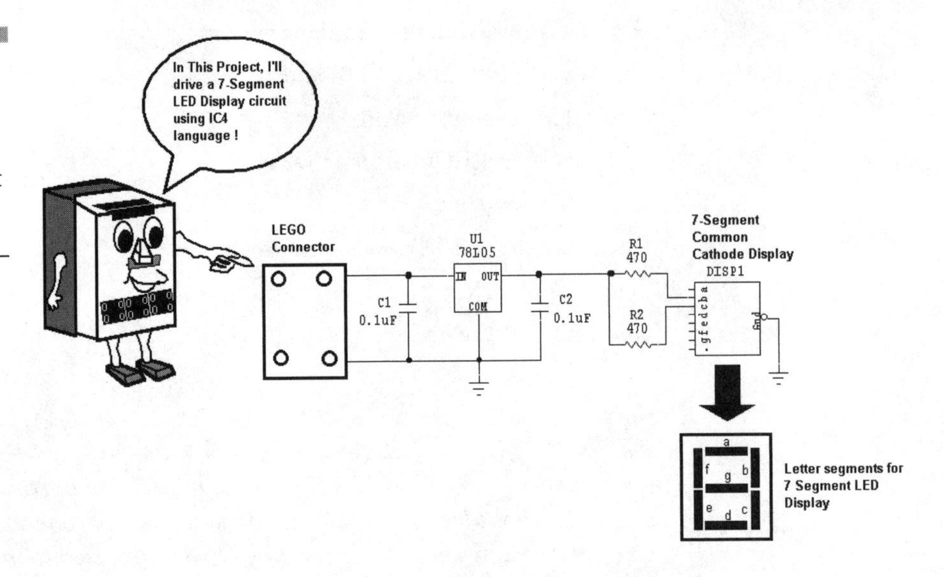

Figure 1-2 Rudy RCX showing the seven-segment LED display circuit for the IC4 lab project

TECH NOTE: *A supply bus rail is quite easy to implement on a solderless breadboard. Take a positive supply voltage (in the case of the RCX P-Brick, the +7V connector stud), and wire it to a central point or strip of holes on the white board. All circuits that need the voltage can then be electrically connected to the DC source using a jumper wire.*

A Voltage Regulator as a Switching Interface Circuit A question that might be lurking in the mind of the reader is, why use a voltage regulator as a switching interface circuit for the RCX P-Brick? In *LEGO Mindstorms Interfacing*, the voltage regulator IC was introduced on page 58 in Chapter 2, "Developing GUIs: Software Control Basics." The circuit shown enabled a *Robot Digital Assistant* (RDA), called a *Voltage Robot* (V-Bot), to output a constant +5V DC source. A *Visual BASIC for Applications*-built (VBA-built) control panel is used to switch the signal on and off with Active X controls.

In testing the circuit for the robotics application, I came to the realization that the IC makes for a simple and accurate switching circuit for small DC electrical loads or devices. Therefore, based on the lab project described in that section, as well as Chapter 4, "Electronic Switching Circuits," here are four reasons for using the voltage regulator IC as a low-cost switch:

- It isolates the +7V DC output voltage of the RCX P-Brick from external load *transients* or electrical noise.
- It enables the RCX P-Brick to easily control digital logic circuits.

- It enables low-current management of external loads using a constant +5V supply bus rail from the voltage regulator IC.
- It enables digital switching circuits to easily control larger-current electrical loads or devices using a bipolar transistor or Power *Metal Oxide Semiconductor* (MOSFET) electronic components.

Although probably more than four reasons exist for using a voltage regulator IC in Mindstorms electronic switching applications, I believe the items listed are on the top of the applications list.

In Chapter 4, page 144, of *LEGO Mindstorms Interfacing*, the voltage regulator is used to provide a constant +5VDC for a bipolar two-transistor audio tone generator. It also assists a Power MOSFET to drive a seven-segment LED display used in the diagnostics tool shown on page 151. Figures 1-3 and 1-4 show the respective circuit schematic applications.

As an exercise in creativity, see if you can think of additional reasons for using the voltage regulator IC as a switching interface circuit for the RCX P-Brick. Start a robotics design notebook and add your list to the pages of the engineering document. After recording this list in the notebook, see if some of the ideas can be developed into practical Mindstorms robot projects or lab experiments. Remember, the only limitation to Mindstorms robotics creation is a lack of imagination!

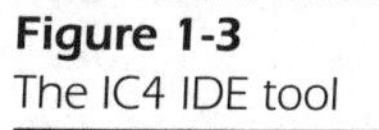

Figure 1-3
The IC4 IDE tool

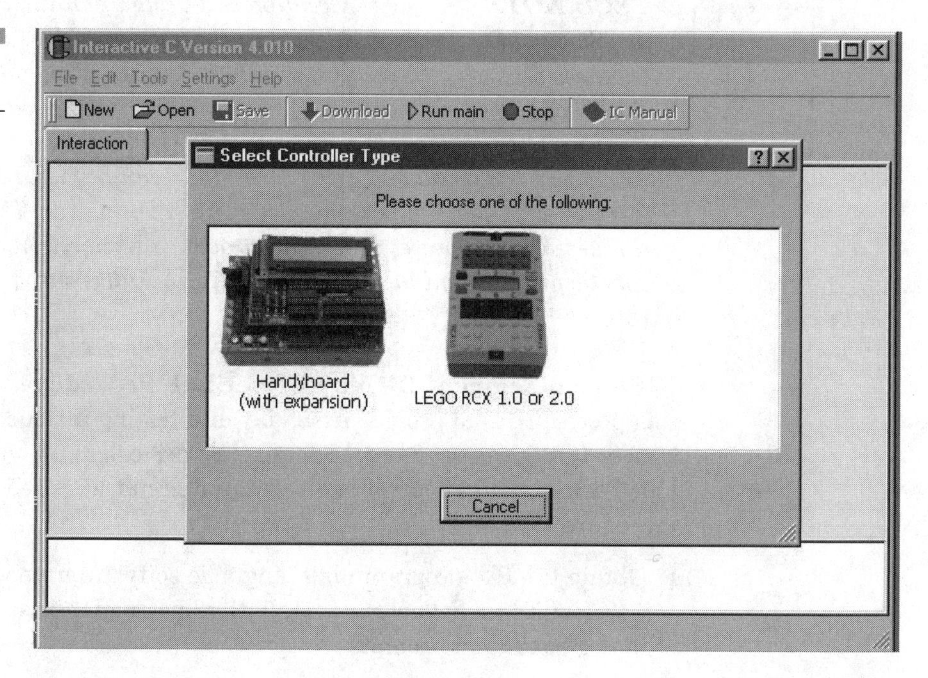

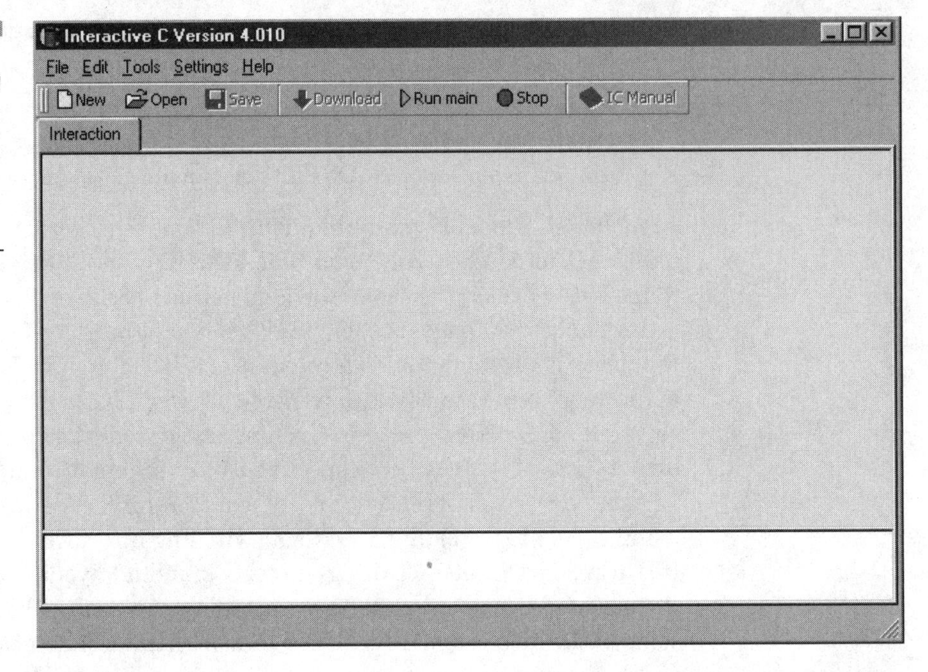

Figure 1-4
The IC4 command prompt text box used for entering functions for direct control of the RCX P-Brick

TECH NOTE: *Creative Creation is a design philosophy that Alexander Slocum, Professor of Mechanical Engineer at MIT, uses to describe engineering problems as opportunities. Passion, Thought Processes, and the Golden Rectangle are just a few of the methods used in developing a Creative Creation design mentality. A PDF presentation by Professor Slocum on Creative Creation can be found at* `http://pergatory.mit.edu/perg/presentations.htm`. *The information discussed in the presentation can provide additional Mindstorms robotic projects and lab experiments for the adventurous amateur roboticist.*

IC4 Seven-Segment Display Driver Lab Procedure The objective of this procedure is to provide assembly and testing instructions for validating an IC4 seven-segment LED display driver circuit. An RCX P-Brick, touch sensor, and 7805 voltage regulator circuit are used in the following procedure:

1. Obtain the IC4 programming language software from `www.kipr.org/ic/download`. Follow the installation instructions using the windows dialog boxes and prompts.

2. Open the IC programming language software and notice the IDE tool displayed on the screen, as shown in Figure 1-3.
3. Turn on the RCX P-Brick and have the infrared tower pointing toward the RCX P-Brick.
4. Choose the *LEGO RCX 1.0* or *2.0* brick, as shown in Figure 1-3. The software will begin to download the firmware and IC4 library to the RCX P-Brick. After the download process is completed, the command prompt text box will be displayed onscreen, as shown in Figure 1-4. The *IC40* logo will be displayed on the LCD screen of the RCX P-Brick.
5. Using the schematic diagram shown in Figure 1-2, wire up the seven-segment LED display driver circuit on the Radio Shack Electronics Lab breadboard or equivalent experimenter's board. The seven-segment LED display should be wired to show the number *1*.
6. Attach a modified LEGO electric wire to the driver circuit and connect the other end to output 1 on the RCX P-Brick.
7. Attach a touch sensor to input 1 on the RCX P-Brick.
8. Type the following function on the command prompt text box, followed by the enter key:

```
fd(1)
```

9. The seven-segment LED display should show the number *1*. To turn off the seven-segment LED display, type the following function on the command prompt text box, followed by the enter key:

```
off(1)
```

   The seven-segment LED display should now be off. Figure 1-5 shows the two functions entered onto the IC4 IDE tool. These two functions will be used to build the final code for driving the seven-segment LED display, with a touch sensor as the input control.

10. On the task bar, click "*New.*" A blank page should be displayed on the IDE tool. Begin typing the following lines of code:

```
void main()
{    while(1){
         if(digital(1)==1){
              printf("1");
              fd(1);
         }
```

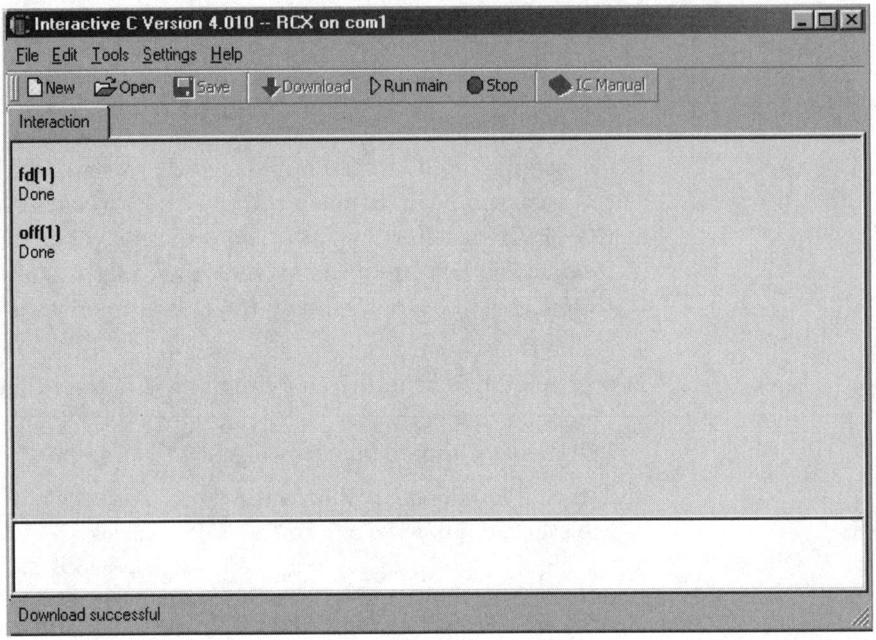

Figure 1-5
The two IC4 functions, fd(1) and off(1), entered onto the IC4 IDE tool

```
            else{
                printf("0");
                off(1);
            }
        }
    }
```

Figure 1-6 shows the code typed on the IC4 IDE editor.

11. Save the code under the filename *ic4_lab1*. Click *Download*, on the task bar of the IDE tool.
12. Once the code is downloaded to the P-Brick, press *Run*.
13. Press and hold the touch sensor. The LCD screen should show a *1* as well as the seven-segment LED display.
14. Release the touch sensor. The LCD screen should show a *1* with the seven-segment LED display turned off.

The code in step 10 allows binary digits *1* and *0* to be displayed when the touch sensor is pressed and released. The seven-segment LED display is turned on only when the RCX P-Brick's LCD screen showed a binary digit *1*.

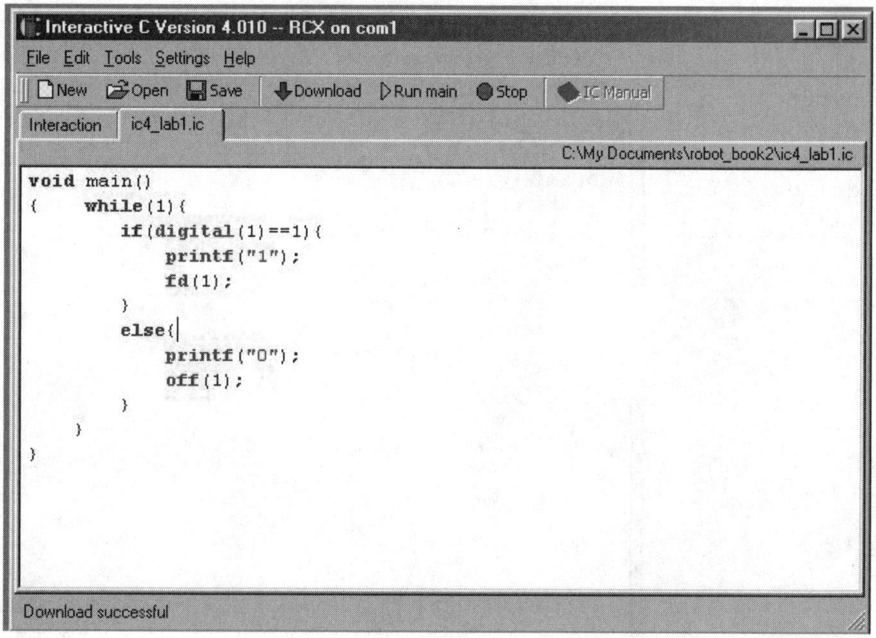

Figure 1-6 The simple motor/LCD control code typed on the IC4 IDE editor

Interactive C Version 4.010 -- RCX on com1

File Edit Tools Settings Help

New Open Save Download Run main Stop IC Manual

Interaction ic4_lab1.ic

C:\My Documents\robot_book2\ic4_lab1.ic

```
void main()
{     while(1){
          if(digital(1)==1){
              printf("1");
              fd(1);
          }
          else{
              printf("0");
              off(1);
          }
      }
}
```

Download successful

How can the text *Hi* and *Lo* be displayed on the RCX P-Brick's LCD screen when the touch sensor is pressed and released? The following IC4 code can be used to display *Hi* and *Lo* on the LCD screen of the P-Brick by using the *printf(string)* function. The following code shows the change made to the *printf(string)* function for displaying the two text words:

```
void main()
{     while(1){
          if(digital(1)==1){
              printf("Hi");
              fd(1);
          }
          else{
              printf("Lo");
              off(1);
          }
      }
}
```

Enter the code using the IC4 IDE editor, as shown in Figure 1-7, and save using the filename of *ic4_lab1_a*. Download the code to the RCX P-Brick and run the program. By pressing and releasing the touch sensor, the text on the LCD should show *Hi* and *Lo*. In addition, the seven-segment display will display binary *1* when the touch sensor is pressed and held.

Figure 1-7 The revised motor/LCD control code being entered using the IC4 IDE editor

Interactive C Version 4.010 -- RCX on com1

File Edit Tools Settings Help

New Open Save Download Run main Stop IC Manual

Interaction ic4_lab1_a.ic

C:\My Documents\robot_book2\ic4_lab1_a.ic

```
void main()
{    while(1){
        if(digital(1)==1){
            printf("Hi");
            fd(1);
        }
        else{
            printf("Lo");
            off(1);
        }
    }
}
```

Download successful

As an additional design challenge, try wiring the seven-segment LED display to show the letter *H* when the programmable brick's LCD displays *Hi*. After completing the wiring, rerun the program and notice the display. With success, the seven-segment LED display should show the letter *H*. If not, recheck the wiring and run the program again.

TECH NOTE: *In the C programming language, one of the most common library functions is called* printf(string).[3] *This is a general output function used to display* strings, *or characters, on the screen. The* printf()*function for IC4 is used to output strings on the RCX P-Bricks' LCD screen.*

[3] Schildt, Herbert. *Teach Yourself C*, 3rd Edition. San Francisco, CA: Osborne-McGraw-Hill, 1997.

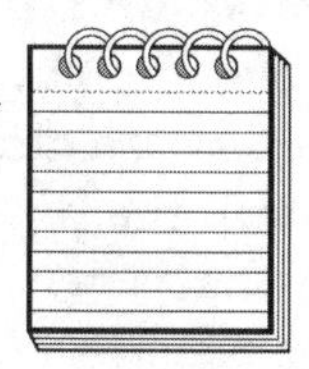

TECH NOTE: *On pages 52 and 53 of the* Basic Electronics Workbook *that accompanies the Radio Shack Learning Lab, Forrest Mims shows the numbers and alpha characters that can be created using a single seven-segment LED display. Also, an unfilled seven-segment display page can be photocopied and used in the design and display of other nonstandard characters.*

Summary of IC4 Lab Project 1 The objective of this lab project was to get you familiar with the IC IDE tool. For those readers who experimented with Chapter 6 in *LEGO Mindstorms Interfacing*, the discussion on entering functions using the command prompt textbox was a review of basic robot control via text messages. Adding a seven-segment LED display and providing a LCD text equivalent were provided to illustrate how the RCX P-Brick can be used in a diagnostic capacity by visually displaying either a direct message or binary data. The seven-segment LED display circuit also showed the effectiveness of the 7805 voltage regulator's capability as an electronic switcher. The two code examples, along with the circuit, will be used in other projects in this book. Therefore, the code and circuit will be set aside for future use in Mindstorms mechatronic projects.

In the next lab project, we will investigate IC4's conditional control capability to operate the direction of a small LEGO electric motor using two touch sensors.

IC4 Lab Project 2

Most mechatronic systems have the capability to reverse or change linear or rotational motion, based on limit or object detection devices that sense *events*. An event is the request of a detection device's switching contacts, by either a single or a double-acting force, initiated by the user (operator) of the machine. One function already discussed was the *fd(int)*, used to turn the small LEGO electric motor in the forward direction. The function *bk(int)* will make the motor turn backwards or in a reverse direction. Therefore, the missing puzzle piece for Lab Project 2 has been found.

The BOM for this project is as follows:

Two touch sensors

One small LEGO electric motor

Three LEGO electric wires

A RCX P-Brick Directional Controller Lab Assembly and Testing Procedure The objective of the lab project is to write code whereby pressing and holding touch sensor 1 will turn the motor in the forward direction. Pressing and holding touch sensor 2 will turn the motor in the reverse or opposite direction. The LCD will show *Fd* and *rev* on the screen as the motor spins in the specified direction. Figure 1-8 shows a pictorial diagram of the components used in controlling the small electric motor and RCX P-Brick's LCD.

To begin the procedure, follow these steps:

1. Attach the two touch sensors to inputs 1 and 2, using the electric wires to the RCX P-Brick.
2. Attach the small electric motor to output A, using an electric wire.
3. Enter the following code into the IDE editor tool. See Figure 1-9 for the code entered into the text editor.

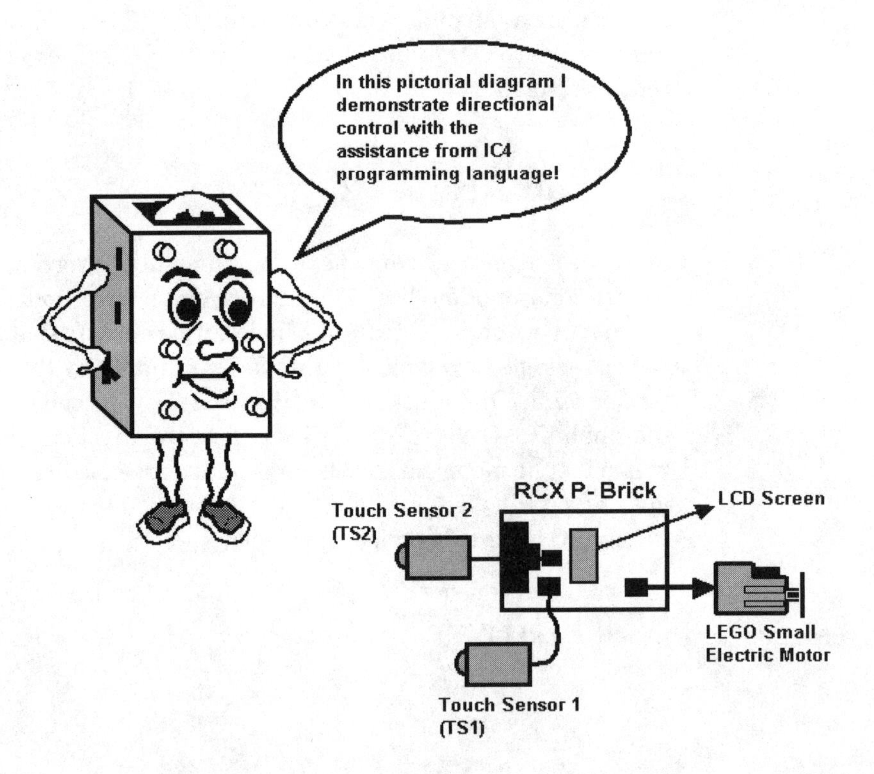

Figure 1-8 Tim Touch Sensor discusses the directional control pictorial diagram

```
void main()
{     while(1){
          if(digital(1)==1){
              printf("Fd");
              fd(1);
          }
          else{
             if(digital(2)==1){
              printf("rev");
              bk(1);
          }
          else{
              printf("-----");
              allbrake();
          }
      }
    }
}
```

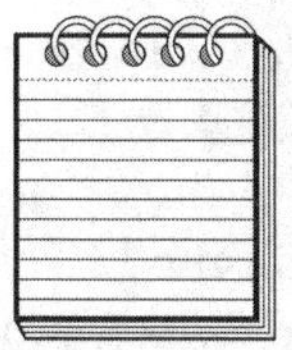

TECH NOTE: *Keeping track of parentheses while writing code can be a software developer's worst nightmare. Fortunately, the IC4 IDE tool has a unique management feature for keeping track of parentheses. By clicking the Tools menu, a* Check parens *feature is available for keeping track of the parentheses used in your code.*

Interactive C Version 4.010 -- RCX on com1

File Edit Tools Settings Help

New Open Save Download Run main Stop IC Manual

Interaction | New | ic4_lab1_b.ic | ic4_lab1.ic

C:\My Documents\robot_book2\ic4_lab1_b.ic

```
void main()
{     while(1){
          if(digital(1)==1){
              printf("Fd");
              fd(1);
          }
          else{
             if(digital(2)==1){
              printf("rev");
              bk(1);
          }
          else{
              printf("-----");
              allbrake();
          }
      }
    }
```

Download successful

Figure 1-9 Directional code for the lab project entered into the text editor

4. Download the code to the RCX P-Brick. Press *Run* on the programmable brick to execute the embedded code. On the LCD screen, five dashes should be visible. If not, recheck the code and repeat this step.
5. Press and hold the touch sensor connected to input 1 on the RCX P-Brick. The LEGO small electric motor attached to output 1 should be rotating with *Fd* displayed on the LCD screen. Release the touch sensor and the motor should stop immediately. The LCD should display five dashes as in step 4.
6. Press and hold the touch sensor connected to input 2 on the RCX P-Brick. The LEGO small electric motor attached to output 1 should spin in the opposite direction of step 5. The word *rev* will be displayed on the LCD screen as well.

Lab Question and Observation: Is the RCX P-Brick's LCD an Output Device? In the lab projects investigation, the LCD was used as a simple text messaging display device. By using the IC4 *printf(string)* function, you can display letters, numbers, and characters on the LCD screen.

I can remember the discussion on output devices and *central processing unit* (CPU) architecture from my digital computer class in community college. The instructor stated that an output device is a "dumb" unit used for displaying data processed by the CPU, which controls the output device. Monitors, terminals, and printers were examples used by the instructor to illustrate devices that mainly display, for operator viewing, processed data produced by the CPU. These items receive commands from control signals flowing through a bundle of wires packaged within the connecting cable.

Therefore, the LCD fits the criteria of being an output device. It has the capability to display data; it is under the control of the microcontroller that resides inside the RCX P-Brick. The connecting wires that send control signals to the various LCD segments are transparent, because they are packaged inside the P-Brick.

Summary of IC4 Lab2 Project The objective of the lab project was to illustrate the ease in which IC4 code can be modified to control the direction of a LEGO small electric motor using two touch sensors. By adding another *if-else* conditional statement, the RCX P-Brick could read the second touch sensor's *event* and produce a reversal motion-control signal to the small electric motor attached at output A on the P-Brick. The LCD became an embedded output device for showing the directional state of the motor by displaying text on the small screen.

This code is quite a useful application in terms of object detection and control for Mindstorms robots and intelligent machines. The touch sensors, in essence, become limit switches for objection detection, allowing the RCX P-Brick to control the direction of the motor.

The next section of this chapter will explore the popular Robolab software, created by Chris Rogers of Tufts University and the National Instruments Corporation.

Robolab

Robolab is a unique embedded software tool for Mindstorms mechatronics because of the combination of LEGO bricks and National Instruments LabView graphical programming language. Chris Rogers, an associate professor in mechanical engineering at Tufts University, created Robolab to introduce engineering concepts to students of all ages—from kindergarten through college. Jeff Kodosky of the National Instruments Corporation invented LabView, with the goal of helping engineers and scientists program engineering and scientific applications using graphical icons in an easy-to-use development environment.[4] The icons look like blocks with an image representing a specific control or function. The blocks are wired together like an electrical circuit, creating the final program known as a *Virtual Instrument* or VI. The VI is connected to the real world using data acquisition cards for obtaining information like voltage, current, temperature, sound, and pressure.

Figure 1-10 shows an example application using the LabView software. The LabView software uses the G programming language, developed by Kodosky to provide the textual code instruction set for processing data obtained from the outside world to the PC or notebook computer. The G programming language is embedded in the background of the virtual instrument presented on the PC or notebook screen. Using the LabView graphical programming language with the RCX P-Brick gives the Mindstorms Robotics invention system greater flexibility and ease when programming mechatronic devices.

In this section, Robolab software will be investigated using the graphical programming language tool to perform some basic control and data collection activities. The following lab projects will demonstrate the fundamental

[4]www.ni.com/company/robolab.htm

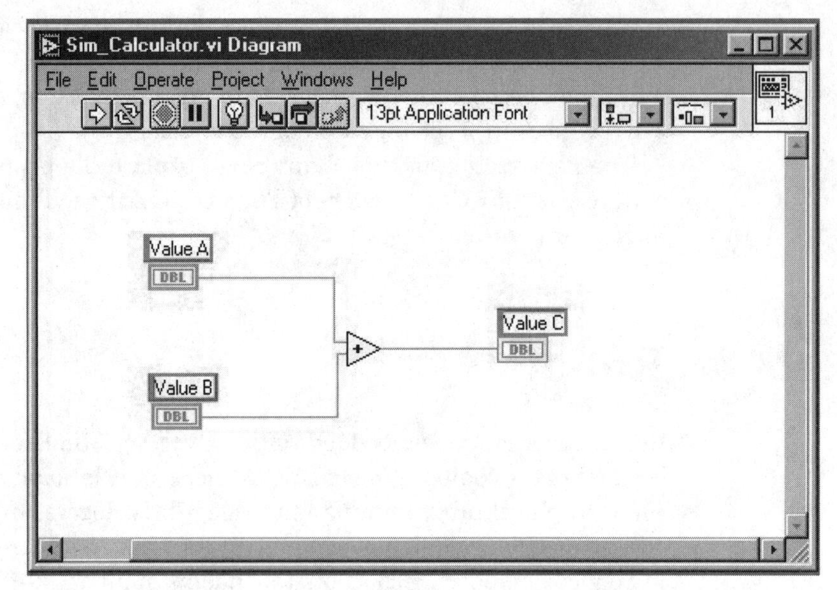

Figure 1-10 An "ADDITION" application example using the LabView software

concepts behind the Robolab software's capability to collect data and control electromechanical devices important to Mindstorms mechatronics.

Robolab Exploratorium Project 1

In this section Robolab data collection techniques will be described using basic laboratory experiments.

Simple Data Collection Mindstorms robots should have the capability to collect data based on their surroundings. This capability enables a true autonomous and intelligent machine to be built using low-cost software and LEGO parts. With LabView's graphical programming language to create virtual instruments for gathering data from physical stimuli (sound, light, and pressure), Robolab allows you to easily create measurement and analysis instruments using an iconic programming environment.

The first lab project will introduce the basic techniques of data collection using the *Download* and *Direct Modes* within the *Invent and Investigator* project area of Robolab. The techniques discussed will be used in subsequent chapters. After completing the following lab projects, the reader will have a basic understanding of programming the RCX P-Brick for data acquisition and control mechatronic applications. The information provided will go to the core of the Robolab software's capability to create mechatronic systems and devices using a graphical programming language IDE.

The BOM is as follows:

One RCX P-Brick

One touch sensor

One infrared tower

One LEGO electric wire

Robolab software

TECH NOTE: *An excellent online reference guide is available for the* Inventor and Investigator *project area of Robolab. If you purchased the Robolab software from Pitsco/LEGO Dacta, this online guide can be printed and used to unlock the data acquisition and controls features of this powerful graphical programming language. The Web site for the Robolab Inventor and Investigator guide is at* `www.lego.com/dacta/Robolab/investigatorprogam.htm`.

Software Assembly and Test Procedure With the software installed on your PC or Mac, and the previous components available, the following lab procedure can be carried out:

1. Upon opening the Robolab software, click the *Investigator* bar on the main menu. See Figure 1-11.

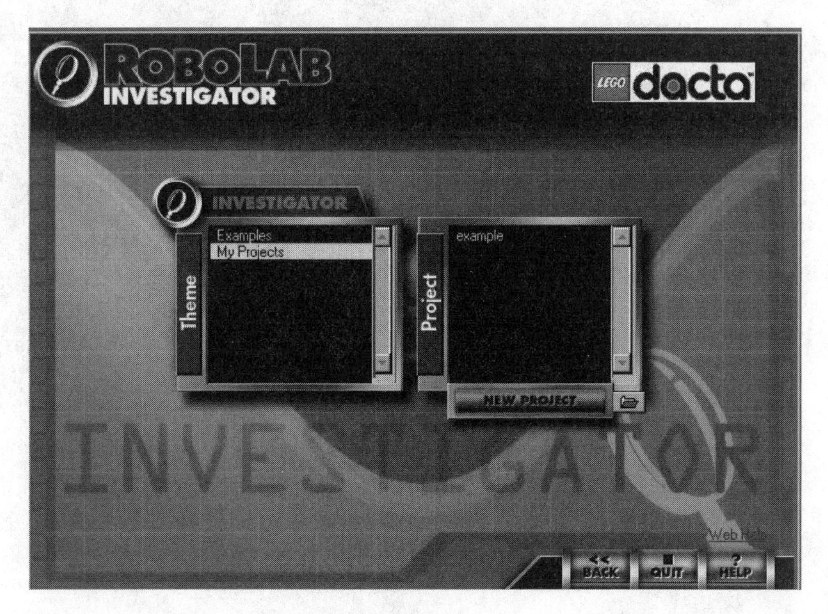

Figure 1-11 Creating a new project using the project dialog box

2. A new screen will appear with two dialog boxes: Theme and Project. Click the *New Project* bar within the Project dialog box. See Figure 1-12.

3. After the project has been created, double-click it to open the file.

The investigator IDE will be displayed on the screen, as shown in Figure 1-13.

As shown in the IDE, the name of the project will be visible along with a simple iconic program. Program level 1 shows a basic Robolab code for collecting data. The first block shows a touch sensor attached to input 1 on the RCX P-Brick. The second block shows the number of data points to be collected. The clock below the sensor shows one second for each data point to be recorded by the RCX P-Brick.

Data Collection Modes Collecting data in Robolab can be done using one of two methods: Download or Direct mode. Download mode enables the data collection program to be stored to the RCX P-Brick during code transmission using the infrared tower. Once the code is embedded within one of the program location slots, it can be executed and the sensor's event will be recorded. After all the data points are collected, the information can be uploaded to the *Upload* area for graphical viewing on a XY graph.

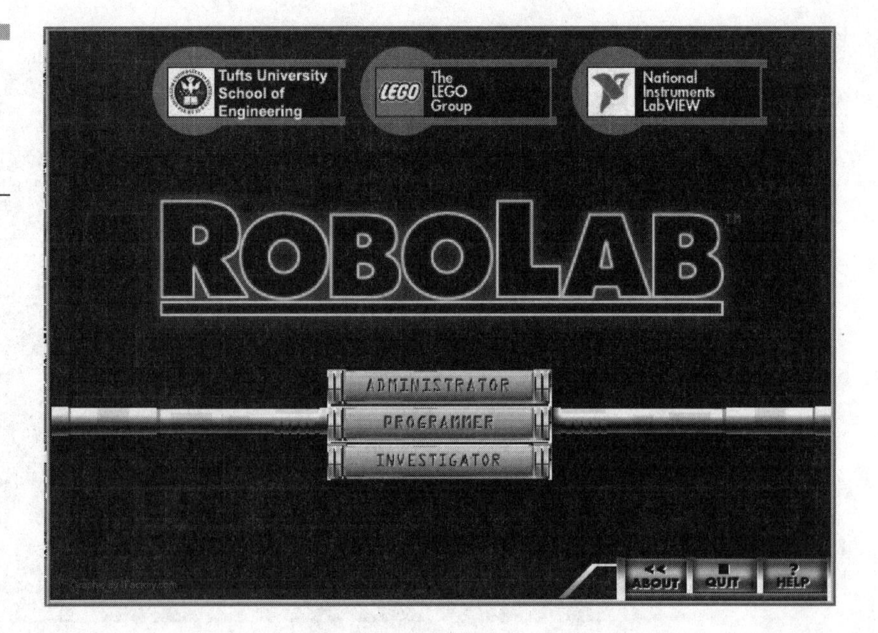

Figure 1-12 Opening the Investigator bar on the main menu

Direct mode enables data to be collected immediately when the sensor's event is requested. The results are displayed in real time on the XY graph. In Figure 1-13, the single arrow identifies the Download mode. The double arrow is used for Direct mode data collection. In this discussion, the continuing lab procedure will be for Download data collection. The lab procedure will walk you through the Download data-collection phase of the project:

1. Attach a touch sensor to input 1 using a LEGO electric wire.
2. Turn on the RCX P-Brick and place it in front of the infrared tower. Have program location 1 ready for data collection code by pressing the Prgm button on the programmable brick until it is displayed on the LCD screen.
3. Click the single arrow for Download mode data collection.
4. Run the data collection code in program location 1 by pressing Run on the P-Brick.
5. Press the touch sensor 10 times, holding the switch for 1 or 2 seconds within the 10-second data collection interval. With each press of the switch, the LCD will display the number of data points collected.

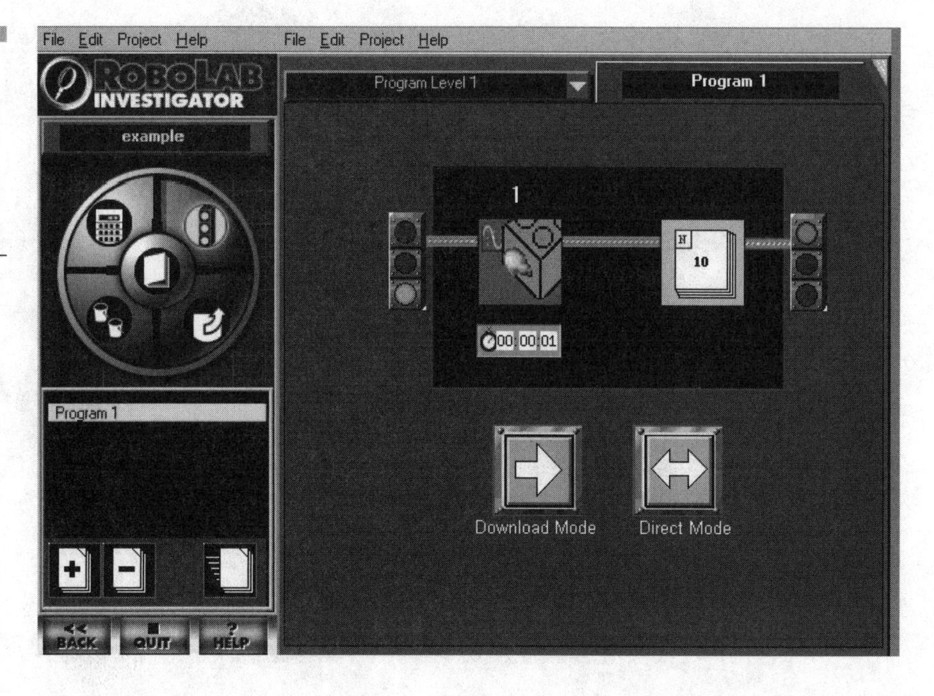

Figure 1-13 Identifying the download mode and direct mode data collection features of Robolab software

6. After 10 data points have been collected, go to the Upload Area to review the results graphically.
7. Click the arrow to review the data. Figure 1-14 shows the results of the number of touch sensor events the author created from the lab project.

Direct Mode Procedure The following procedural steps will walk you through using the Direct mode data collection.

1. Click the double arrows under the Program Level 1 block diagram. Notice the green LED turning on inside the infrared tower. The RCX P-Brick's LCD screen will be active via the bar graph and the transmit-receive icon will be displayed.
2. A XY graph will be displayed onscreen. Push and release the touch sensor. Notice that the data collected is immediately displayed on the graph.
3. Continue pressing and releasing the touch sensor until all 10 data points are captured and plotted on the graph. Figure 1-15 shows the data plotted by the author.

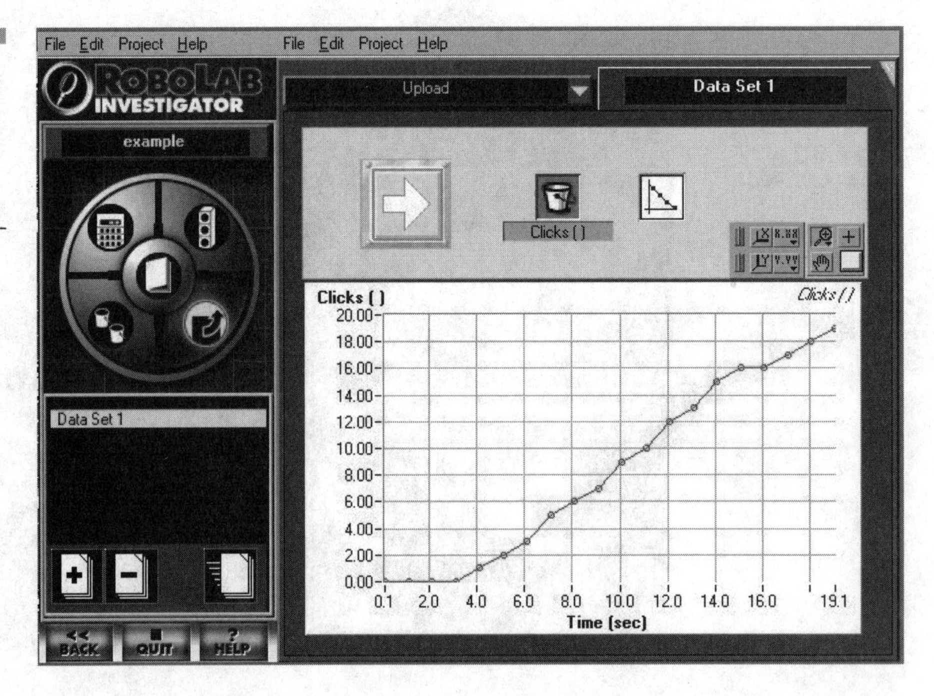

Figure 1-14 The results of the number of touch sensor events created from the lab project.

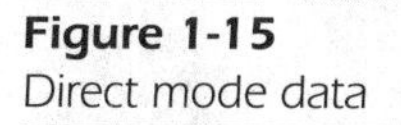

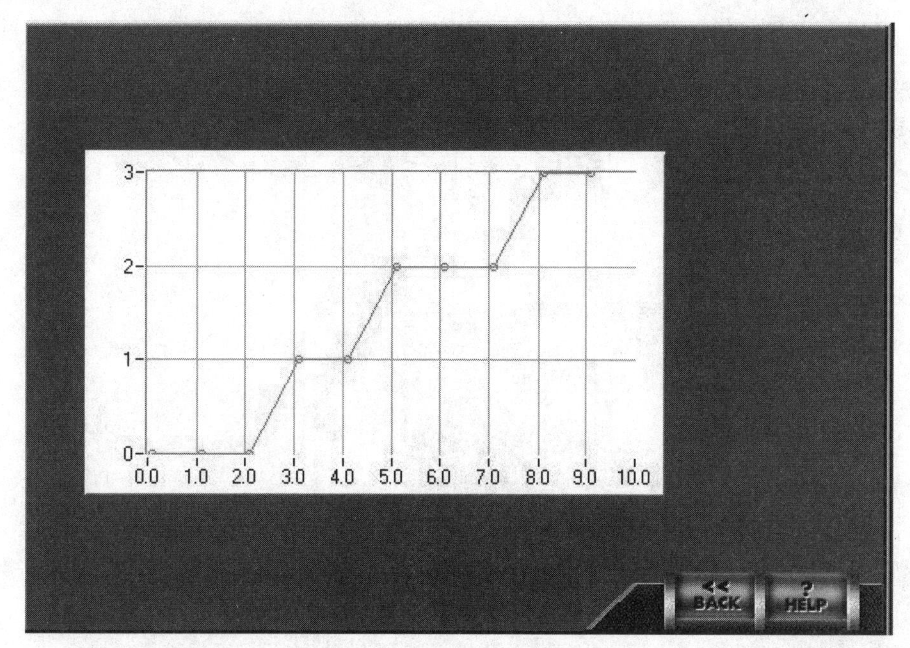

Figure 1-15
Direct mode data

The number of data points can be changed with one click to the block diagram and by selecting a new timing value. A dropdown box will appear with several data point collection values; click the one with the question marks (?) inside it (see Figure 1-16). This particular block diagram allows the developer to type in the data points he or she wants to collect. As an experiment, try changing the value to 30. Run lab procedure steps 7 through 11 (Download mode) and steps 1 through 3 (Direct mode) again, comparing this new XY graph with the old data plots.

Data Collect Mode Summary As seen in the last two lab procedures, Robolab allows the data collect function to be implemented easily for any Mindstorms robot or intelligent machine. Download and Direct modes of data collect provides two methods for gathering data and displaying it. The next lab exercise will walk you through programming the RCX P-Brick for a basic machine control function using the LabView feature of the software.

Basic Machine Control Function The Investigator IDE is a powerful tool that enables small subsystems to collect data that can be explored. The first three levels of preprogrammed software enable experimentation in the areas of collecting data with sensors as well as motor control. The next two

Figure 1-16 Dropdown box with data collection Options

levels of preprogrammed code provide a gateway into developing complex sensory and control applications for Mindstorms mechatronics-based robots and intelligent machines using VI block diagrams.

This brings us to the discussion of how to write basic machine control functions using VI programming techniques. As discussed in Chapter 4, page 59, of *LEGO Mindstorms Interfacing*, state machines provide a graphical method of describing how hardware functions work with the embedded software layer of a mechatronics device. By using state diagrams, machine control functions can be designed effectively, using the least amount of time.

The first step in developing a machine control function for a LEGO Mindstorms mechatronics creation is to draw a state machine on paper. For example, let's say the target robotics application requires an output actuation of 2 seconds, activated after the input-sensing device is requested. Using the two main elements of a circle or an ellipse as the *state* and an arrow as the *transition* throughout the graphics diagram, Figure 1-17 shows the state machine for the mechatronic control function. The state machine will then be converted into a VI block diagram using Robolab's Data Logging and Motors control function palette shown in Figure 1-18.

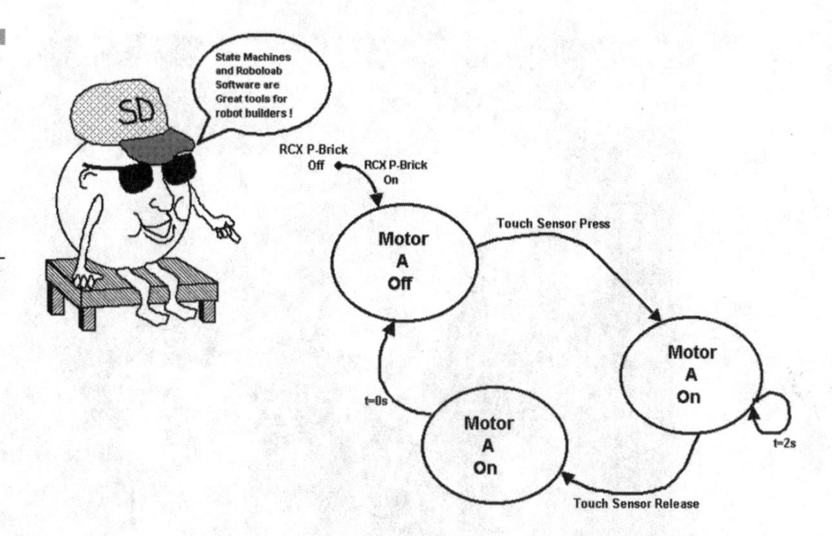

Figure 1-17 Sam State and the basic motor control state machine (mechatronic control function)

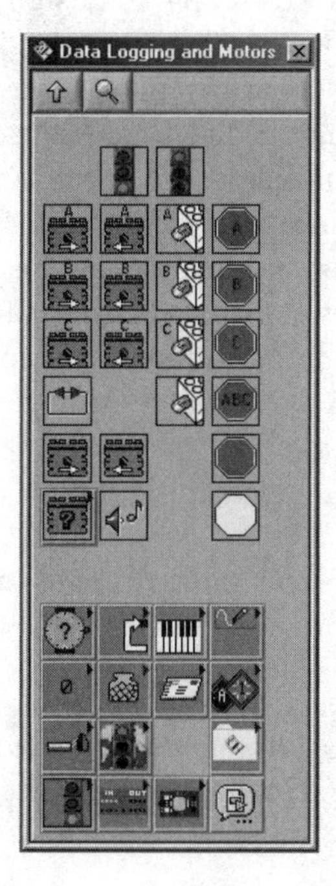

Figure 1-18 Robolab's Data Logging and Motors fontrol function palette

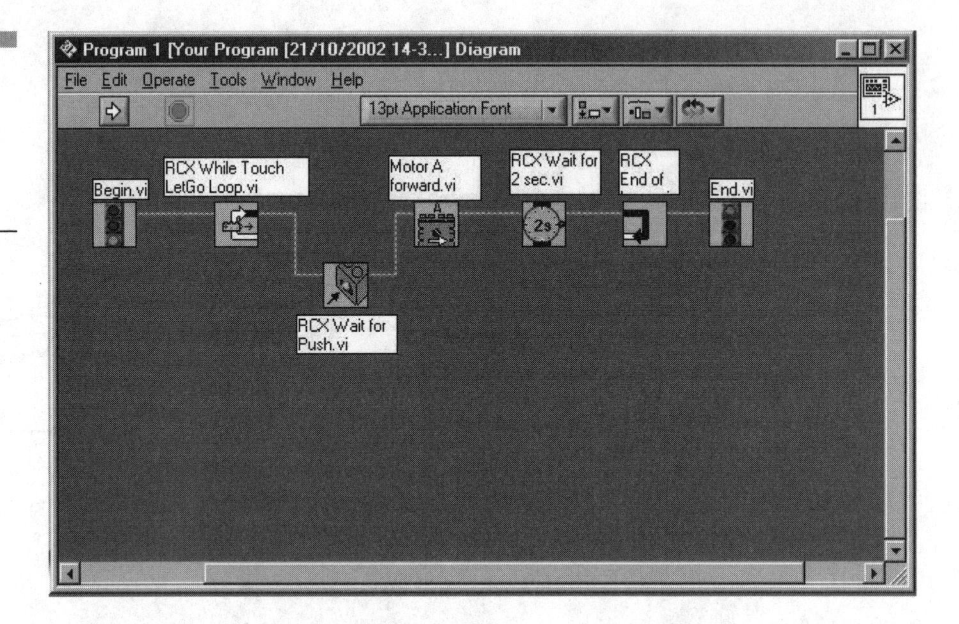

Figure 1-19 The completed Robolab VI block diagram for the basic motor control function

The purpose of the conversion process of paper design to Robolab code is to identify the virtual I/O of the state machine with physical electrical components. Therefore, the *Motor A* on/off states in the state machine are handled by the small LEGO electric motor. The touch sensor is linked to the *press / release* transitional events, as shown in the state machine diagram. The time delays, denoted by $t = 2$s and $t = 0$, have a Robolab software counterpart of a *2s* timer. A stop sign icon in Robolab has the same definition as the virtual state *Motor A Off* shown on the diagram. To ensure that the state-transition process of a *timed delay-on* motor control continues with a touch sensor press/release event, Robolab allows the usage of a *While* loop to be placed on the VI block diagram.

By using these VI icons and wiring them together, a functional block diagram is created and ready to be downloaded to the RCX P-Brick. Figure 1-19 shows the completed Robolab VI block diagram for the basic motor control function.

TECH NOTE: *Program Levels 4 and 5 allow you to explore the LabView programming environment. Highlights of the these two levels include the following:*

- *Advanced programming levels for the* Invent *feature of Robolab are available, using LabView's data logging and motors control functions.*

- *The levels use the LabView VI programming language exclusively.*
- *Systems engineering methods for* partitioning *(separating) and* decomposition *(breaking into small levels) can be easily implemented on mechatronic-based devices.*

The final Robolab Exploratorium project will allow you to experiment with this basic motor control application.

Robolab Exploratorium Project 2

The following lab project will explore how to program Robolab to perform a simple machine control function.

Basic Machine Control Function A simple machine control function of a timed delay-on will be explored using the Robolab software *Invent-Investigate* programming feature. A time delay on function is where the electric load connected to an electronic controller will remain on for a predetermine time after the input electrical signal is removed.

The BOM is as follows:

One RCX P-Brick

One touch sensor

One infrared tower

Two LEGO electric wires

One small LEGO electric motor

Robolab software

Software Assembly and Test Procedure To conduct the procedure, follow these steps:

1. With the touch sensor attached to input 1 of the RCX P-Brick using one LEGO electric wire, attach the small LEGO electric motor to output A using the second LEGO electric wire. See Figure 1-20.
2. Build the Robolab basic motor control function by placing the While Touch loop, Wait for Touch, Motor A forward, Wait for 2secs, Stop A, and End Of VI icons onto the VI block diagram window. These icons are found in the Data Logging and Motors control function palette (see Figure 1-21).
3. Turn on the RCX P-Brick. Download the basic motor control function (built in the previous step) into program slot 3 using the infrared tower.

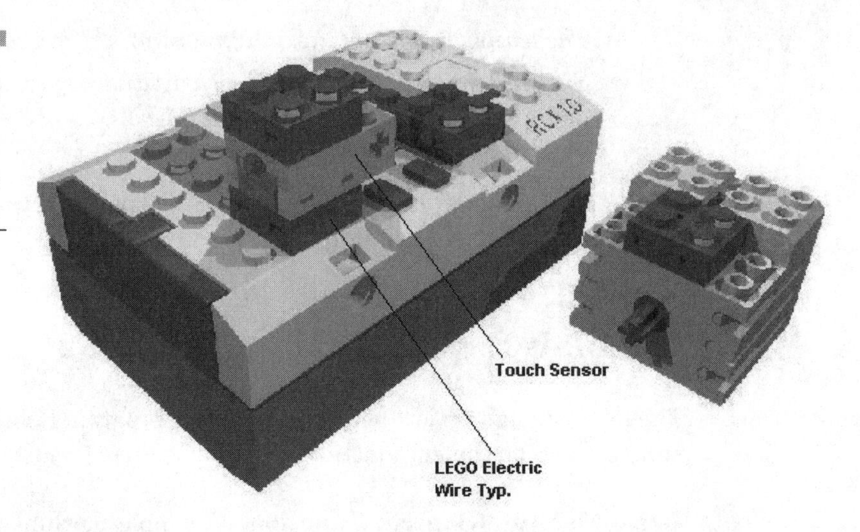

Figure 1-20 Attaching a touch sensor, an electric motor, and an electric wire to an RCX P-Brick

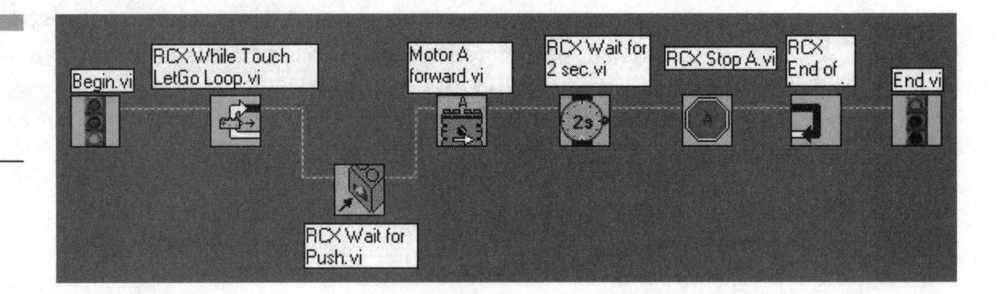

Figure 1-21 Robolab basic motor control function VI

4. Press Run on the RCX P-Brick. Then press and release the touch sensor while observing the motor. Did the motor stay on for 2 seconds and then stop? Yes___ No___.
5. Change the timed delay-on to 10 seconds and repeat step 4. Using a stopwatch, measure the actual time the motor stays on. Did the motor stay on for 10 seconds and then stop? Yes___ No___.
6. Stop the program and turn off the RCX P-Brick.

TECH NOTE: *To show labels on the VI block diagram, right-click each motor control icon and move the cursor to Visible Item on the pulldown menu and select Label.*

Basic Machine Control Function Lab Summary The basic machine control function lab project demonstrated how Robolab software can be used to operate a small LEGO electric motor with a touch sensor. This lab also demonstrates how a VI block diagram can be built and downloaded to the RCX P-Brick for basic I/O control of a Mindstorms mechatronics device. Additional Robolab software lab projects will be discussed in Chapter 2, Electronics Hardware Add-Ons.

Java (RCXTools ver 1.4)

RCXTools is basically a Java-based Windows IDE for downloading programs onto the RCX P-Brick as well as for controlling the outputs directly and reading input data. The tools were developed by Tim Rinkens and can be found at `http://rcxtools.sourceforge.net`. RCXtools have two IDEs: Direct and Download. They are batch files (with the *.bat* extension). This allows for bulk processing of the supporting Java source files, applets, and *Application Programming Interfacing* (APIs) required for RCX P-Brick I/O control applications.

In addition, a text box for testing message appearance is available on the RCX P-Brick's LCD within the Direct mode IDE. The virtual LCD on the IDE will display how the text message will appear once the information is entered. Direct mode IDE enables one to plan robot movements, audible alerts, and input sensor values to use in developing code for a mechatronics application.

Download Mode IDE

The Download mode IDE enables a amateur roboticist to compile code and send it to the RCX P-Brick. The leJOS firmware required for writing Java-based mechatronics applications can also be downloaded to the P-Brick using this convenient IDE tool. If you are like me, DOS prompts and changing directories to obtain files and execute applications via command lines are too much to remember as well as type. I would rather spend my time debugging the code instead of the mistyped command lines. Because of its Windows-based visual appearance, this tool provides convenient code testing without worrying about correct pathways and setting directories necessary to compile the source file when using DOS commands.

The following lab projects will provide a quick overview of two RCX tools for mechatronics applications. The activities will give the amateur roboticist a mini-tour into using an IDE tool for developing and testing Java-based robots and machines. The objective of the following lab projects is to illustrate another free software tool that can assist one in Mindstorms mechatronics embedded systems development.

RCXTools Lab Project

This lab project will investigate how to use both Direct mode and Download IDEs for virtual screen control of a small LEGO electric fan. The amateur roboticist should be able to send the leJOS firmware to the RCX P-Brick using the Download IDE. One should be able to control the speed of the LEGO electric fan using a slide control on the Direct mode tool.

The BOM is as follows:

LEGO bricks for constructing an electric fan model

One small LEGO electric motor

One LEGO electric wire

One RCX P-Brick

The leJOS software package

The RCXTools version 1.4 software

Java 2 Platform Standard Edition (J2SE) version 1.3.1_01 software

An oscilloscope (optional)

Infrared tower

Software and LEGO Model Assembly Test Procedure The procedure is as follows:

1. Go to the Sun Microsystems Web site (`http://java.sun.com/j2se`) to download and install J2SE version 1.3.1_01 software onto your notebook or PC's hard drive. Figure 1-22 shows the home page for J2SE software.

2. Go to Jose Soloranzo's web site (`http://lejos.sourceforge.net/download.html`) to download and install leJOS 2.0 software onto your notebook or PC's hard drive. Figure 1-23 shows the download page for leJOS software.

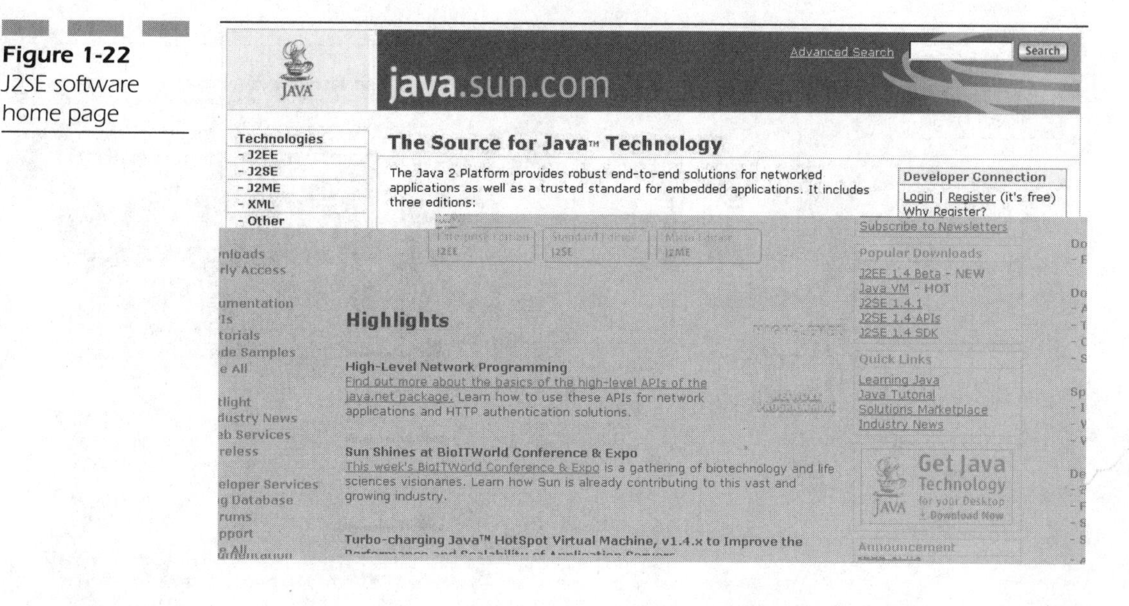

Figure 1-22 J2SE software home page

Figure 1-23 leJOS Software download page

Download Area

Linux/OSX/Solaris leJOS

Download version 2.0.0
Download API docs
View Unix Install Notes

Win32 leJOS

Download version 2.0.0
Download API docs
View Win32 Installation Notes

Download Area at Sourceforge

Other releases, notes, dates and sizes are available at the SourceForge download area.

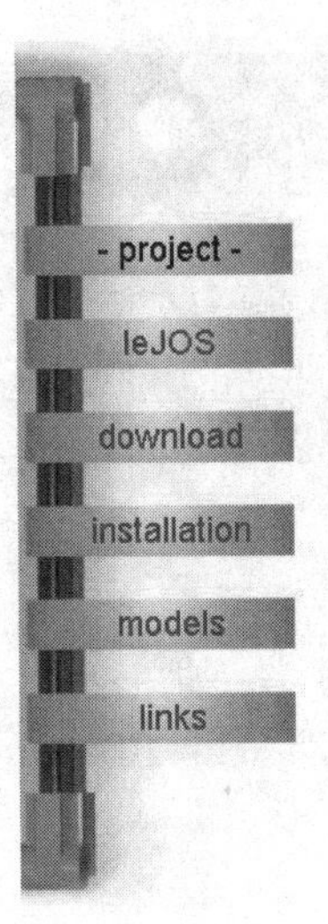

RCXDownload and RCXDirectMode

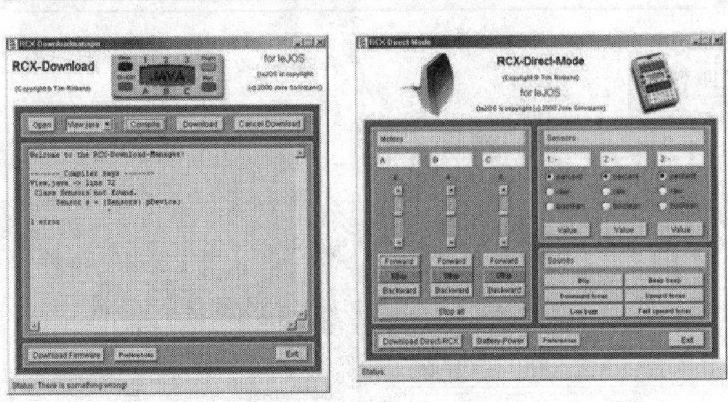

RCXDownload and RCXDirectMode both have software which is applecable with Java Runtime leJOS for the Lego Mindstorms RCX. Because of its ability not to need line commands on the one hand and the transparency of both programs on the other hand, it makes it easy to start in object orientated programming.
RCXDownload automatically sets the JDK-, leJOS- and ClassPaths, compiles the chosen Java-Source, shows the compiler messages and is able to link and load both, the compiled classes and the leJOS-firmware.
RCXDirectMode directly controls the RCX-Brick. Its transparent GUI covers the complicated communication with the RCX. It needs just one mouse-click to start the motors, to control their speed, to check out the sensors, to play an assortment of sounds and to confirm the level of remaining battery power.

Figure 1-24 RCXTools software download page

3. Go to Tim Rinkens' Web site (`http://rcxtools.sourceforge.net/`) to download and install the RCXTools IDE software onto your notebook or PC's hard drive. Figure 1-24 shows the download page for RCXTools software.

With the main software components residing on your notebook or PC's hard drive, the leJOS firmware is ready to be installed to the RCX P-Brick.

The leJOS Software Install Procedure To complete the procedure, follow these steps:

1. Go to the *RCXTools_1_4* directory created in step 3. Open the folder and double-click the *RCXDownload.bat* file. After a few seconds, the IDE tool should be on your notebook or PC's video screen.
2. Click Preferences and check the pathway and COM port for your system. Make the appropriate changes and click Accept.
3. Turn on the RCX P-Brick and have the infrared tower in front of it.
4. To send the leJOS firmware to the RCX P-Brick, click the Download Firmware button. A dialog box will appear on the IDE tool asking if you

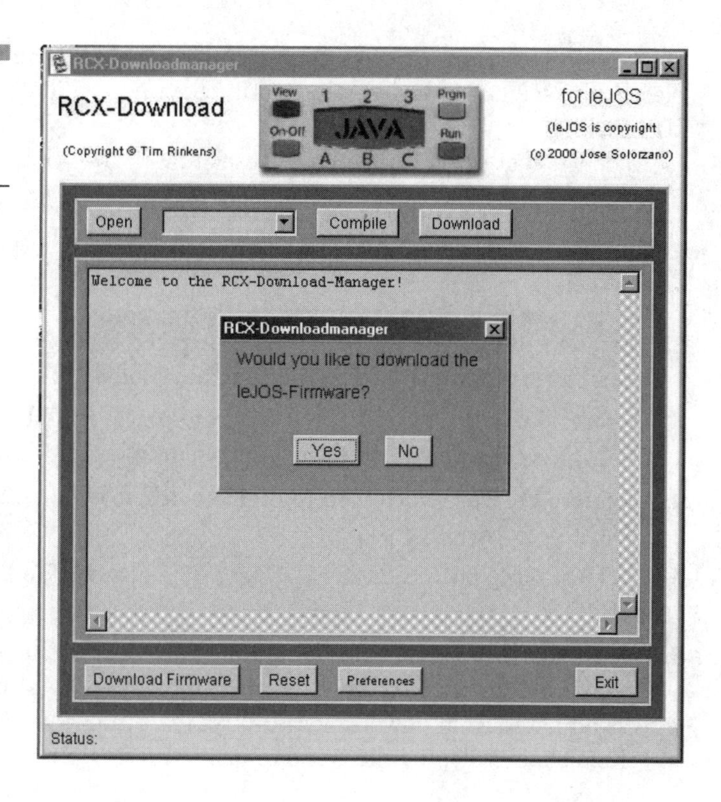

Figure 1-25 Firmware Download Confirmation

would like to download the firmware. Click Yes. Figure 1-25 shows the dialog box for firmware download confirmation.

5. After downloading the firmware to the RCX P-Brick, close the Download RCXTool.

The following steps are for the construction and control of the LEGO electric fan model using the Direct Mode RCXTools IDE.

LEGO Electric Fan Build and Test Procedure A LEGO electric fan can be used to familiarize you with the Direct Mode IDE tool control functions. You might ask the question, why an electric fan? I have been doing volunteer work at my daughter's elementary school discussing simple machines. The project I'm having the fourth-grade students work on is to build a small LEGO fan using gears. This program will allow the students to understand how the mechanics of the simple machine works, through hands-on construction and observation. I have added an electric motor and a RCX P-Brick to create a *smart fan*, demonstrating to the fourth-grade class how electronics can assist in simple machine design. Therefore, the electric fan seemed like a suitable testing device to validate the Direct Mode IDE tool.

To do the test procedure, follow these steps:

1. Build a small LEGO electric fan like the one shown in Figure 1-26 or spin your own design. Attach the electric motor to output A of the RCX P-Brick.
2. Turn on the RCX P-Brick. Have the infrared tower in front of the P-Brick for direct control of the electric fan model.
3. Download the Direct RCX Java program to the RCX P-Brick by clicking the button. The download time will take a few seconds to complete.
4. Click the Forward button on the IDE tool. The electric fan blades should start turning.
5. Click the Stop button on the IDE tool. The electric fan blades should stop.
6. Adjust the speed of the fan using the slide control A on the IDE tool. The fan blades should turn slow at 0 and fast at 7. As the slide control A is adjusted from power-levels 0 through 7, the RCX P-Brick's LCD screen should display the same information as well.

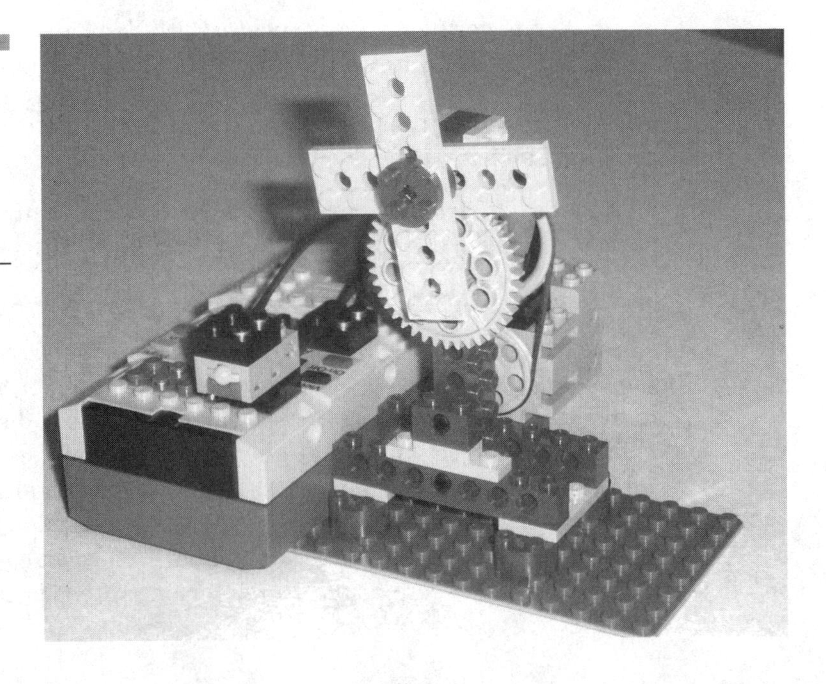

Figure 1-26 The small LEGO electric fan used in the Direct Mode Java RCXTools lab project

Pulse Width Modulation Lab This purpose of this lab is to investigate the output signal of the RCX P-Brick using an oscilloscope. The output speed control is based upon a *Pulse Width Modulation* (PWM) signal driving the LEGO electric motor. By using a pulse width modulated signal with a fixed frequency, the on/off time of the periodic waveform can be varied, thereby changing the amount of current that flows through the electric motor's windings. The following lab will demonstrate this output signal control of the LEGO electric motor, with the aid of the Direct mode IDE. By attaching an oscilloscope across the motor, a PWM signal can be observed.

PWM Signal Observation Test Procedure The test procedure is as follows:

1. Modify a LEGO electric wire in order to attach the RCX P-Brick's output to an oscilloscope. Figure 1-27 shows the author's modified LEGO electric wire used to observe the PWM signal produced by the P-Brick's output A.
2. Set the oscilloscope up for a *volts per division* (volts/div) of about 2V and the time base of 2 *milliseconds per division* (ms/div). Set the output for DC measurement.
3. Click the Forward button of the Direct Mode IDE tool. The blades on the electric fan should be rotating.
4. Adjust the slide control for output A on the Direct mode IDE tool to a power level of 0. The oscilloscope should be displaying a series of pulses. Figure 1-28 shows the pulse train signal, captured by the author using a Fluke Model 97 Scopemeter.

Figure 1-27 The author's modified LEGO electric wire used to observe the PWM signal produced by the P-Brick's output A

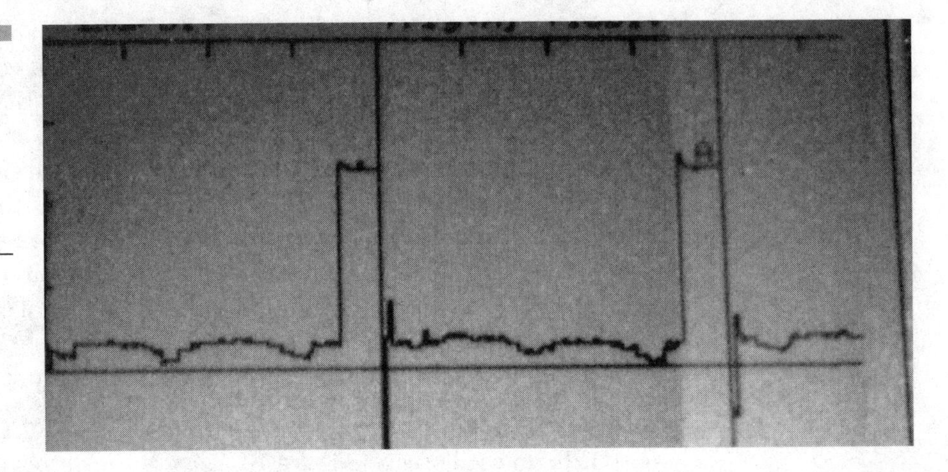

Figure 1-28 The pulse waveform signal, captured by the author using a Fluke Model 97 Scopemeter

5. Change the slide control for output A on the Direct mode IDE tool and notice the waveform varying in width. At a power level of 7, the output signal should be a straight line with a minimum amount of ripple.
6. Click the Stop button on the IDE tool and turn off the RCX P-Brick.

Java RCXTools Version 1.4 Summary

Direct mode and Download IDE tools provide a convenient way for the amateur roboticist to load and control the RCX P-Brick using the Java programming language. According to Tim Rinkens, Java RCXTools developer, "RCX Download and RCX Direct mode both have software which is applicable with the Java Runtime leJOS for the Lego Mindstorms RCX. Because of its ability not to need line commands on the one hand and the transparency of both programs on the other hand, it makes it easy to start in object-oriented programming."[5] Therefore, the previous lab projects demonstrate the ease in which it is possible to download and control a Mindstorms mechatronic device using these Java-based tools.

[5]From the RCXDownload and RCXDirectMode web site by Tim Rinkens, `http://rcx-tools.sourceforge.net/e_home.html`.

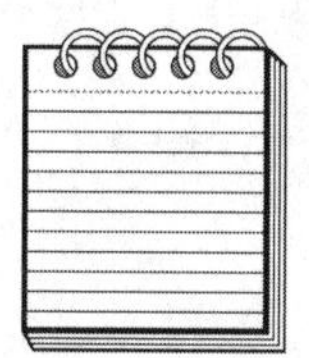

TECH NOTE: The Direct mode IDE can assist in the design and development of robotic or intelligent machine movements by turning the appropriate outputs on or off in a sequence using the controls of the tool. By switching the RCX outputs in a preplanned manner, the actual autonomous motion of the Mindstorms mechatronic device can be designed.

NQC (BricxCC)

In my first book, *LEGO Mindstorms Interfacing,* I discussed writing robotics application code using the NQC programming language. The example code used in the discussion was to develop an *AND* logic using NQC.

I also discussed how to write code from scratch, using a programming-style block diagram shown on page 205. BricxCC IDE comes packaged with some fun, practical, and powerful tools for controlling Mindstorms-based robots. This section will explain an embedded-tools application within the BricxCC IDE that can be used by the amateur roboticist for direct control of a Mindstorms-based robot or intelligent machine.

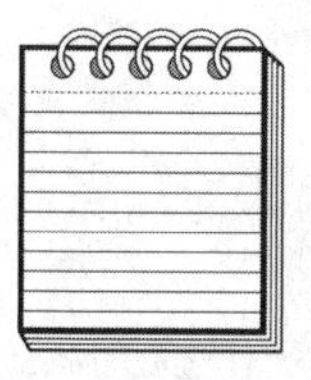

TECH NOTE: *John Hansen has a new version of BricxCC known as 3.3.6.5. This new version of the software supports programming the RCX (all versions), Scout, Cybermaster, and Spybot P-Bricks, all in one convenient package. It also supports programming the Scout, RCX2, and Spybot, using The LEGO Company's MindScript™ and* LEGO Assembly *(LASM™) languages via the RCX 2.0 SDK. Visit the Web site* `http://members.aol.com/johnbinder/bricxcc.htm` *to download this latest version of IDE software.*

Direct Controller Tool

The Direct Controller is a self-contained testing tool for operating the RCX P-Brick. It provides the amateur roboticist with the ability to experiment with autonomous motions for Mindstorms robots and intelligent machines. The controller has direct control of all three outputs, as well as monitoring the inputs in either Boolean, Raw, or Percentage-sense modes. NQC Global variables can be examined and tasks can be initiated through the use of the controller.

The following lab project will explore the direct control of the RCX P-Brick's outputs. Think of this lab as a basic introduction to *telerobotics*. Telerobotics uses a controller, camera, and human operator in order to directly control a robot or intelligent machine.

The BOM is as follows:

BricxCC version 3.3.6.5 software

Two LEGO electric wires

RCX P-Brick

Constructopedia manual version 1.5

Assorted bricks from Mindstorms RIS kit

Infrared tower

Oscilloscope (optional)

One 10 KΩ resistor, 1/4 watt (optional)

Radio Shack Electronics Learning Lab breadboard or equivalent

Software and LEGO Model Assembly Test Procedure The following lab procedure will allow you to control a multiaxis robot of your own design:

1. After installing the software onto your desktop or PC's hard drive, open the BricxCC IDE.
2. Click Tools, located at the main menu bar.
3. Click Direct Control from the pulldown menu. The Direct Controller GUI will appear on the screen.
4. Build a two-axis robot such as InventorBot in the *Constructopedia Manual* (1.5 version), pages 57-80. Or spin your own robot design. Make sure to use outputs A and C for BricxCC IDE control.
5. Place the infrared tower in front of the robot, with RCX P-Brick turned on.
6. Under Motors Control, click the A-forward button with power level set at 5. See Figure 1-29. The mechanical assembly driven by the LEGO electric motor from output A should be active.
7. Click the red button to stop the motor.
8. Click the A-forward button again, but instead of stopping it immediately, select the yellow button to float the output for a gradual slowing of the LEGO electric motor.
9. Repeat steps 6 through 9 for output C.

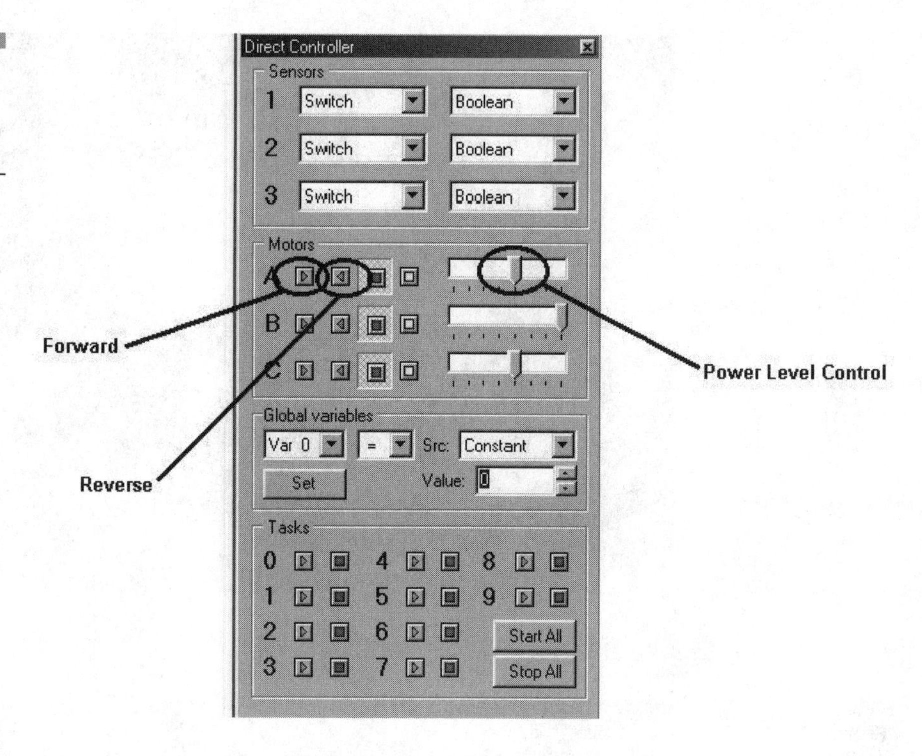

Figure 1-29 A-forward button with power level set at 5

If an oscilloscope is available, the following test procedure can be used to view the PWM signal discussed in Java RCXTools section:

1. Remove the small electric motor from Output A. Using the Radio Shack Electronics Learning Lab breadboard or equivalent, place a 10 KΩ resistor onto the electrical wiring prototyping device.
2. Connect the modified LEGO electric wire (as shown in Figure 1-30) to output A, the oscilloscope, and the 10 KΩ resistor.
3. Set the oscilloscope with the following measuring scales: 2 V/div for voltage, and 5 ms/div for the timebase measurement scale.
4. Place the infrared tower in front of the RCX P-Brick. Turn on the P-Brick.
5. Click the A-forward button on the Direct Controller IDE and adjust the power level using the slide control for 0. The LEGO motor that is wired to output A should activate, displaying a series of pulses as shown in Figure 1-31.

Figure 1-30 Connecting the modified LEGO electric wire to output A, the oscilloscope, and the 10 KΩ resistor

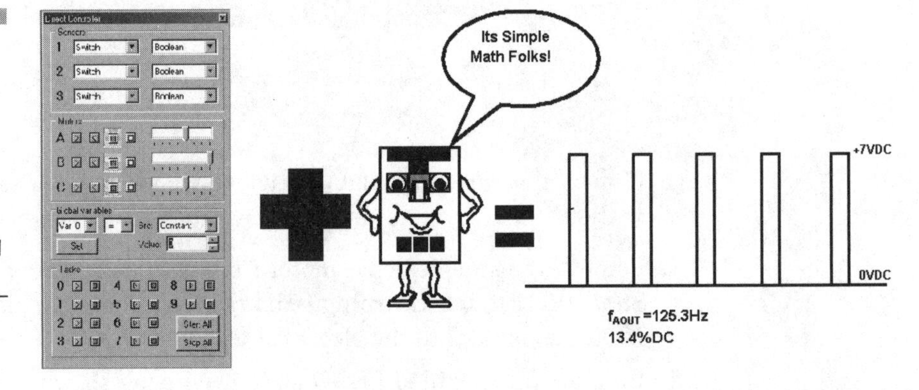

Figure 1-31 The Direct Controller tool plus Rudy RCX's Output A can produce a pulse width modulated signal with a fixed frequency.

6. Play with the PWM output signal of the RCX P-Brick by adjusting the slide control. Notice with a power level of 7 no pulses occur, but a steady DC voltage signal is present. The measured voltage is about +7V DC.
7. Click the A-reverse button on the Direct Controller IDE and adjust the power level using the slide control for 0. The output signal will be inverted on the oscilloscope, showing a negative voltage.

8. Repeat step 7, noticing the negative DC voltage present with power level 7 at output A. The measured voltage is about -7V DC.
9. Turn off the RCX P-Brick.

Direct Controller Tool Summary The following lab project provided another IDE tool for controlling a Mindstorms-based robot, using the free BricxCC software. This tool is analogous to the Java RCXtool demonstrated earlier in this chapter. The BricxCC Direct Controller IDE provides another software tool for preplanning autonomous robot motions.

Another novel application of the BricxCC Direct Controller is to act as a virtual control for an electronic audible oscillator circuit.

A Multimedia Control for an Electronic Oscillator Circuit

The RCX P-Brick controls the speed of a small LEGO electric motor by PWM. The previous lab project, as well as the Java RCXTool experiment, demonstrated the embedded speed-control feature. Besides controlling the speed of a small electric motor, the PWM output signal generated by the RCX P-Brick can be used with a simple electronic amplifier to produce a multimedia-controlled audible oscillator circuit.

Multimedia is a multidisciplinary technology that combines visual, audio medias for the purpose of disseminating information. The BricxCC IDE is a software tool used with a computer that has text and images for directly controlling robots or developing embedded robotics code applications. The power-level slide control provides a software interface to the physical RCX P-Brick's outputs. By adjusting the power level, the output signal's pulse width varies. The observance of the varying pulse width is associated with its *duty cycle*, also referred to as DC.

The Duty Cycle Equation The duty cycle of a squarewave or pulse train is the mathematical relationship between a squarewave, pulse train *Time On* (t_{ON}), or pulse width (p_w) to its total time (t_{Total}). The mathematical equation for duty cycle is represented as

$$DC = \left[\frac{t_{ON}}{(t_{ON} + t_{OFF})}\right] \times 100$$

where t_{ON} equals p_W
and $(t_{ON} + t_{OFF})$ equals total time.

Duty cycle is expressed as a percentage because of the ratio related to the whole output voltage signal produced by the frequency generator source. For example, the waveform displayed in Figure 1-31 at power level 0 has a duty cycle of 13.4 percent. Therefore, the speed of the motor will be 13.4 percent of the total 7V DC signal applied to its windings. A duty cycle of 100 percent relates to the full 7V DC output signal from the RCX P-Brick being applied to the windings of the LEGO small electric motor.

The measured frequency of 125.3 Hz never changes, because this electrical parameter has no effect on controlling the speed of the motor. The low frequency generated by the RCX P-Brick's internal circuit is used because of the nonelectromagnetic interference with other consumer electronics around the home.

Build a Multimedia-Controlled Electronic Oscillator The following lab project will demonstrate the mechanics of DC and virtual controls using the RCX P-Brick and an electronic amplifier circuit. The end product will be a multimedia-controlled electronic oscillator.

The BOM is as follows:

- One 2N3904 NPN transistor
- One audio transformer (500 Ω centered tapped primary, 8 Ω secondary winding)
- One 470 microfarad 10V DC electrolytic capacitor
- One 10 KΩ, 1/4 watt resistor
- One 8 Ω speaker
- A variable-output DC power supply
- Radio Shack Electronics Learning Lab breadboard or equivalent
- Infrared Tower
- Oscilloscope

Software and LEGO Model Assembly Test Procedure The following lab procedure will allow the amateur roboticist to build and investigate a multimedia-controlled electronic oscillator:

1. Breadboard the electronic amplifier circuit shown in Figure 1-32 using the Radio Shack Electronics Learning Lab Breadboard or equivalent.
2. Adjust the output voltage of the variable DC power supply for +6V DC.

3. Attach the +6V DC and ground connections of the DC power supply to the corresponding *battery* and *GND* connections of the electronic amplifier in Figure 1-32.
4. Connect the modified LEGO electric wire used in the direct control lab project to the electronic amplifier circuit built in step 1.
5. Connect the other end of the modified LEGO electric wire to output A of the RCX P-Brick.
6. Place the infrared tower in front of the RCX P-Brick.
7. Turn on the RCX P-Brick.
8. With the direct control tool opened earlier, adjust the power-level slide control of output A for a scale reading of 0.
9. Click the *A-forward* button on the IDE window. A 125.3 Hz tone should be emitted from the 8 Ω speaker of the electronic amplifier circuit.
10. Adjust the power-level slide control for a DC of 50 percent (midscale) and listen for the 125.3 Hz sound emitted from the 8 Ω speaker. The tone should be louder than what was heard in step 9.

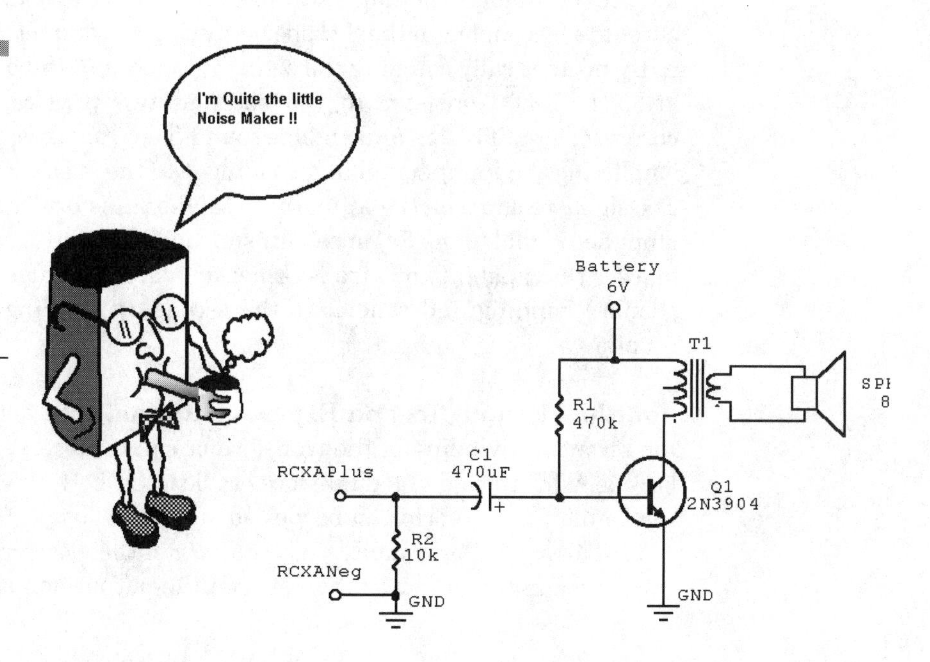

Figure 1-32 The electronic amplifier circuit used in the multimedia-controlled electronic oscillator lab project, with Tom Transistor

11. Adjust the power-level slide control for a scale reading of 7 (DC of 100 percent). An initial loud click should be heard once upon initiating this power level.
12. To visually see the RCX P-Brick-produced PWM signals driving the electronic amplifier, attach an oscilloscope across the 10 KΩ resistor and repeat steps 8 through 11. The oscilloscope will allow visual inspection of the power level and the duty cycle associated with each adjustment made with the slide control.
13. Turn off the RCX P-Brick.

Multimedia-Controlled Electronic Oscillator Summary The lab project illustrated the concept of using virtual controls via the direct controller of the BricxCC IDE to assist in creating a multimedia-controlled electronic oscillator. This Mindstorm RIS application is not the norm of traditional robotics projects, because no mechanical assemblies are attached to the RCX P-Brick.

The objective of the experiment is to show how a homebrew audible tone generator can be built using relatively inexpensive electronic parts from an amateur roboticist's junk box. The simple electronics amplifier used in the lab project enables the pulse train signal produced by the RCX P-Brick's output to be coupled to the 8 Ω speaker using the audio transformer.

By magnetically coupling the signal injected into the base circuit of the 2N3904 NPN transistor to the speaker, two physical effects can be observed: amplification and audible sound. The transistor is responsible for amplifying the input signal and sending it to the speaker for audio sound. The signal residing on the primary side of the transformer sends this stored amplified signal to the 8 Ω speaker using magnetic coupling. *Mutual inductance* is the capability of a transformer to take the voltage or current on its primary winding and send it to the secondary winding using magnetic coupling.

Additional Thoughts and Experimentation The amplified signal on the secondary winding of the transformer can be observed using an oscilloscope. With the aid of a dual-trace oscilloscope, both the input signal and the amplified waveform can be viewed simultaneously.

This activity allows you to view the *gain* of the electronic amplifier. The gain is the mathematical ratio between the output voltage and the input voltage.

The electronic amplifier circuit used in the multimedia lab project's amplification gain can be experimented with by changing the 2N3904 transistor with a higher *beta,* as well as by replacing the 470 KΩ resistor with a different resistance value. The beta of a transistor is the ratio between induced changes in collector current compared to the applied change of base current.

This electronic amplifier circuit can also be simulated using Electronics Workbench or CircuitMaker software to obtain predata prior to breadboarding. The actual circuit data can be compared to the simulation, thereby allowing the software model to be modified for a true representation of its physical counterpart.

The next chapter will continue the discussion of using electronics hardware add-ons, with aid of the Basic Stamp and the Palm Computer to enhance the development of LEGO Mindstorms mechatronic devices.

CHAPTER 2

Electronic Hardware Add-Ons

To enhance a mechatronics-based device or machine using either the *Robot Command Explorer* (RCX) or Scout *Programmable Brick* (P-Brick) requires a basic understanding of electronics *input/output* (I/O) interfacing techniques. Chapters 3 and 4 of *LEGO Mindstorms Interfacing* discussed this topic using several lab projects. The reason for understanding interfacing techniques is to allow the amateur roboticist to modify the special functions of his or her Mindstorms-based robot using electronic circuits. Electronic circuits enhance the P-Brick's hardware architecture by connecting hardwired application-specific subcomponents to its embedded software layer. The P-Brick's embedded computer program will use the electronic circuit's I/O for data management and control of the target robot built by the amateur roboticist. The Hately/Pirhbai Method is a structural approach that can assist the amateur roboticist in capturing requirements for the development of the electrical/electronics system of the Mindstorms robot. This system's engineering technique can be used to define the function of operation of these hardware *add-ons*, and the interaction they have with the target P-Brick. The main objective of the electronic circuits or hardwired add-ons connected to the RCX or Scout P-Brick-based robot or intelligent machine is to improve the specific function or behavior of the mechatronics system. A library of circuit blocks will be used for the purpose of Mindstorms mechatronic enhancement development.

The objective of this chapter is to investigate how electronic hardware add-ons can assist the amateur roboticist in developing circuit blocks and controls. The goal is to create unique mechatronics devices and machines using the Mindstorms P-Bricks as the *Central Processing Unit* (CPU) controller. The Basic Stamp, 68HC11 *Engineering Evaluation Board* (EVB), and Palm computer will be examined. To explore the use of these development tools in building hardware add-ons, the following topics will be discussed:

What is a circuit block?

Robotics control prototyping

Engineering evaluation boards

Basic stamp interfacing

Palm diagnostics controller.

What Is a Circuit Block?

Electronics can enhance a mechatronics-based device or intelligent machine, without using software. The RCX or Scout's P-Brick internal *hardware architecture* is comprised of a microcontroller, power supply, solid-state motor driver, an LCD driver, and an electrical input circuit. Figure 2-1 shows a typical block diagram of the RCX or Scout's hardware architecture.

The hardware architecture is the structural organization and the function of an electronic product. The P-Brick's internal devices mentioned earlier provide the foundation and operation of the Mindstorms controller. The internal elements that provide electrical functions for the Mindstorms P-Brick are known as *circuit blocks*.

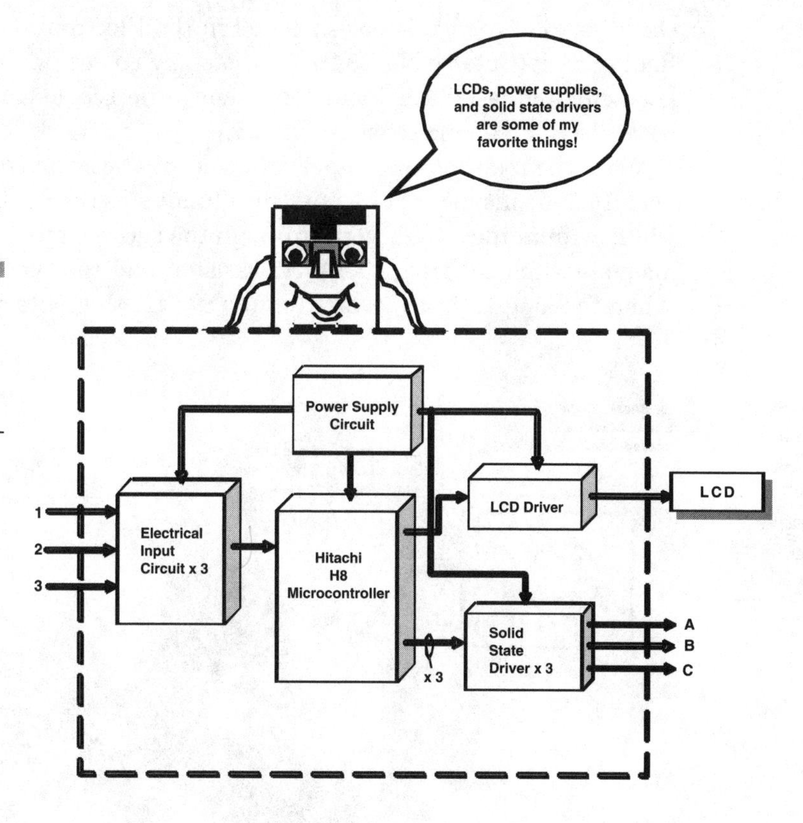

Figure 2-1 A typical block diagram of a P-Brick's hardware architecture.

The circuit block is an essential component because it provides a hardwired *application-specific* function to an electronic product. An application-specific function is an object committed to doing a dedicated operation. The object can be a *small applet* (small software application) or an electrical/electronic circuit. The circuit block is not only connected internally to the microcontroller, but can be externally wired to the mechatronics-programmable controller as well. Figure 2-2 shows the concept of externally hardwiring a circuit block to a P-Brick. To understand the role of a circuit block in relation to a mechatronics programmable controller, let's look at a Mindstorms *Proof of Concept* (POC) prototype example.

Creating a Smart Switch Concept

On page 26 of the workbook included in the Electronic Sensors Lab from Radio Shack (Catalog No. 28-278), written by Forrest Mims, a simple magnet-switch is described. Figure 2-3 shows the schematic diagram of the basic electromagnetic-sensing switch.

When the reed-sensor contacts of the magnetic switch detect a magnetic field from the magnet, they close. This illuminates the red LED. The current will flow from the +9V battery through the series circuit, consisting of the magnet-switch contacts, the 470Ω resistor, and the red LED, to ground. When the magnet is removed from the switch, the contacts will open, and

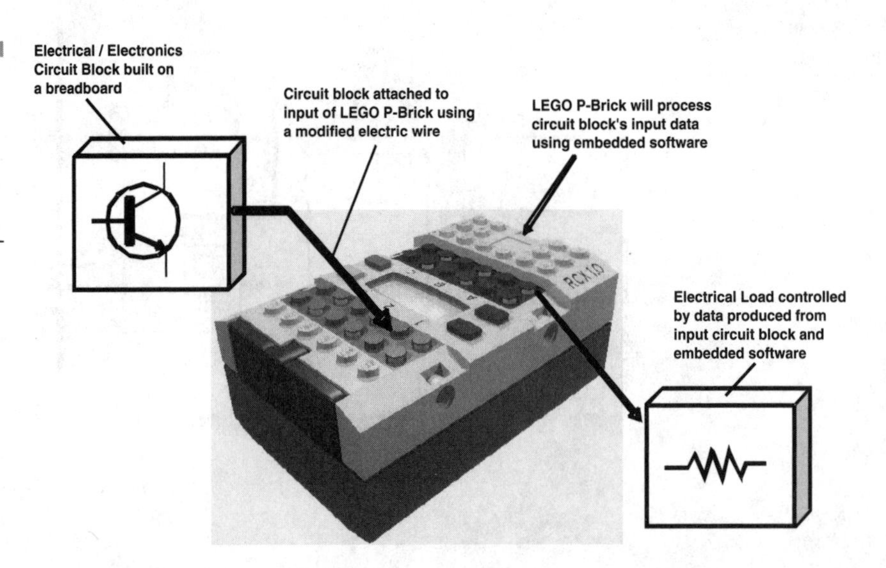

Figure 2-2 Concept of externally hardwiring a circuit block to a P-Brick.

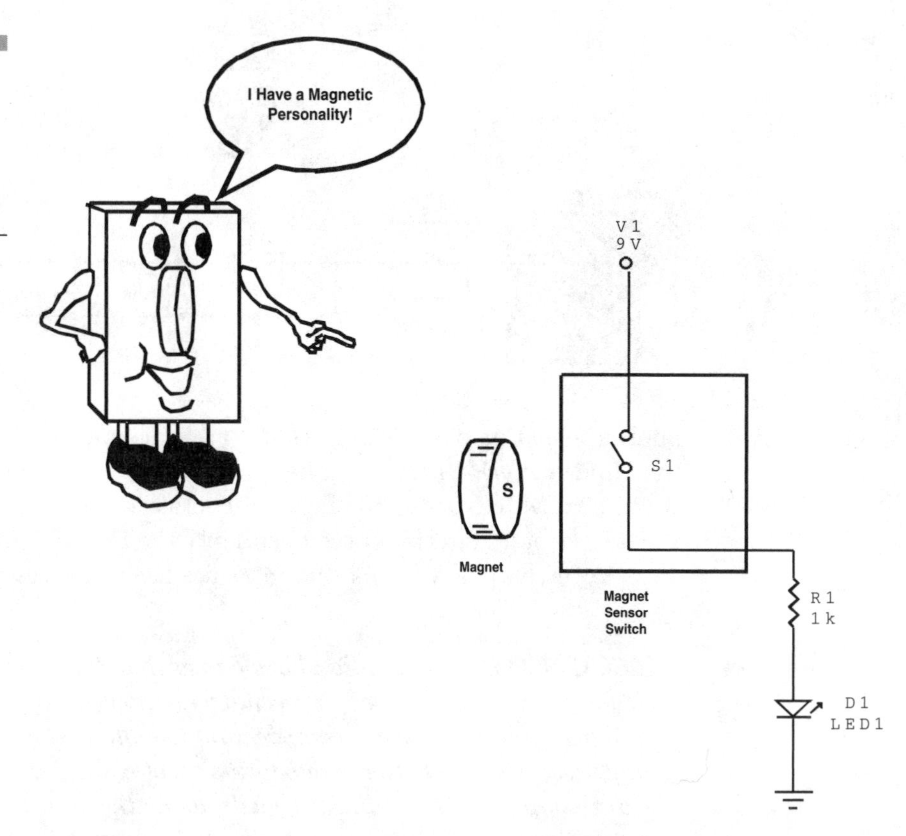

Figure 2-3 Schematic diagram of the base electro-magnetic-sensing switch.

electric current will stop flowing through the series control circuit. This basic DC control function can be hardwired to the RCX or Scout P-Brick as an input circuit block, thereby providing a programmable on/off switching output control-delay function for the red LED. To accomplish this programmable feature for the red LED, the magnet-switch must be wired to the input of the RCX or Scout P-Brick. The 470Ω resistor and red LED circuit need to be wired to the output of the P-Brick. Figure 2-4 shows the modified schematic diagram, with the input circuit block connected to the P-Brick.

Time delay software for switching the output of the P-Brick on or off would be coded and downloaded to the P-Brick. Depending on the type of time delay software embedded within the P-Brick, the output switching control of the red LED will be initiated by the magnet-switch detecting a magnetic field. A modified electric wire attached from the output of the LEGO P-Brick to the LED circuit would complete the output circuit block. After successful testing of the software and hardware, the I/O circuit block can then be placed in an electronic file for *book shelving,* or storing, the hardwired devices for future Mindstorms mechatronics-based systems

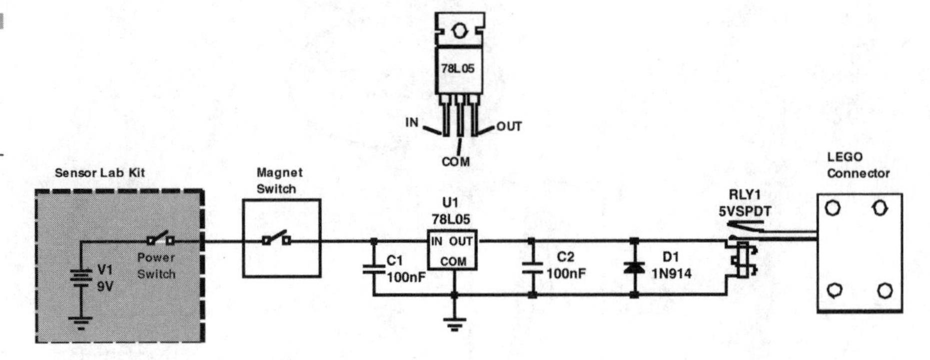

Figure 2-4 Input circuit block for magnet sensor switch

applications. Complex electronics I/O interfacing circuits can be added to the Mindstorms P-Brick using the same hardwired techniques discussed in the Smart Switch concept application. Therefore, the circuit block is an electrical/electronics interface used to enhance the P-Brick's input sensing and output controls operation for mechatronics-based devices and systems.

TECH NOTE: *The Radio Shack Sensor Lab kit (Catalog No. 28-278) is a great prototyping tool for mechatronics development because of the ample supply of electronic components, a small breadboarding area, and several sensor cards. Mindstorms robot sensory-detection designs can be implemented and tested easily, using the multitude of sensing devices (rotation, touch, magnetic, probe, induction coil, magnet, phototransistor, thermistor, and photoresistor) that come with the lab kit.*

Magnet-Switch Lab Project

The Magnet-switch lab project will demonstrate the operation of the magnet-detection control circuit shown in Figure 2-4. By building and testing this basic DC circuit, the amateur roboticist will understand how to incorporate the magnetic detection and control functions of the circuit to a RCX or Scout P-Brick. It is highly recommended that the Radio Shack Electronic Sensor Lab be purchased for ease in constructing the circuit described in the Assembly and Test procedure section of the lab project.

The *Bill of Materials* (BOM) is as follows:

One 5V electromechanical relay

One 7805 linear voltage regulator

Two 100 nF nonpolarized capacitors

One Radio Shack sensor lab (Catalog No. 28-278)

One RCX P-Brick

One LEGO electric motor

One modified LEGO electric wire

One pre-cut 22 AWG solid wire

Constructopedia Manual version 1.5

Objective of Lab Project

The objective of this project is to build a magnet-switch circuit block and use it to control a Pathfinder mobile robot. When a magnet is placed in front of the switch, the robot should move forward for 10 seconds and stop. The two modes of active use from the RCX P-Brick will be the default setting for the P-Brick. Figure 2-5 shows the pictorial setup diagram for the lab project.

Assembly and Test Procedure

The steps for this procedure are as follows:

1. Build the enhanced magnet-switch circuit using the Radio Shack sensor lab, as shown in Figure 2-4.
2. Turn on the power switch of the sensor lab and place a magnet in front of the switch. The internal contacts of the 5V electromechanical relay should close. If not the, turn off the power switch, recheck the wiring, and repeat this step.
3. Build the Pathfinder robot on pages 6-7 in the *Constructopedia Manual* version 1.5, or the equivalent mobile-based machine of your choice.
4. Attach the robot to the magnet-switch circuit using a modified LEGO electric wire. Figure 2-5 shows the setup for this laboratory project.
5. Build the software program (using Robolab, RCX code, *Interactive C Version 4* [IC4] or *Not Quite C* [NQC]) that will move the robot forward upon the sensor switch detecting a magnet. The forward motion of Pathfinder should continue for 10 seconds and then stop.
6. Download the code into the robot and run the program.

7. Press the *Run* button on the P-Brick. Place the magnet in front of the magnet-switch. Did the robot move forward?

   Yes___ No___

8. Did Pathfinder move forward for 10 seconds and stop?

   Yes___ No___

TECH NOTE: *If the robot did not move forward, check the software and repeat this step.*

9. Turn off the power to the sensor lab and Pathfinder.

Magnet-Switch Lab Project Summary

This laboratory project was developed to illustrate the concept of building an electronic circuit and attaching it to the LEGO P-Brick to enhance its input sensory capability. The circuit block allows the amateur roboticist the ability to customize the LEGO RCX for an application-specific robot, using basic or complex electronic circuits for detection and motion control. One final note on this lab project is the choice of programming languages in Step 5 of the Assembly and Test Procedure. The robotics designer has a choice of

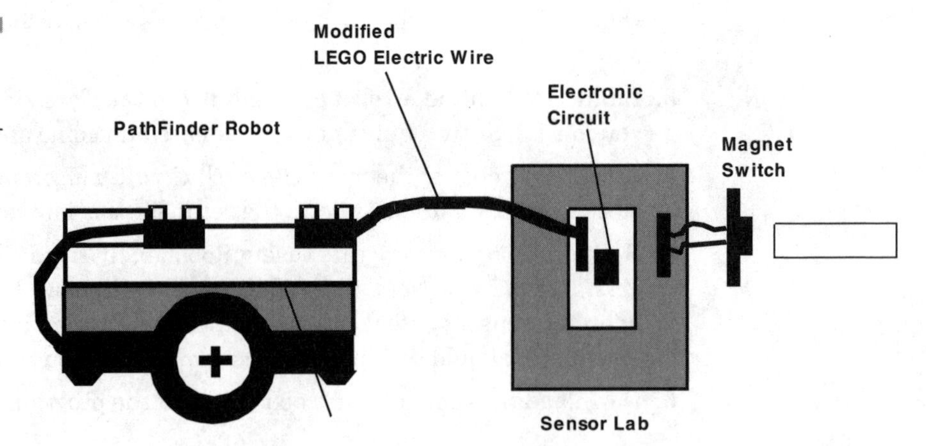

Figure 2-5 Setup of magnet-switch lab project

programming languages when testing his or her mechatronics-based machine. The author left the choice of software to reader.

Robotics Control Prototyping

In the previous discussion, the circuit block concept was presented through a hands-on laboratory project, using the Radio Shack Sensor Lab. The intent behind this kit is to allow the electronics hobbyist to experiment with sensors and circuits on a breadboard-spring terminal platform. By using the pre-cut wire that comes with the kit, the electronics hobbyist can complete the 95 projects in the workbook. As illustrated in the Magnet-switch Lab Project, the sensor lab can be used as a robotics-control prototyping station.

The idea behind prototyping is to build a functional device that resembles the actual product the consumer will buy. The robots built by the LEGO Mindstorms enthusiast are definitely not for consumer purchase, but for education and robot-construction entertainment. Therefore, quick-building techniques in electronics and mechanical fabricating will enhance the enjoyment of the building experience tremendously. Prototyping allows the designer to check function, feel, and look of the represented device or system. The prototype becomes an important design tool for the designer, allowing improvements to be made before the product is ready for final production.

In mechatronics projects, prototyping is critical to the success of the product, because of the amount of capital investment made in its development. For the LEGO Mindstorms enthusiast, the investment in purchasing the Robotics Invention System is quite small, compared to the hundreds of thousands of dollars required for industry-based prototypes. In building experimental robots, the designer should have tools that allow quick changes to be made to the product and tools that assist in the troubleshooting or modification of the mechatronics-based device or machine.

Therefore, the sensor lab meets the requirements mentioned in developing robots quickly. In addition to the sensor lab, the Radio Shack Electronics Learning Lab (Catalog Number 28-280) is another suitable choice for *Rapid Robot Control Prototyping* (RCP). RCP is suitable because the large breadboard area facilitates quickly building and testing electronic circuits. With the spring terminals, additional electrical and electromechanical loads, like light bulbs and small DC motors, can easily be added to the target electronic circuit. By modifying a LEGO electric wire, the amateur roboticist can easily attach these devices to the P-Brick. Figure 2-6 shows how the sensor lab can assist in the developing of a new detection circuit for the LEGO P-Brick.

Velcro™ is an excellent prototyping material to use in robotics mechanical construction when adding non-LEGO pieces to the LEGO P-Brick or other brick components. The *hook and loop* adhesion of Velcro provides a rigid foundation for adding non-LEGO pieces to Mindstorms-based mechatronic devices and machines. Again, the idea behind robot control prototyping is constructing mechatronically-sound designs that can be built quickly, debugged, and tested easily.

Using some of the products discussed in the previous paragraphs should give the amateur roboticist development tools to validate his or her autonomous robots and mechatronic-based machines quite easily and inexpensively. In future chapters in this book, additional prototyping techniques in electronics and mechanics will be explored, thereby allowing LEGO Mindstorms enthusiasts an opportunity to enhance their robot-building experience.

Another advance robot control prototyping tool that allows amateur roboticists to enhance their LEGO Mindstorms mechatronics devices and intelligent machines is the Engineering Evaluation Board. The following topic will explain what an Engineering Evaluation Board is and how LEGO Mindstorms enthusiasts can use it to develop their mechatronics-based robots and machines.

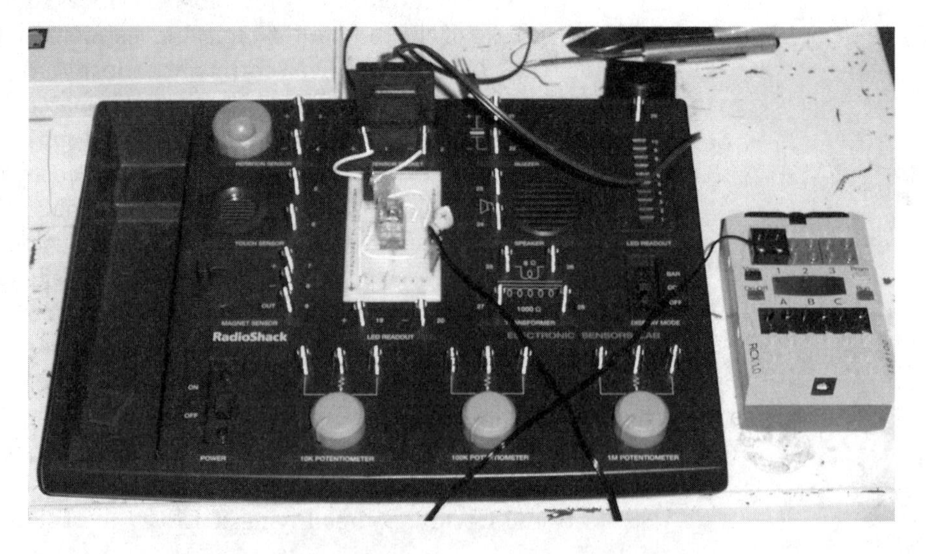

Figure 2-6 Developing a new detection circuit using the Radio Shack sensor lab

TECH NOTE: *There is a technical paper, written by three robotics engineers from Carnegie Mellon University, that describes work on high quality rapid prototyping. The authors explain how to use servomotors, LEGO bricks, and electronics to achieve satisfying results. The paper can be downloaded from the following Web site:* `www.cs.cmu.edu/~reshko/Publications/prototyping.pdf`.

Engineering Evaluation Boards

Another electronics hardware add-on that can assist in developing LEGO Mindstorms robots is the *Engineering Evaluation Board* (EVB). An EVB is a specialized prototype board used to educate the developer on the core component package on it. Sensors, integrated circuits, passive devices, and electromechanical devices are generally packaged on a printed circuit board, with supporting components that allow the developer to experiment with the specialized function or electrical feature of the component under evaluation. The EVB usually has a small breadboard area where additional components can be added to the target device under investigation. Several semiconductor companies like *Texas Instruments* (TI) and Motorola are known for providing EVBs to engineers in order to evaluate their electronic components. Some of the common electronic components packaged on EVBs are microcontrollers, switching regulators, sensors, and high-intensity LEDs.

In Figure 2-7(a), the EVB can be set up such that its output would be wired to one of the LEGO P-Brick's inputs. In Figure 2-7(b), there is a reversal in device control; The LEGO P-Brick drives the input of the EVB. Further analysis of Figure 2-7(a) shows the output of the EVB wired to an electromechanical relay. The switching contacts are wired in series, with a 1K resistor providing the digital input signal that the LEGO P-Brick reads. One of the P-Brick's outputs is attached to electromechanical drive assembly or small LEGO electric light.

Figure 2-7(b) works in reverse of the block diagram shown in figure 2-7(a). The EVB reads the digital output signal produced by the LEGO P-Brick and provides the appropriate driving signal response to the electromechanical drive assembly or small LEGO electric light. Microcontroller EVBs are commonly used in embedded controller applications, thereby making suitable electronics hardware add-ons for LEGO mechatronics projects. The example EVB to be discussed in this section of the

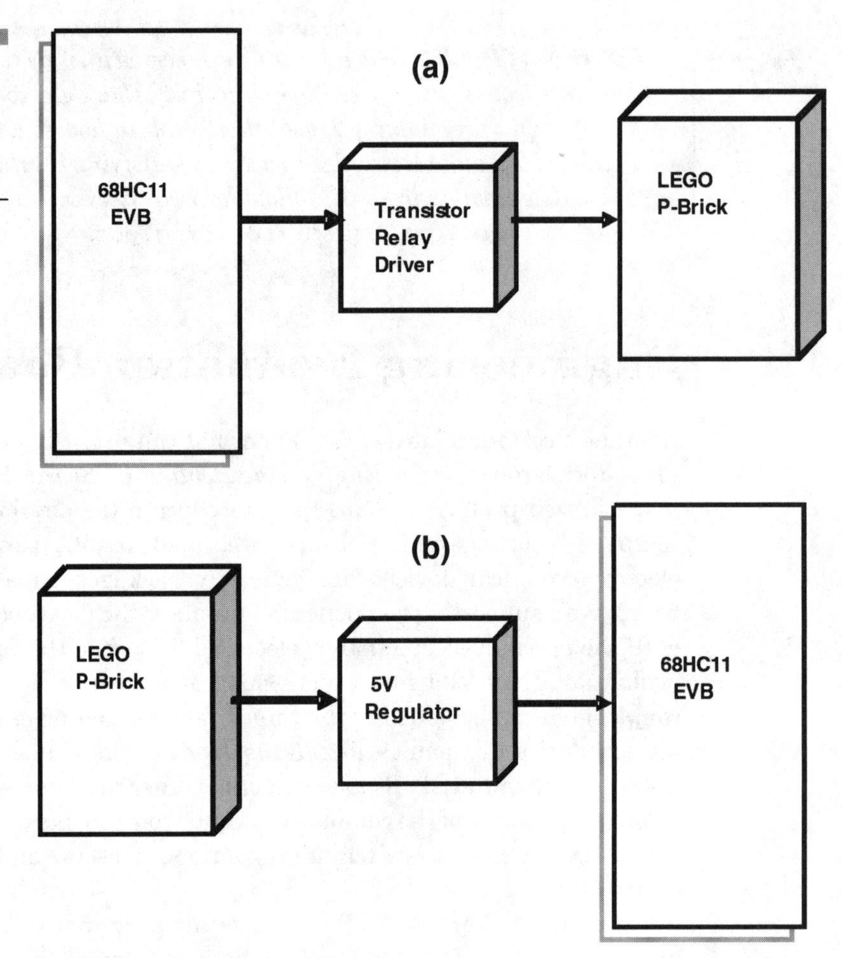

Figure 2-7 Wiring interface concepts for EVB and the LEGO P-Brick

chapter is a 68HC11-based unit. Figure 2-8 shows a picture of the author's EVB.

The 68HC11 microcontroller is a popular programming device used in many undergraduate electrical and computer engineering university courses. The purpose of this example is not to teach the 68HC11 microcontroller's architecture, but discuss how this device can assist in LEGO Mindstorms mechatronic applications through the EVB. Therefore, the I/O associated with the 68HC11 microcontroller is seamlessly integrated to the outside world, using the EVB printed circuit board as a wiring media to the LEGO P-Brick's inputs and outputs.

Start with Figure 2-7(a) as the origin for discussing the EVB as an electronic hardware add-on.

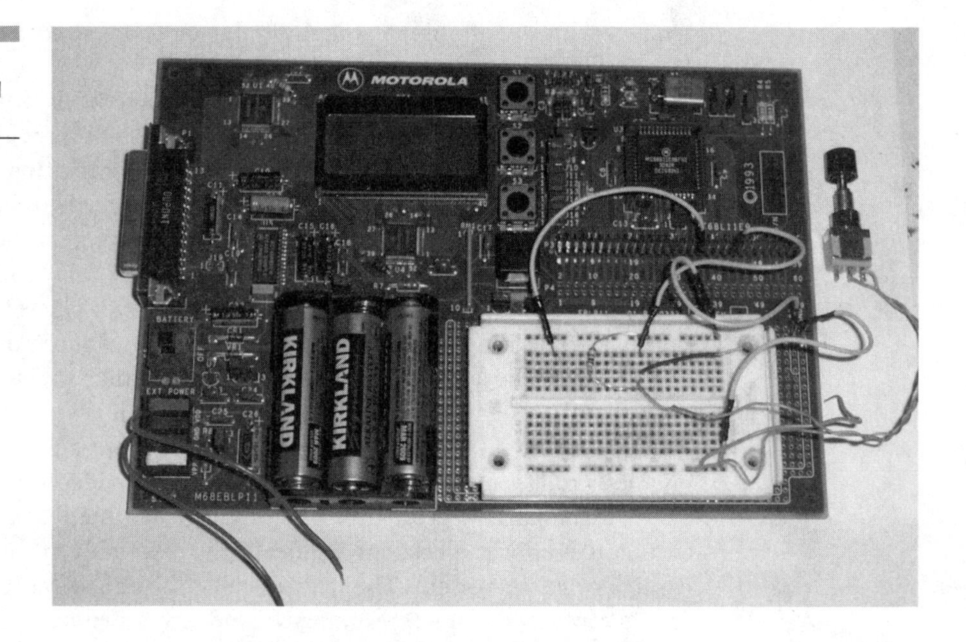

Figure 2-8
A 68HC11-based EVB.

The EVB Driving a LEGO Mindstorms P-Brick

In Figure 2-7(a) we see a simple block diagram, showing the EVB driving a LEGO Mindstorms P-Brick. Often in microcontroller training courses, an I/O lab experiment will consist of an EVB programming design problem requiring a microcontroller-based development board to activate an LED and to represent an electrical or electromechanical load of an industry-specific application. The LED could represent an electric lamp, a DC motor, or an eletromechanical actuator. The LEGO Mindstorms P-Brick, along with a mechanical drive assembly, gives the EVB-based application more of an industrial representation—an engineering model, if you will.

Therefore, the student has a better perspective of how the microcontroller can be used in an industrial application, without having greasy mechanical parts to contend with. The engineering model, consisting of the EVB wired to the LEGO Mindstorms-based electromechanical drive assembly, is a good example of mechatronics at work. An electromechanical switch can easily be hardwired to one of the EVB's input ports, thereby providing a seamless I/O distributed control from the microcontroller-based development board to the LEGO P-Brick's input. Separate software would be written for each programmable unit: Assembly or C for the EVB and RCX; NQC, IC4, Java, or Robolab for the LEGO P-Brick.

In order to the make the EVB's output compatible with input of the LEGO P-Brick input, an interface circuit must be provided. Figure 2-9 shows the complete interface circuit for accomplishing this function.

Upon the 68HC11 *Port-pin E0* receiving a 0V (logic 0) input signal, the output port *B1* will be +5V (logic 1), thereby turning on the Q1 transistor. Current will flow from *VDD* (+5V) through the 5V relay coil, via the collector-emitter circuit of the NPN transistor, to ground. The relay contacts will close, allowing the 1K resistor to be electrically connected to the internal 10K resistor. These two resistors will form a voltage divider circuit, allowing approximately 0.45V to be read by the Hitachi H8 microcontroller. This voltage will then be interpreted as an active input signal enabled by the microcontroller's user-defined program. The embedded software written in an RCX-compatible programming language will activate the designated output, driving the electromechanical drive assembly of the mechatronics-based device or system. The 1N4001 Si (Silicon), wired across the +5V relay's contacts, is for suppressing the inductive switching transients (voltage spikes) of the coil. The assembly language software used in the 68HC11 EVB is shown in Listing 2-1.

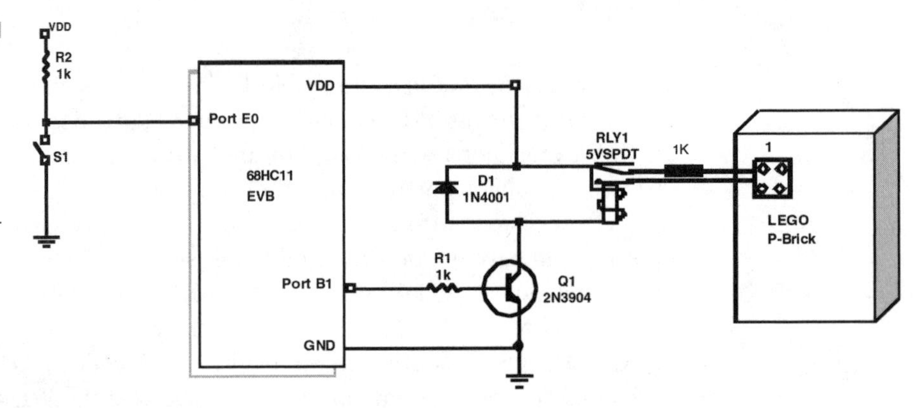

Figure 2-9 An EVB driving a LEGO P-Brick with a NPN transistor relay driver interface circuit

Listing 2-1

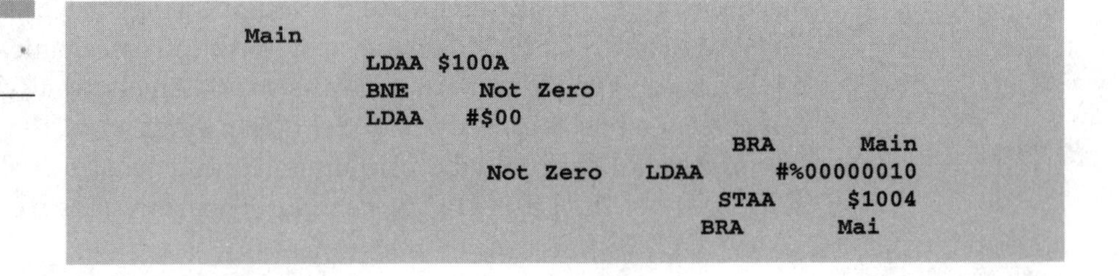

```
Main
        LDAA $100A
        BNE       Not Zero
        LDAA    #$00
                                  BRA         Main
              Not Zero    LDAA          #%00000010
                                 STAA        $1004
                               BRA         Mai
```

The assembly language code works by loading accumulator *A* at address *$100A*, which is port *B*, with the immediate hex value of #$00. This digital logic value resets accumulator *A*, thereby cleaning up the latched bits in its data register.

If bit 7 of the port *E* register is not zero, then jump to accumulator *A* and set bit 1 of the port *B* register to 1 and latch it. Read the value at port E, looping back at the beginning of the program (Main) to poll another logic input signal. A rocker or self-latching electromechanical switch should be used in order to read a latched 0 or +5V digital signal for a correct EVB output switching response.

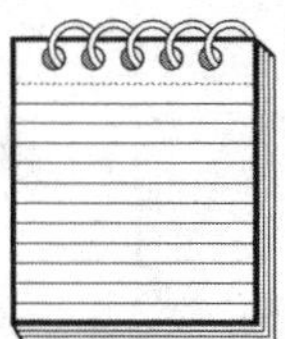

TECH NOTE: *Typing 68HC11 within the Web-search browser textbox will provide several links on programming, software, and how to build a 68HC11 EVB from scratch.*

LEGO P-Brick Driving the 68HC11 EVB

In Figure 2-10 the block diagram shows how the LEGO P-Brick can drive the 68HC11 EVB. The expanded circuit schematic is shown in Figure 2-11.

The key to having the LEGO P-Brick drive the EVB is to step down the output voltage of the P-Brick to a digital-logic voltage level conducive for microcontroller controls applications. The output voltage reduction can be accomplished using a +5V linear voltage regulator circuit. The voltage reg-

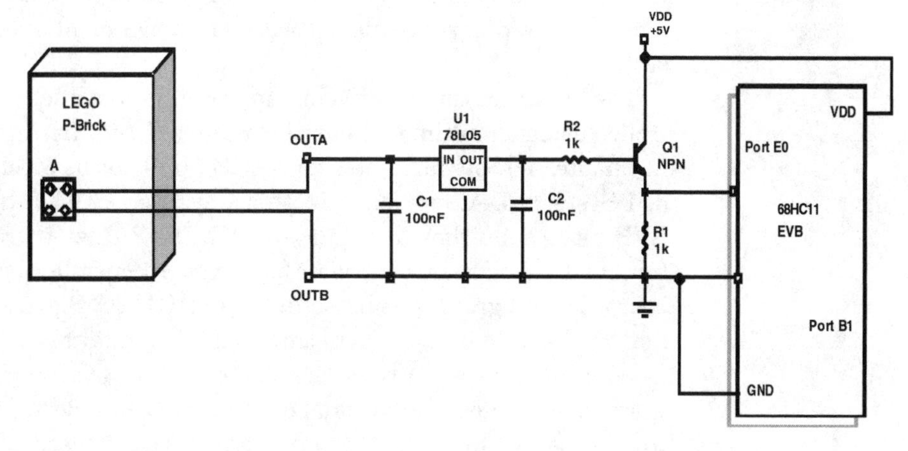

Figure 2.10
A LEGO P-Brick driving a +5V voltage regulator-NPN emitter follower interface circuit.

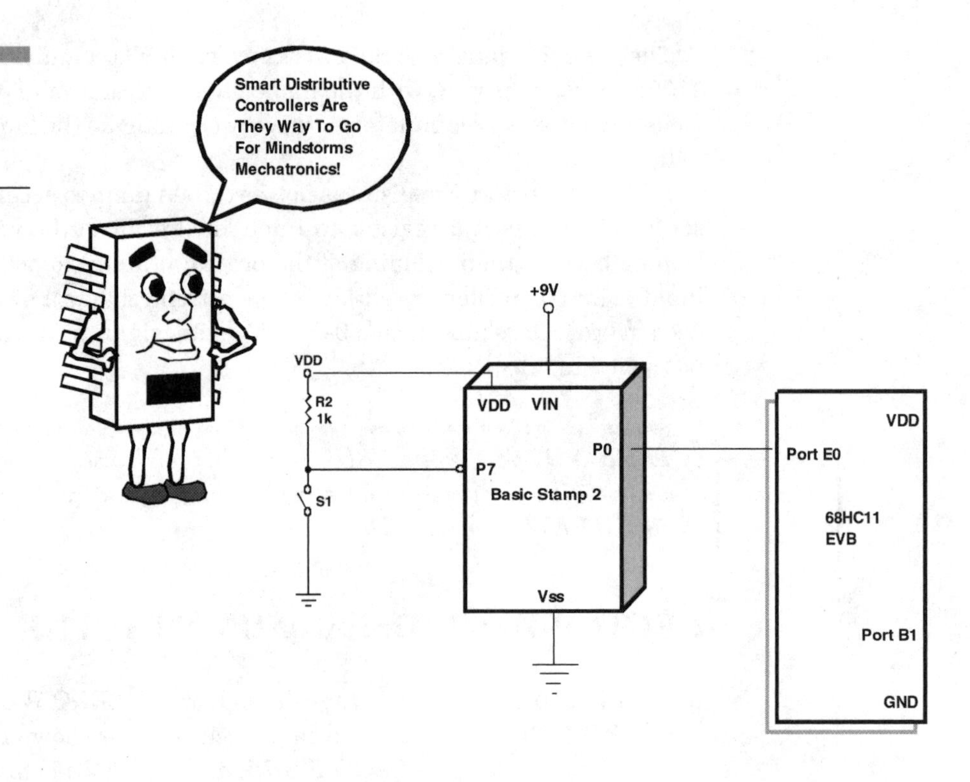

Figure 2-11 Basic Stamp 2 hardwired to a 68HC11 EVB

ulator in conjunction with the NPN transistor act as a DC to DC converter that reduces the output produced by the LEGO P-Brick from +7V to +5V. The NPN transistor is an emitter follower that provides a little amplification in current that is required for the EVB to respond. The output voltage received by port *E0* is the same as the input signal received at the NPN resistor-based circuit. The LEGO P-Brick's output can be controlled via a compatible programming software language mentioned in the previous section.

The appeal of this interfacing approach is that the P-Brick can be controlled using a multimedia-based *Graphical User Interface* (GUI) discussed in Chapter 1, "Software Tools for LEGO Mindstorms Mechatronics Embedded Systems Development." In this way, the LEGO P-Brick controls the EVB, using a multimedia-managed GUI. The P-Brick therefore becomes the front end or upstream controller for the microcontroller-based development board. The program code listed in the 68HC11 EVB driver application section can be used in this mechatronics implementation as well.

When using the EVB as the electrical interface controller to the real world, a multitude of additional outputs can be used through the microcontroller's ports. This is another interesting feature associated with using the

EVB. The output ports of the 68HC11 can be selected by writing an 8-bit binary word to their assigned addresses. Selection of outputs can be set with a *1* to the appropriate register bits. The outputs can be switched on or off with the aid of the LEGO P-Brick driving three of the inputs to the EVB. This gives the amateur roboticist a wide range of mechatronic control functions to experiment with.

The 68HC11 EVB allows for microcontroller experimentation in various robotics and mechatronics applications, such as an electromechanical switch tester, wire continuity validation, rotation speed measurements, security detection devices, and multimedia GUI interface controllers. The key to developing some or all of these applications is as follows. Select one of the driver interface concepts discussed. Design and build the appropriate intermediate circuit block. Write simple Assembly or C code. Finally, execute and debug the EVB program.

For the amateur roboticist who feels adventurous, the Orgler Electronics Web site, <http://space.tin.it/computer/lorgler/sw-e.htm>, has a circuit schematic for the 68HC11 microcontroller that can be built from scratch. The bootstrap circuit schematic is ideal for experimenting with the LEGO P-Brick, due to the ease of programming compared to the Expanded Mode configuration. Figure 2-12 shows the bootstrap circuit schematic used in building a 68HC11 EVB.

68HC11 EVB Driver Lab Project

This lab project will demonstrate how a 68HC11 EVB can be used as a Front End Controller for driving a LEGO P-Brick.

The BOM is as follows:

One 68HC11 EVB or equivalent demonstrator

Three 1 KΩ resistors, 1/4 W

One latching electric switch

One 1N4001 Si (silicon) diode

One 2N3904 NPN transistor

One 5V electromechanical relay

One modified LEGO electric wire

One LEGO P-Brick

One +5V DC power supply, 300 *milliampere* (mA)

One Radio Shack electronics learning lab or equivalent breadboarding system

Figure 2-12 Bootstrap circuit for the 68HC11 microcontroller

Assembly and Test Procedure The steps of the procedure are as follows:

1. Build the EVB controller, using the circuit in Figure 2-9(a). Attach V_{DD} (Drain-Supply voltage) to +5V DC of the 300mA power supply.
2. Build and assemble the code in Listing 1. Check for typing errors before assembling the code.
3. Compile and assemble the code using an assembler.
4. Download the assembled code to the 68HC11 EVB.
5. Build the transistor-electromechanical relay driver circuit using a Radio Shack Electronics Learning lab breadboard or equivalent. Make sure the correct pins of the EVB are wired to the base, collector, and electromechanical relay circuits.
6. Apply the +5V DC power supply to the 68HC11 EVB circuit.

7. Flip the electric switch to the off position. The contacts of the 5V DC electromechanical relay should close. This switched state can be verified with a *Digital Multi-Meter* (DMM), set to ohms. After placing the test leads from the DMM across the electromechanical relay's contacts, 0 ohms of resistance should be displayed on the electrical measurement tool. If not, turn off the power supply. Check the assembly code and wiring of the circuit. Repeat this step after the software and hardware has been confirmed.
8. Place the connector from the modified LEGO electric wire to input 1 of the LEGO P-Brick.
9. Write a simple program (using RCX code, NQC, IC4, or Robolab) for the LEGO P-Brick to perform a simple turn on/off switching application. Program the LEGO P-Brick to beep upon input 1 receiving the corresponding digital logic value.
10. Download the program into the LEGO P-Brick, using the infrared tower. Once the program has been successfully downloaded to the P-Brick, the control software can be executed.
11. Flip the electric switch on the EVB to make the LEGO P-Brick beep. If the P-Brick does not beep, check the modified electric wire and the embedded control code of the LEGO P-Brick. Repeat this step after the problem has been debugged.
12. Toggle the switch several times. Notice the position of the electric switch (0 or 1) and the response of the LEGO P-Brick.
13. Turn of the +5V DC power supply and the LEGO P-Brick.
14. If a Scout P-Brick is available, program it using *Stand Alone Code* (SAC), as discussed on page 10 of *LEGO Mindstorms Interfacing*. Replace the LEGO P-Brick with it. Repeat steps 10 and 11. Did the Scout P-Brick work correctly under the control of the EVB?
15. Replace the Scout P-Brick with LEGO P-Brick. Leave the project together for the next lab project.

Summary of the EVB Controller Lab

This lab project was designed to illustrate how an engineering evaluation board can assist the amateur roboticist in developing an off-board controller, using an advanced microcontroller like the 68HC11. By writing some simple assembly language code and building a transistor relay driver interface circuit, the LEGO Mindstorms P-Brick can be used as distributive smart controller for an electromechanical assembly.

Whenever a logic low signal is present at port pin *E0* of the 68HC11 microcontroller, it will process the digital signal and activate port pin *B1*. The 4.99V output voltage is capable of turning on the transistor-electromechanical relay that provides a 0.45V signal via the 1 KΩ resistor. The LEGO P-Brick's software and Hitachi H8 microcontroller will interpret this low voltage signal as a digital logic *Hi*, thereby allowing the P-Brick to beep.

The Scout P-Brick can also be used in controlling electromechanical assemblies by programming the programmable unit using SAC. The same output response (an electronic beep) can be provided using the Scout's internal speaker.

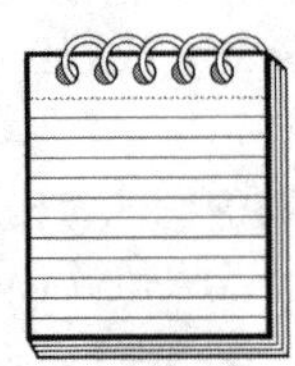

TECH NOTE: *The Scout can easily be used as a surrogate P-Brick for the lab projects. By using SAC or its own assembly language, the Scout P-Brick can be use in advance robotics projects quite easily.*

Basic Stamp Interfacing

The Basic Stamp is an EVB for evaluating the PIC165C7 microcontroller. The BS1 or BS2 is a small computer board that includes the PIC165C57 microcontroller, a 24LC16B 2K *Electrically Erasable Programmable Read Only Memory* (EEPROM) chip, a 20 Mhz ceramic resonator, a +5V voltage regulator, and 16 I/O. Like the 68HC11 EVB, the Basic Stamp can be used to drive a LEGO P-Brick as well as be driven by the P-Brick. The same electronics circuit drivers used in the 68HC11 EVB discussion can be reused with the Basic Stamp as well. Programming the Basic Stamp does not require assembly language like the 68HC11. It requires a higher-level language called *PBASIC*. PBASIC is combination of *BASIC* (Beginners All-purpose Symbolic Instruction Code) and Parallax, Inc.-defined programming instructions.

The PIC165C57 microcontroller (made by Microchip Technology, Inc.) and custom-programmed by Parallax, Inc. provides a unique chip, one ready for LEGO Mindstorms mechatronics projects. The PBASIC language is Parallax, Inc., firmware that turns the microcontroller into *interpreter* chips. An interpreter is software that reads and processes the program code one line at a time.

Since the microcontroller is an interpretive device, the Basic Stamp PIC has the entire PBASIC programming language embedded into the internal program memory (*One Time Programmable Read Only Memory* [OTP ROM]). Therefore, replacing a damaged PIC165C7 microcontroller with another one requires burning the firmware (the interpretive programming language) into the replacement chip. Two options are available to the amateur roboticist if the microcontroller becomes damaged: Purchase either the chip or a new Basic Stamp from Parallax, Inc.

The Basic Stamp can control a LEGO P-Brick directly, or it can control another EVB controller that will drive the P-Brick. The Basic Stamp can drive the 68HC11 without an interface driver circuit because each output delivers +5V at 20mA. This volt-ampere output is directly compatible with the 68HC11 EVB, making for some interesting LEGO Mindstorms-based mechatronics projects.

One question someone might ask is, why the need for two embedded-based development systems to control the LEGO P-Brick? In Gordon McComb's *The Robot Builder's Bonanza*, second edition, he states in the Introduction:

> There is plenty of room for growth, with a lot of discoveries yet to be made—perhaps more so than in any other high-tech discipline.

The key word is *discovery.* By exploring new techniques of controlling the LEGO P-Brick, the amateur roboticist is growing in his/her understanding of the field of robotics. The unique aspect of this two EVB microcontroller-based controller is that it allows laboratory experiments in computer programming, systems architecturing, and interface driver circuits development investigations to be carried out with the aid of the LEGO P-Brick. By programming in the different computer language formats, the amateur roboticist begins to learn about software design and the development of embedded-based controllers. Finally, creativity and innovation are at the forefront in experimentation, as unique ways of controlling mechatronics-based machines can be explored through the LEGO Mindstorms P-Bricks.

Two Embedded Controllers Lab Project

The objective of this lab project is to explore the use of the 68HC11 EVB and the Basic Stamp 2 to control a LEGO Mindstorms P-Brick through hands-on experimentation.

The BOM is as follows:

One 68HC11 EVB or equivalent

One Basic Stamp 2

Basic Stamp Windows Editor version 1.3

One LEGO P-Brick

One 1N4001 Si (dilicon) diode

One 2N3904 NPN transistor

One 5V electromechanical relay

One modified LEGO electric wire

One LEGO P-Brick

One +5V DC power supply, 300 mA

One 9V Battery

One Radio Shack Electronics Learning Lab or equivalent breadboarding system

One latching electric switch

Three 1 KΩ resistors, 1/4 W

Assembly and Test Procedures To conduct the procedure, follow these steps:

1. Using the lab project setup circuit shown in Figure 2-9, replace the digital switch circuit with the Basic Stamp 2 controller to port pin *E0*. See Figure 2-10. The output *P0* of the Basic Stamp will connect to the port pin *E0*. Apply power to the Basic Stamp 2 controller using a 9V Battery or +5V DC power supply.
2. Using the Basic Stamp 2 Windows Text Editor or DOS equivalent version, enter the program shown in Listing 2.
3. Download Listing 2 into the Basic Stamp, using the serial cable connection from the PC or notebook's serial port to the embedded controller.

4. Connect a DMM to pin B1 of the 68HC11 EVB.
5. Press the digital switch connected to the Basic Stamp 2 controller. Observe the voltage reading at pin B1 of the 68HC11 EVB. Is the output voltage approximately +5V DC?

   Yes___ No____.

   If there is no voltage reading at this port pin, recheck the wiring and software. Repeat this step after the problem has been identified and corrected.
6. Set the two embedded controllers aside for the Final Assembly and Test Procedure section of this laboratory project.

With the assembly language program in 68HC11 EVB running and the Basic Stamp 2 software executing, the two embedded controllers are ready to be tested. The following procedures will outline the attachment to the LEGO P-Brick and the two embedded controllers.

Final Assembly and Test Procedure The steps for this last procedure are as follows:

1. Remove the DMM connected to port pin B1 of the 68HC11 EVB and attach the LEGO P-Brick to it, using the modified electric wire. Use input 1 of the LEGO P-Brick as the termination point for the two embedded controllers.
2. Attach a small motor or light bulb to output A of the LEGO P-Brick.
3. Write code (using RCX programming language, Robolab, NQC, or IC4) whereby output A turns on when input 1 is active. Download the code to the P-Brick.
4. Press the digital switch and notice the motor attached to output A of the LEGO P-Brick. Does the motor turn on when the digital switch is pressed?

   Yes ___ No ___.

   If no, check that the correct input of the LEGO P-Brick is attached to the transistor relay driver circuit of the 68HC11 EVB. Verify that the software downloaded to the P-Brick is correct. Repeat this step after the problem has been identified and corrected.
5. Play with the two embedded controllers by pressing/releasing the digital switch, observing the response of the LEGO motor.

6. Replace the LEGO P-Brick with the Scout P-Brick, using SAC (see Step 13 under the 68HC11 EVB Lab Project section within this chapter).
7. Turn off the two embedded controllers and the LEGO P-Brick.

Summary of the Two Embedded Controllers Lab Project The two embedded controllers lab project illustrates, at a low level, the concept of *Smart Distributed Systems* (SDS). Complex industrial based systems use SDS as a means of controlling smart actuators remotely using embedded controllers and hardwired or communication protocols.

In robots, the various sensors used to obtain environmental data for electromechanical or electrohydraulic control are remotely located throughout the body structure of the robot. A main controller would be used to obtain sensor data, and process the information before sending it to the smart actuator.

The two embedded controllers lab project demonstrates this concept. The Basic Stamp controller obtains the data and sends the logic information to the main controller (68HC11 EVB). The EVB is responsible for reading the digital logic event of the Basic Stamp, processing it, and sending an input control signal to the LEGO P-Brick. With the small electric motor attached to the output of the P-Brick, it becomes a smart actuator. Once the flexible hardware architecture is built, expandability is based on writing additional lines of code and hardwiring more sensors and electromechanical devices to the appropriate embedded controllers and the P-Brick. The Basic Stamp 2, the 68HC11 EVB, and the LEGO P-Brick can be swapped around, which

Listing 2-2

```
Dpin      CON    0       ' Basic Stamp Driver output  P0
SWpin     CON    7       ' Resistor-switch connected to pin 7
DlyVal    CON    255     ' Delay  Value
ALow      CON    0       ' Active Low switch state
SWrate    CON    255     ' Switch cycles between autorepeats
btnWk     CON    byte    ' Workspace for BUTTON instruction
btnWk=0

Loop:
BUTTON SWpin, ALow, DlyVal, SWrate, btnWK, Dpin, noPress ' Goto noPress
TOGGLE  Dpin                                          ' Toggle pin 7
noPress:  GOTO Loop                                        ' Repeat
```

allows for different hardwire and software controls to be explored by the amateur roboticist.

TECH NOTE: *The Basic Stamp Windows Editor ver. 1.3 can be attained from Parallax, Inc.'s Web site for free: Their Web site address is* `www.parallaxinc.com/html_pages/downloads/software/software_basic_stamp.asp`. *In addition to this development tool, other software downloads are free as well.*

Palm Diagnostics Computer

The handheld remote control that is packaged with the LEGO Mindstorms Ultimate Accessory allows the amateur roboticist to quickly operate a robot without writing code once the mechatronics-based machine is built. The handheld remote control also provides a convenient method in troubleshooting the LEGO P-Brick by its ability to quickly check out the outputs and the tone generator via turning on motors, lights, and an audible sound. Unfortunately, the inputs cannot be read due to limited embedded software resources of the handheld remote control.

The Palm computer can assist the amateur roboticist with advanced diagnostics techniques. Activating individual tasks within an embedded software program, reading sensor data, and activating motors, electric lights, and tone generators using the outputs; all of these techniques are available using the Palm computer. The Palm computer has a lot of computing power in a small package, which gives the amateur scientist an alternate means of operating and troubleshooting LEGO Mindstorms-based robots and mechatronics. The infrared port, through software, can send and receive commands to the transceiver of the LEGO P-Brick, to operate and troubleshoot the robots wirelessly. Activating *tasks,* or sub-programs, embedded within the core robotics software allows individual sections of code to be tested and verified, without the need of a notebook or PC. This portability is a handy tool, especially during robotics competitions, because of the quick and easy *go-no-go* testing that can be performed on the robot.

Wireless Diagnostic Controller Concept

Here is another useful approach to using the Palm computer. Use the handheld device as a wireless front end for driving a Basic Stamp or 68HC11 EVB, through the LEGO P-Brick. In the previous section, a discussion on how to create a Smart Distributed Systems was explained by the two embedded controllers lab project. The approach of hardwiring one embedded controller to another can be reused with the wireless front-end driver for the Basic Stamp. Figure 2-13 shows the setup for the Wireless Diagnostic Controller concept.

The Palm P-Brick Library

With the basic concept of how the Palm computer can be used as a wireless diagnostics controller for LEGO Mindstorms-based robots and mechatronics now understood, the question asked by the amateur roboticist is, where to obtain the handheld software? The answer is to go to `www.harbaum.org/till/palm/pbrick/`. This Web page has amble supply of *Application Programming Interface* (API) libraries for the Palm computer. The "PBrickLib" is an open-source API, allowing the public to download it for free. The program is installed on the Palm by *HotSyncing* the library to the Palm computer.

With the programming utility loaded into the handheld device, a host application package must be installed onto the programming unit as well. The disadvantage to the API library is it provides only Palm-to-LEGO

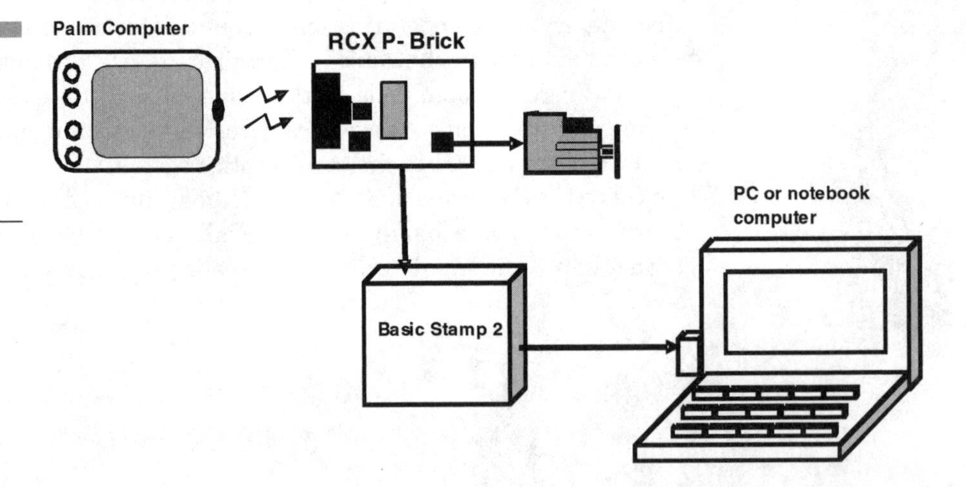

Figure 2-13 Setup for the Wireless Diagnostic Controller Concept.

P-Brick data transfers. The PBRemote.prc program works like the LEGO remote control, using one-way communication to the P-Brick, provided by the *PBrickLib* API functionality. The infrared tower is not required to send/receive commands and data to the Palm computer. Simply place the handheld unit in front of the LEGO P-Brick to start operating the yellow P-Brick. The *RCXLib 0.1* API provides bi-directional control of the Palm computer and LEGO P-Brick. The disadvantage of the *RCXLib 0.1* is that the cradle is needed in order to send/receive data from the LEGO P-Brick.

This experimental diagnostics controller will be explored in Chapter 3, *Instrumentation*, using a hands-on laboratory experimentation approach to building robots and mechatronics.

TECH NOTE: *To make the Palm diagnostics controller an integral part of a robot or mechatronics-based machine, the handheld computer can be mounted on the LEGO P-Brick with standard Legos. See the following Web site,* www.beanos.com/~tsoutij/legopalm.php, *for construction details.*

Chapter 2 Summary

In this chapter, the concept of adding external electronic controls (like circuit blocks, embedded controllers or handheld computers) was illustrated with several laboratory projects. The idea behind explaining the various external controls, using electronics, is to give the amateur roboticist several experimental tools to choose from when developing his or her robotic and mechatronics-based machines and devices. Some of the electronic controls investigated were serendipitous, because of the distributive approach of controlling the LEGO P-Brick. Again, robotics is such a wide-open technical discipline that all concepts are acceptable and possible when using the LEGO Mindstorms Robotics Invention System.

Web Site Resources

Here is a list of the Web site resources highlighted in this chapter.

Carnegie Mellon University robotics rapid prototyping method:

www.cs.cmu.edu/~reshko/Publications/prototyping.pdf

Orgler Electronics Web site for building a 68HC11 EVB:

`<http://space.tin.it/computer/lorgler/sw-e.htm>`

The Basic Stamp Windows Editor version 1.3:

`<www.parallaxinc.com/html_pages/downloads/software/software_basic_stamp.asp>`

The Palm Programmable Brick Library:

`www.harbaum.org/till/palm/pbrick/`

The LEGO Palm Computer:

`www.beanos.com/~tsoutij/legopalm.php`

CHAPTER 3

Instrumentation and Robots

Mechanical assemblies, electromechanical motor drives, and electronic circuit interfaces are necessary components needed in robotics design. Understanding the mechanical, electrical, and software function and performance of the mechatronics-based machine is equally important to the amateur roboticist. Mechanical test fixtures and electronics instrumentation are necessary to design an analysis tool for validating a robotics device or system, because the collected information gives guidance to what area needs repair or improvement.

Most test fixtures and electronic instrumentation are external to the mechatronics-based *Unit Under Test* (UUT). With the LEGO *Programmable Brick* (P-Brick), the amateur roboticist has the ability to embed data-logging features using software. The data can be accessed using external hand-built or off-the-shelf measuring tools.

This chapter will discuss the importance of using an experimental analog DC voltmeter, Digital voltmeter, oscilloscopes, and function generators in mechatronics design of robots and intelligent machines. The following topics will be discussed in this chapter:

- Basics of the Analog DC voltmeter and the Digital voltmeter
- Function Generators
- Palm Diagnostics and Robots.

A core robotics assembly will be described, along with the target internal or external instrumentation used in obtaining functional and performance data. The main objective in this section is to explore several measurement tools that can assist the amateur roboticist in his or her LEGO Mindstorms mechatronics development.

Why Robotics-Based Instrumentation?

The problem with modern scientific instrumentation is the *black box* term commonly used to describe the equipment. Today's black-box test instruments are highly effective in making mechanical and electrical measurements and collecting data, allowing novices to perform advanced engineering and scientific experiments. Black boxes are opaque (their inner workings are often hidden and not understood by the user) and they are plain in appearance, making it difficult for the user to feel a sense of personal connection with the measuring activity. Electronics and computational technologies have accelerated the development of engineering- scientific measuring instruments because of virtual controls that allow ease in building new testing devices for engineers and scientists. The LEGO Mindstorms *Robotic Invention System* (RIS) can add a new dimension to instrumentation development, because robotics-based sensing devices can be seamlessly embedded within the target-measuring unit. *Robot Digital Assistants* (RDA) can be incorporated within the electromechanical, electronic, and software components that make up the unique measuring instrument.

Therefore, a new mechatronics-based invention opportunity is available to the amateur roboticist by exploring the integration of robots and measuring instruments. The following topics outlined in this chapter will illustrate this robot-instrumentation integration concept with laboratory projects and experiments.

Basics of Analog DC voltmeters and the Digital Voltmeter

A discussion on the basics of a DC Voltmeter will be explained in this section. A laboratory project will also be presented to illustrate the concept of robotic-instrumentation as well.

DC Voltmeter Basics

An electrical ammeter movement responds to DC or AC current in the wire-wound moving coil. By adding a resistor in series with the movement, a simple DC voltmeter is constructed. Figure 3-1 shows a circuit schematic of a DC voltmeter.

The series resistor, known as the *multiplier*, is packaged inside the voltmeter case. Because the DC voltmeter has high resistance, it must be connected in parallel to measure the potential difference across two points in a circuit. By connecting the voltmeter in parallel with the circuit under test, its high resistance will not affect the voltage reading being measured. Therefore, the *IR* drop (current × resistance) will be greater with the series $R_{MULTIPLIER}$ (resistor-multiplier) - *rm* (meter resistance) than the circuit being tested, giving a true voltage reading being measured.

Figure 3-1
Circuit Schematic of a DC Voltmeter.

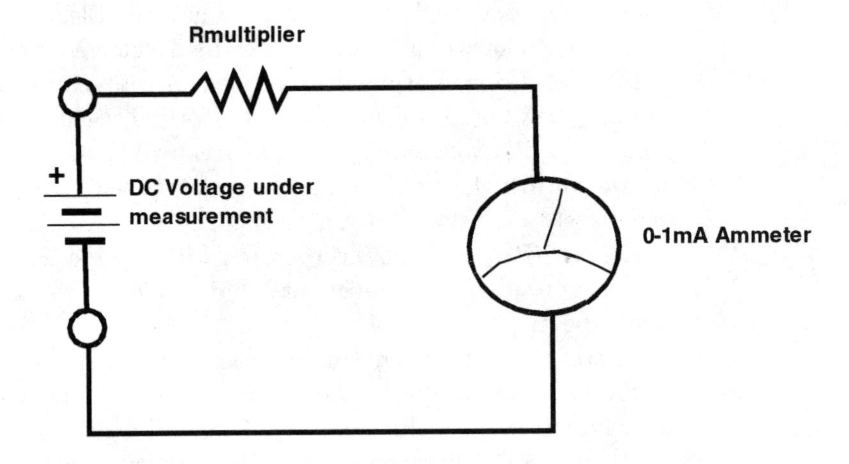

Determining $R_{MULTIPLIER}$

The $R_{MULTIPLIER}$ establishes the full-scale voltage range that the DCDC ammeter movement can deflect without being damaged. By using Thevenin's Voltage Theorem (the sum of the individual voltage drops in a closed loop series circuit equals zero), $R_{MULTIPLIER}$ can be determined. Therefore, mathematically:

$$\Sigma V_T - V_{RMULTI} - V_{rm} = 0$$

Using full-scale in the equation to solve for R_{MULTI}:

$$\Sigma(V_{Full\text{-}Scale} / I_{Full\text{-}Scale}) - (V_{RMULTI} / I_{Full\text{-}Scale}) - (V_{rm} / I_{Full\text{-}Scale}) = 0$$

Solving for $R_{MULTIPLIER}$ algebraically:

$$(V_{RMULTI} / I_{Full\text{-}Scale}) = (V_{Full\text{-}Scale} / I_{Full\text{-}Scale}) - (V_{rm} / I_{Full\text{-}Scale})$$

Simplifying the fractions:

$$R_{MULTIPLIER} = (V_{Full\text{-}Scale} / I_{Full\text{-}Scale}) - rm$$

In order to calculate $R_{MULTIPLIER}$, the three electrical parameters associated with the DC voltmeter circuit in Figure 3-1 must be known. When purchasing analog DCDC ammeters, $V_{Full\text{-}Scale}$, $I_{Full\text{-}Scale}$, and *rm* must be known. If the resistance of the internal ammeter is not known, a simple equation can help determine the electrical parameter value.

Determining *rm* of a DCDC Ammeter (Analysis Example)

On the Radio Shack's Electronics Learning Lab kit, the full-scale current, $I_{Full\text{-}Scale}$, is shown on the milliammeter (See Figure 3-2). In designing a DCDC voltmeter, a target full-scale voltage or $V_{Full\text{-}Scale}$ value is determined by the electrical designer. In the circuit schematic of Figure 3-1, the desired $V_{Full\text{-}Scale}$ value is 10V. The $I_{Full\text{-}Scale}$ is 1 *milliampere* (mA), as shown on the milliammeter's data sheet or the electrical measuring meter. The internal milliammeter resistance is not known and therefore needs to be determined. By measuring the electrical passive part's resistance *rm* directly, with an ohmmeter, and using a simple math ratio, the full *rm* can be determined. The mathematical ratio is:

$$[(1/2) \times \text{rm}] / (1/2 \times I_{Full\text{-}Scale}) = \text{rm} / I_{Full\text{-}Scale}$$

By solving *rm* algebraically,

$$\text{rm} = [I_{Full\text{-}Scale} \times (1/2) \times \text{rm}] / I_{Full\text{-}Scale\ reading}]$$

Determining (½) x rm and ½ x $I_{Full\text{-}Scale}$ is easily done by connecting an ohmmeter across the milliammeter's coil windings. A small DC current is introduced into the milliammeter from the ohmmeter, causing the needle to deflect. The reading from the meter's scale, along with the resistance value displayed on the ohmmeter, is used in the above equation to determine the *rm* value. With this value known, the $R_{MULTIPLIER}$ for the DC voltmeter can be calculated.

TECH NOTE: *A computer-aided tool to assist in creating state machines can be downloaded free of charge, from the CARMS web site www.tc.umn.edu/~puk/carms.htm. CARMS, which stands for Computer-Aided Rate Modeling and Simulation, is an integrated tool for modeling and simulating time dependent prediction-oriented problems.*

Figure 3-2
Determining of a DCDC Ammeter.

A Robotics-Actuated DC Voltmeter Lab Project

With an understanding of DC voltmeter basics established, a robotics-controlled measuring device can be built. Figure 3-3 shows the circuit schematic of the *Robotics-Actuated Voltmeter* (RAV).

The function of the RAV is simple: upon a *touch sensor* (TS) being pressed, the robot will move forward at a preset speed-level, and stop at a specified time. The robot will turn on the voltmeter, taking a measurement of the target DC source for a specified amount of time.

The following lab project will allow the amateur roboticist to explore the robotics DC measuring system shown in Figure 3-3.

The *Bill of Materials* (BOM) is as follows:

One 0–1mA DC meter

One 10KΩ, 1/4 W resistor

One 7805, +5V linear regulator *integrated circuit* (IC)

Two 100nF (0.1μF) capacitors

Two 1KΩ, 1/4 W resistors

One 1N4001 si diode

One 5V electromechanical relay

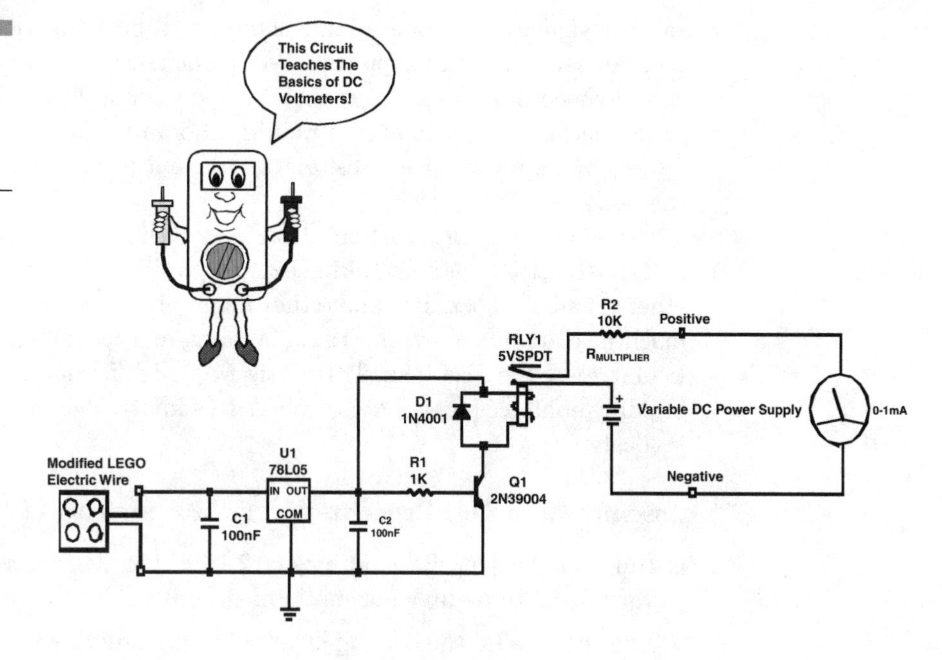

Figure 3-3 Circuit schematic for the Robotics-Actuated DC Voltmeter

One modified LEGO electric wire

One DC power supply

One touch sensor

Three standard LEGO electric wires

One NPN transistor, 2N3904 or equivalent

Constructopedia Manual version 1.5

LEGO Mindstorms RIS

NQC with BricxCC software

One infrared remote control

One *digital multimeter* (DMM) DC voltmeter

Philosophical Design Notes on Building *Electronic Beetle* (E-Beetle) Bot The electromechanical driving base instructions, found on pages 12-16 in the *Constructopedia Manual* version 1.5, and the legs assembly instructions, shown on pages 18–20, were combined to make an E-Beetle Bot. In building robots, the concept of taking design sections from one unit and adding them to another is the actual practice used by most robot designers. Robots are built using small building blocks like electromechanical drives for propulsion, 3-D geometric structures, and a series of simple machines like gears, wheels and axles, and levers. By taking all of these mechanical parts and combining them in a multitude of configurations, unique robot structures, platforms, and body forms can be built by the designer.

E-Beetle Bot's construction, described earlier, used this simple design and construction method. Taking existing validated mechanical designs by other amateur roboticists and either modifying or combining the core submachine elements allows for rapid building of a robot that will meet a particular mechatronics task. This design-by-modification approach to robot building enhances the amateur roboticist's knowledge in machine mechanics design as well.

Assembly and Test Procedure The procedure is as follows:

1. Build the Driving Base on pages 12-16 in the *Constructopedia Manual* version 1.5, or equivalent mechanical units.
2. Add the legs for the driving base, using the building instructions on pages 18-20 in the *Constructopedia Manual* to create E-Beetle Bot. The completed E-Beetle Bot is shown in Figure 3-1.
3. Test the E-Beetle's movement, by pressing up and down on the arrow buttons *A* and *C* on the infrared remote control. If there are any binding mechanical movements of the gears or the motor, make the necessary changes before proceeding to the next steps.
4. If the E-Beetle mechanical movement works correctly, turn off the LEGO P-Brick and compile and download the following *Not Quite C* (NQC) programming code, shown in Listing 2, using the BricxCC IDE tool. Store the program in location 1. Figure 3-2 shows the code, as typed in the BricxCC IDE tool.
5. Test the software partially, by running the program in location 1. Press and release the touch sensor and watch the bot *hobble* forward. The movement should continue for 10 seconds and stop. Did the E-Beetle bot move forward for 10 seconds?

   Yes___ No____

   If not, check your listing in the BricxCC IDE tool against that shown in List 2 for typos. In addition, check to see if the code was downloaded to program slot 1. If everything checks out, repeat this step to validate correct bot movement-response before proceeding to the next step.

6. Build the circuit shown in Figure 3-3. Attach it to E-Beetle Bot's output *B*, using the modified LEGO electric wire as shown in Figure 3-4(a). Note the placement and orientation of the modified electric wire connector onto the LEGO P-Brick. Figure 3-4(b) shows how to make an extension wire, using another standard LEGO electric wire. Note the orientation between connectors.
7. Attach a touch sensor to input 1, using a standard LEGO electric wire.
8. Check the circuit wiring for the *Transistor Relay Driver* (TDR) controlled DC Voltmeter before running the embedded software code.
9. Attach a DMM or DC voltmeter across the base and emitter leads of the 2N3094 NPN transistor. Adjust the meter for the appropriate DC measurement scale.
10. Activate E-Beetle Bot and run the NQC code stored in *P1* location.
11. Press and release the touch sensor and watch the bot move forward for 10 seconds. After 10 seconds, the DC voltmeter or DMM should read around 0.78V. The reading will be displayed for 10 seconds. A click from the electromechanical relay will be heard as well. If not, recheck the wiring and repeat this step before proceeding.
12. Provide a 3.9V DC source for the RDA DC voltmeter to measure by attaching a DC power supply to the mechatronics-based instrument.

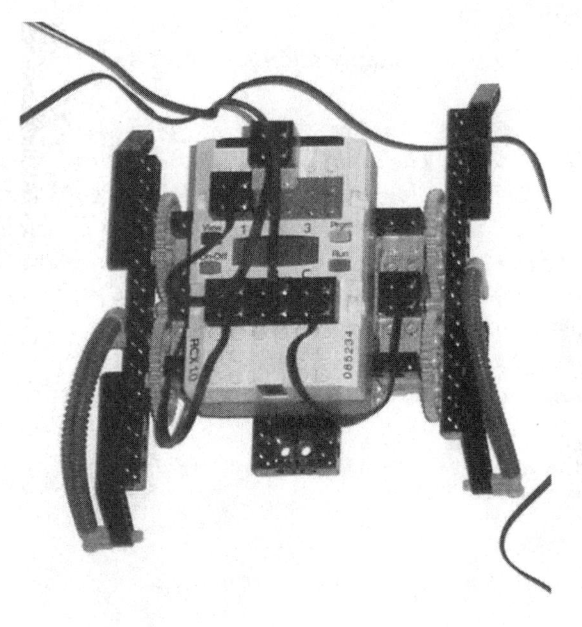

Figure 3-4(a) Placement and orientation of the modified electric connector onto the LEGO P-Brick

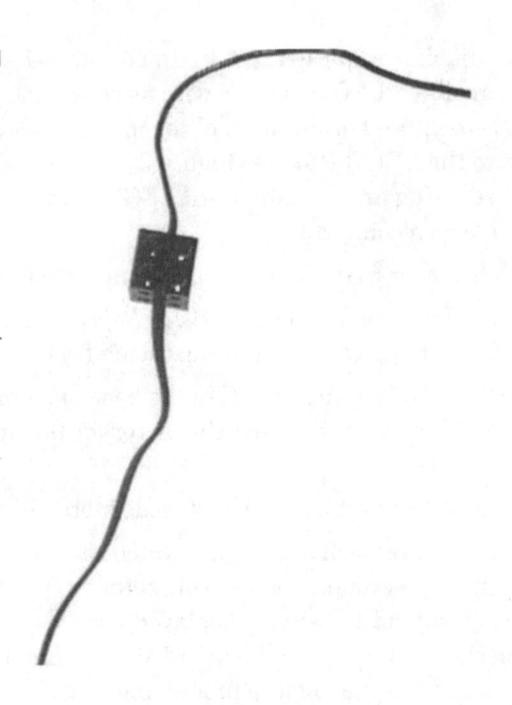

Figure 3-4(b) An extension wire using two standard LEGO electric wires. Note the orientation of between the connectors.

13. Run the program and watch the needle deflection on the DC meter. Record the measured value _____V. The reading on the 0-1mA meter should be around 0.4. This value is equal to 4VDC.
14. Change the 3.9V DC source and repeat step 13. Record the measured DC value ____V.
15. Turn off the power supply and E-Beetle bot.

TECH NOTE: *The amateur roboticist can produce different walking motions for E-Beetle Bot by raising or lowering the legs of the robot. With one leg up and the other one down (offset position), the bot will emulate the movement of a beetle or insect.*

```
// E_Beetle.nqc

// source RIS 1.5
// target LEGO P-Brick

#define Motor_A  OUT_A
#define Motor_B  OUT_B
#define Motor_C  OUT_C

task main ()
{
```

```
SetSensor (SENSOR_1, SENSOR_TOUCH);
while (true)
{
  until (SENSOR_1==1);
  OnFwd (Motor_A + Motor_C);
  OnFor (Motor_A + Motor_C,  1000);
  Off (Motor_A + Motor_C);
  OnFwd (Motor_B);
  OnFor (Motor_B,  1000);
  Off (Motor_B);

 }

}
```

Summary of the Robotics-Actuated DC Voltmeter Lab Project

This project illustrated how a simple DC measurement circuit can be upgraded with the aid of a mechatronics based driver system. By using a simple DC series circuit controlled by a transistor relay driver, an RDA-actuated DC voltmeter was created. The voltage 0.78V, measured across the base-emitter leads of the transistor, is known as the base-emitter junction voltage. The base-emitter junction voltage is symbolized in most electronics text books as V_{BE}.

The NQC code used in the lab project works by the following line-code description.

```
#define Motor_A  OUT_A
#define Motor_B  OUT_B
#define Motor_C  OUT_C
```

These three lines of code use the constants of *OUT_A* , *OUT_B* and *OUT_C* being defined with *Motor_A*, *Motor_B*, and *Motor_C*. The next line of code under *task Main ()*, Set Sensor (SENSOR_1, SENSOR_TOUCH), associates input 1 of LEGO P-Brick the with the touch sensor device. In order for the code to continue to monitor the touch sensor press-event being triggered, the *while (true)* will allow the program to loop continuously, waiting for a false condition to occur. To make E-Beetle Bot move forward, the touch sensor's digital event value must equal 1, detected by the line code instruction *until (Sensor_1 ==1)*. The forward motion for E-Beetle bot is provided by the *OnFwd(Motor_A + Motor_C)* line code instruction. The time delay for keeping the bot in motion for 10 seconds is accomplished with the *OnFor(Motor_A + Motor_C, 1000)* instruction. The time is in milliseconds. To stop E-Beetle bot after the 10 second movement, execution of the *Off(Motor_A + Motor_C)* NQC statement is used. The output *B* is controlled in the same manner as the A and C ports on the LEGO P-Brick.

The Digital Voltmeter

The *Digital Voltmeter* (DVM) has a numerical display for showing the measured quantity, instead of using an analog meter. The DVM requires converting the analog voltage to be measured into an accurate digital equivalent value. An *Analog to Digital Converter* (ADC) is used for converting, or *digitizing*, the applied voltage signal into a binary equivalent value. Digitizing is the process of converting a decimal-based signal to its equivalent binary value, using a weighted place-holder method. Digitizing also helps in determining the number of units or increments that comprise the measured analog voltage as well. Once the data is present in its digital form, the information can be displayed using several popular types of displays. Examples such as *Light Emitting Diodes* (LEDs) or *Liquid Crystal Displays* (LCDs) are used to provide a visual numerical value to the user.

Digital voltmeters are much easier to use because they reduce the human error that occurs in reading the different scales on an analog meter. The DVM works by the input analog voltage being converted into a digital form, using the ADC. The ADC needs an external clock that generates timing pulses, and another electronic circuit that counts the squarewave signals on/off switching transitions.

A special ADC IC chip, known as the ICL735, is used in this pulse counting activity and drives a 4½ digital display. The IC counts the timing pulses from the external clock circuit, and the digital equivalent value from its internal ADC. The analog voltage applied to the DVM is referenced to the IC's internal standard voltage value. A comparison is made to the ADC's binary value and the internal equivalent digitized data. The result from this comparison activity produces a new equivalent binary value that drives the digital display unit.

Figure 3-5(a) shows the block diagram of a typical DVM. The laboratory project will illustrate the concept of analog-to-digital conversion by building a digital bargraph voltmeter.

A Robotic-Actuated, Digital Bar Graph, DC Voltmeter Laboratory Project

This laboratory project will illustrate the concept of analog-to-digital DC voltage-measurement conversion, using a robotic-actuated digital bar graph DC voltmeter. The *transistor relay driver* (TRD) circuit of E-Beetle

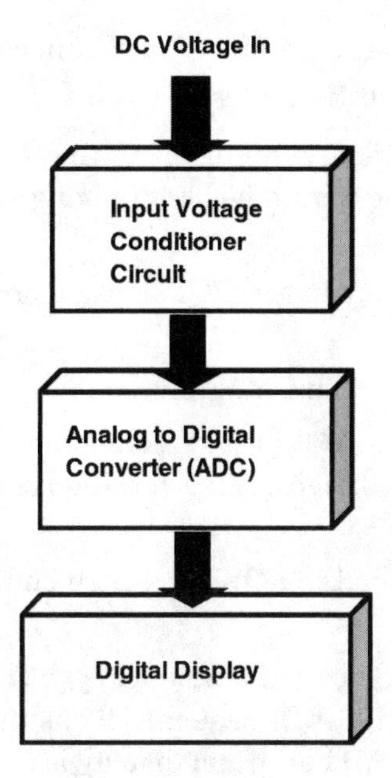

Figure 3-5(a)
Block diagram of a typical DVM

bot will be used to take DC voltage measurements and display them on a LED bar graph. The same robotic behavior will be the enabling feature for automatic DC voltage-measurement actuation. Therefore, the robotic DC voltage-measurement unit in the previous laboratory project will be used in this instrumentation experiment.

The BOM is as follows:

E-Beetle Bot TDR controls

One TLS272 Dual op-amp IC or equivalent

One 10 KΩ potentiometer

One 10 KΩ, 1/4W resistor

One 100 KΩ, 1/4W resistor

One 1 MΩ, 1/4W resistor

One Radio Shack sensor lab kit breadboarding system

One variable DC power supply

One DMM

Digital DC LED Bar Graph Voltmeter Circuit Assembly and Test
The following procedure outlines the circuit assembly and test of the digital DC LED bar graph voltmeter. These assembly and test steps will vali-

date the circuit's ability to measure a DC voltage range of 0-1.5V, and display the results on the LED bar graph display.

1. Wire the circuit shown, in Figure 3-5(b), using the Radio Shack Sensor Lab's breadboard. Check the wiring before applying power to the electronic circuit.
2. Put the LED bargraph display in the *DOT* mode, using the *DISPLAY MODE* switch. See Figure 3-6.
3. Attach a black test lead from the DMM to ground.
4. Make a Data Table as shown in Table 3-1, using Excel spreadsheet software, to record the DC voltages at the specified circuit nodes shown in Figure 3-5.
5. Apply power to the circuit via the *POWER* switch on the Sensor Lab kit.
6. Rotate the knob on the 10KΩ potentiometer slowly. The LED bar graph display should turn on the 10 LEDs sequentially as the knob is adjusted clockwise. If the LED bar graph display is not responding to the 10KΩ potentiometer adjustment, turn off the power to the Sensor Lab kit and recheck the wiring. After the problem has been corrected, repeat this step.
7. Turn the 10KΩ potentiometer knob, until the bar graph display *1* is lit.
8. Turn on the DMM. Adjust the voltage range scale for 1V, and place the red test lead on point *A,* as shown on the circuit diagram, and record the voltage. Pt A:_____V

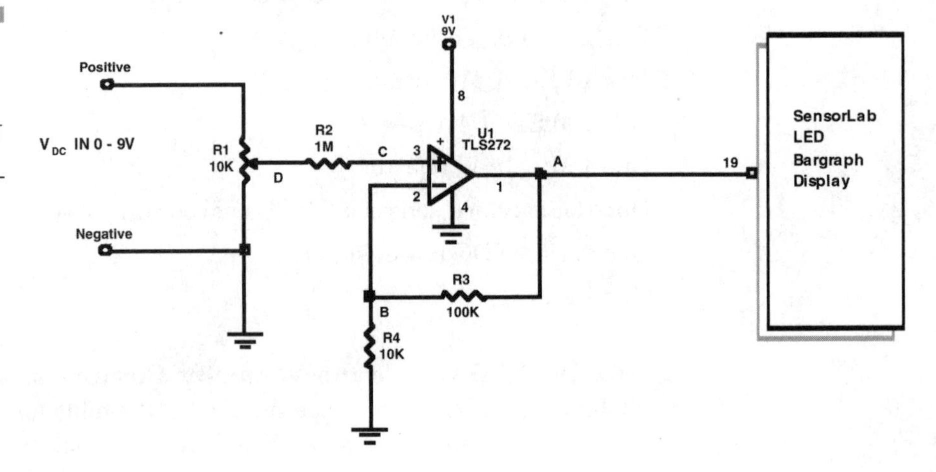

Figure 3-5(b) Circuit schematic diagram of the digital DC LED bar graph voltmeter

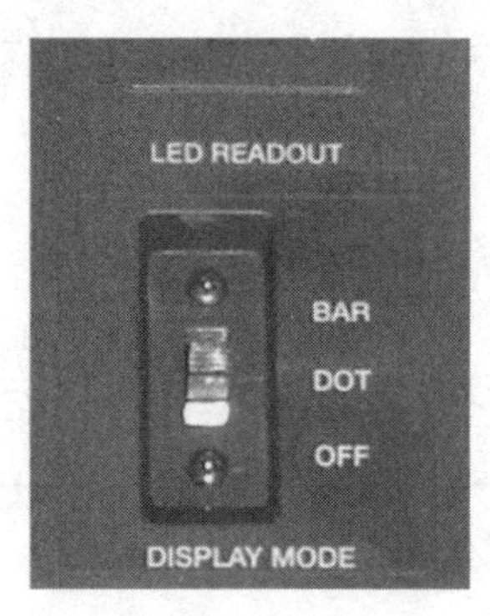

Figure 3-6 Selecting the LED bar graph display mode using the mode switch on the Radio Shack sensor lab

Table 3-1 Circuit Data Table

Measuring Point	Voltage	LED Position ON
A		
B		
D		
D		

9. Place the red test lead on point *B*, as shown on the circuit diagram, and record the voltage. Pt B:_____V
10. Place the red test lead on point *C*, as shown on the circuit diagram, and record the voltage. Pt C:_____V
11. Place the red test lead on point *D*, as shown on the circuit diagram and record the voltage. Pt D:_____V

    Compare your results with the author's recorded data, shown in Table 3-2. Your results might be slightly different due to the Sensor Lab kit's 9V battery strength.
12. Enter the data recorded in steps 8 through 11 onto the Excel spreadsheet table.
13. Repeat steps 8 through 11, changing the LED positions of the bar graph display using the 10KΩ potentiometer. Create a new spreadsheet, with tables capturing the measuring points' A to D voltages.

Table 3-2

The author's recorded data from the Digital DC LED Bar Graph Voltmeter

Measuring Point	Voltage	LED Position ON
A	217mV	1
B	19.8mV	
D	17mV	
D	19.8mV	

The DC LED bar graph voltmeter circuit is now ready for the final assembly and test.

Final Assembly and Test Procedure The following procedure will outline attaching E-Beetle Bot to the digital DC LED bar graph voltmeter. It will also outline testing the complete robotics-controlled measuring system.

1. Attach E-Beetle bot to the digital DC LED bar graph voltmeter, using the circuit diagram shown in Figure 3-7. Figure 3-8 shows each subcircuit used in the final assembly of the digital DC LED bar graph laboratory project.
2. Remove the +9V connection from the circuit.
3. Using a variable DC power supply and a DMM, adjust the output voltage to 1.5V. Turn off the power supply.
4. Attach the variable DC power supply across the 10K potentiometer. Turn the knob of the potentiometer to full counterclockwise.
5. Turn on the power to the Sensor lab kit and the variable DC power supply.
6. Set the display to read 605 *millivolts* (mV), using one of the Excel spreadsheet tables created in the Circuit Assembly and Test Procedure section of this laboratory project.
7. To test the complete control circuit, press the *B* up-arrow on the LEGO infrared remote control. The position *4* LED should be turned on. If not, turn off both power supplies and recheck the wiring. After the problem has been corrected, repeat this step.
8. Run program 1 and observe E-Beetle Bot's forward motion. After 10 seconds, the bot should stop moving and LED *4* on the Radio Shack Sensor Lab's bar graph display should be on. The LED will stay on for 10 seconds, then turn off.

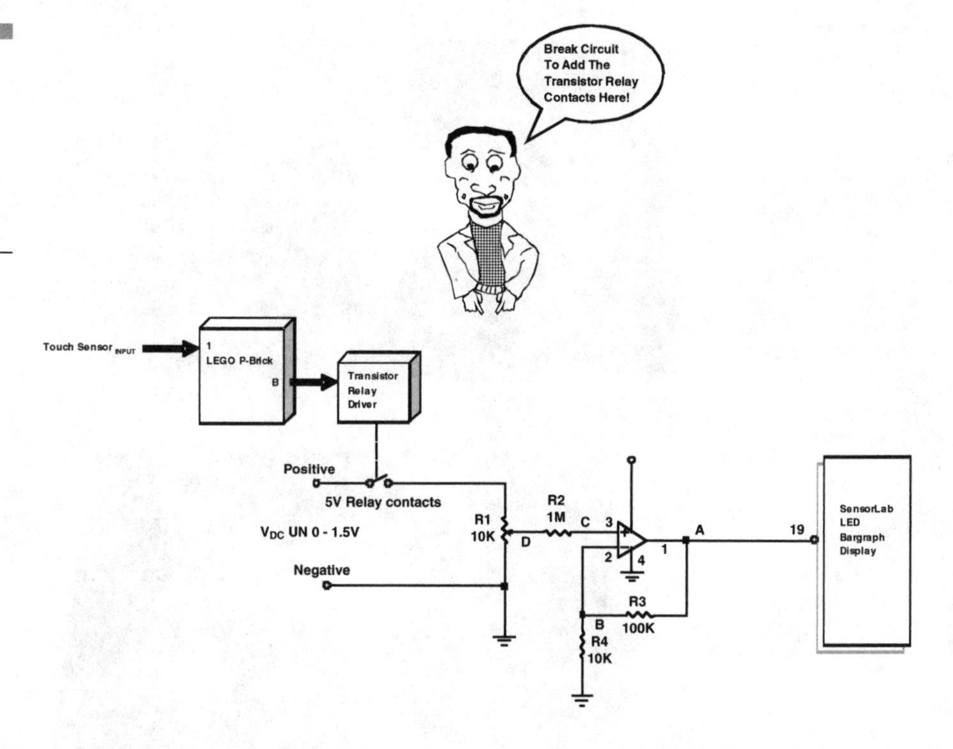

Figure 3-7 Circuit diagram used to attach E-Beetle Bot to the Digital DC LED bar graph voltmeter

Summary of the Digital DC LED Bar Graph Voltmeter Lab Project

The robot-based Digital DC LED Bar graph Voltmeter Lab Project was designed to illustrate how an analog voltage can be visually represented digitally, using a LED bar graph. A quick analysis of the bar graph circuit shows the *operational amplifier* (op-amp) having an amplification gain of 10. This is significant in amplifying an input signal of a few millivolts and producing an output voltage capable of driving an LED bar graph.

The 10KΩ potentiometer provided an input voltage adjustment-control, allowing the output voltage to be displayed in an easy to read visual digital scale. The 1MΩ series resistor is used to limit the current flowing into the non-inverting pin of the TLS272 dual op-amp to a level measured in microamperes (μA). The input voltage applied to the non-inverting amplifier can be in the operating voltage range of 0-9V DC. E-Beetle Bot will display the appropriate LED on the bar graph, by adjusting the input potentiometer to the data recorded on the Excel spreadsheet tables discussed in the laboratory project.

A reasonable applied input voltage to use, without being concerned of exceeding the maximum LED reading, is 1.5V DC. Having this voltage

Figure 3-8
Subcircuits used in the final assembly of the Digital DC LED bar graph voltmeter: (a) the Transistor Relay Driver (TRD) circuit, (b) the Digital DC LED bar graph circuit, and (c) the complete assembly

(a)

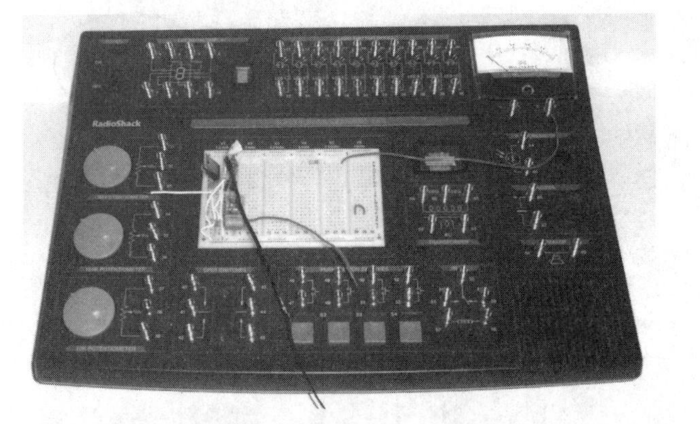

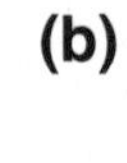

divided into discrete *millivolt* (mV) steps will ensure the amateur roboticist will not go beyond the tenth LED on the bar graph display.

Change the *DISPLAY MODE* to bar graph to create another visual effect with the LED bargraph. The LEDs will move in an upward-fill/downward-unfill motion, as the analog input voltage is increasing/decreasing in value. Replacing the 10KΩ resistor with a 1KΩ resistor can change the sensitivity of the op-amp's voltage gain.

Now that a basic understanding of analog and digital voltmeters is under our belts, let's continue the venture into instrumentation and robots with a discussion on function generators.

TECH NOTE: *To learn more about the ICL7135 IC, go to Digikey's Web site,* `www.digikey.com`*. With the home page open, type in the part number inside the search text box and click GO. When the parametric table page is open, click on the Maxim data sheet to download the PDF document.*

Function Generators

A function generator is a signal source that can produce different waveforms. All electrical waveforms can be expressed as mathematical functions of time. For example, the sine wave can be expressed mathematically in the form:

$$f(t) = a \times sin(bt),$$

where a is the amplitude of the waveform, t is the time, and b is the frequency of varying signal. Square waves, sawtooth waves, and all other periodic signals can be expressed as mathematical functions of time.

Most function generators can produce sine waves, sawtooth waves, and square waves. Some function generators can produce a series or sequence of pulses. The primary use of the function generator is to assist in the design and testing of electronic devices, such as audio amplifiers and digital circuits.

Building a Square and Triangle Wave Generator with a LEGO P-Brick

In Chapter 1, "Software Tools for LEGO Mindstorms Mechatronics Embedded Systems Development," the section titled *A Multimedia Control for an Electronic Oscillator Circuit* explained how a *Pulse Width Modulation* (PWM) signal is used to control the power level and speed of a LEGO DC motor. The 125 KHz squarewave signal provides the switching frequency for *electrically* slowing the small electromechanical rotating component attached to one of the three available outputs of the LEGO P-Brick. Therefore, the LEGO P-Brick can act as a squarewave generator for audio applications or for driving digital electronic switching circuits.

How can a triangle wave be created using a LEGO P-Brick? The square wave signal generated by the LEGO P-Brick can be converted into a triangle waveform. Filter the on/off pulses, thereby creating a switching *ramp* waveform that is in sync with the rising and falling edges of the PWM signal generated from the P-Brick.

This square wave filtering method is known as *integration*. Integration is the total accumulation of a quantity over a particular interval. Another way to define integration is the sum of all the points within a specified area. Therefore, the injected squarewave signal that the integrator sees will sum up all of the points within the t_{ON} pulse area and produce a proportional signal. This signal is the ramp or triangle waveform. The mathematical form for symbolizing integration is shown as:

$$y = \int_{o}^{n} dx(t) = x(t)$$

where $\int$ is the integral symbol and *0-n* are the data points within the area of the function *dx(t)*. After the integration process is completed, the final function is $y = x(t)$. By substituting numeric values between 0-n for t, the value of y can be evaluated. It is quite easy to perform this integration process electronically. Figure 3-9 shows the circuit blocks required to integrate a squarewave into a triangle signal.

The LEGO P-Brick's PWM signal can be used to produce a triangle waveform using an op-amp, a resistor, and a capacitor. Figure 3-10 shows a circuit diagram of a triangle waveform generator.

The values of R and C where selected whereby their time constant τ (pronounced "tau") is 10 times as the period of the LEGO P-Brick's PWM signal. To demonstrate this concept mathematically:

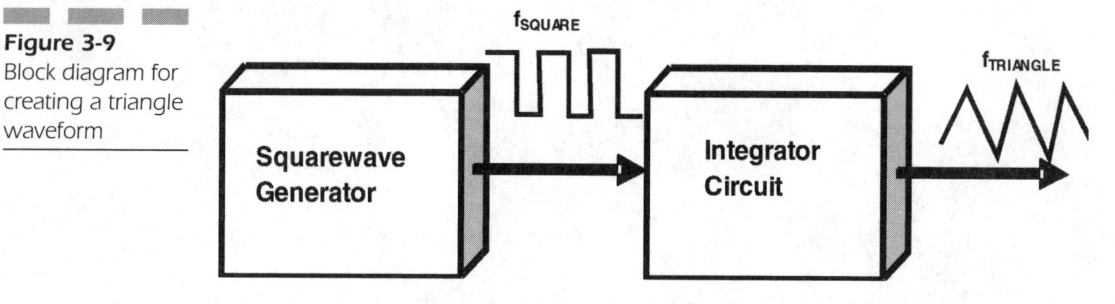

Figure 3-9
Block diagram for creating a triangle waveform

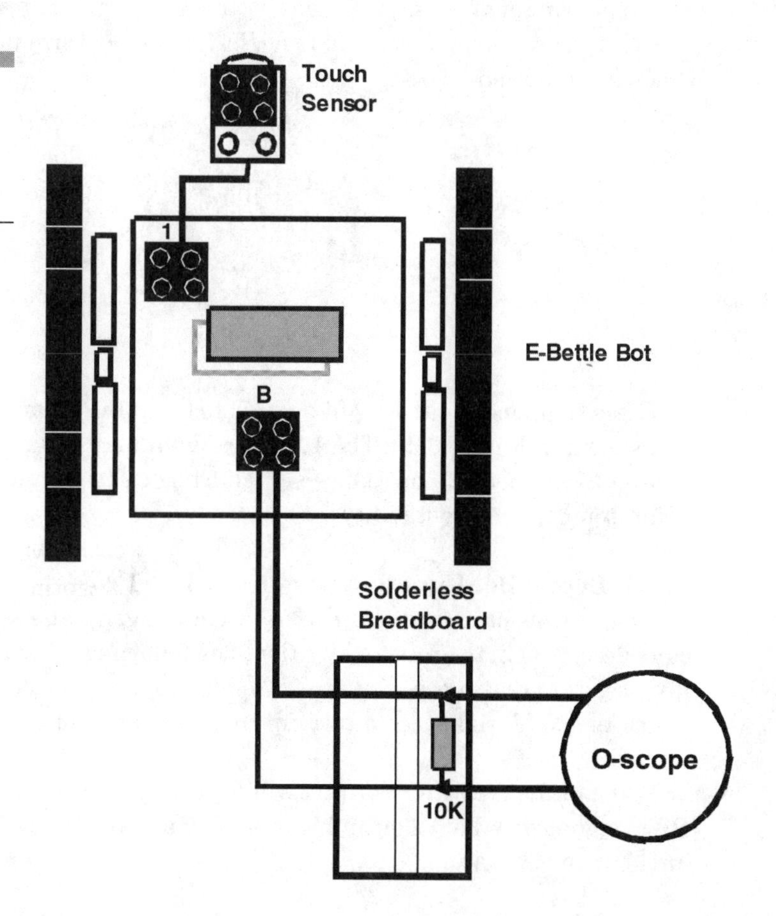

Figure 3-10
An operational amplifier (Op-Amp) integrator circuit

$$\text{LEGO P} - \text{Brick period} = \frac{1}{\text{frequency}}$$

$$\text{LEGO P} - \text{Brick period} = \frac{1}{125\text{Hz}}$$

$$\text{LEGO P} - \text{Brick period} = 8\text{mS}$$

Next, to calculate the value of , multiply the 8 mS value by 10. The time constant value equals 80mS. The RC combination required to manage an 80mS integration period is determined using the following equation:

$$t = 10 \times R \times C$$

The value of C to be used in the op-amp integration design of 100nF, therefore, has the value of R. R can be determined with the following equation:

$$R = \frac{t}{10C}$$

$$R = \frac{80mS}{10 \times 100nF}$$

$$R = 80K\Omega$$

These component values will be used to build the op-amp integrator circuit shown in Figure 3-10. The 100KΩ resistor across the 100nF capacitor is used to compensate or reduce signal drift due to circuit *bias* (operating point) temperature instability.

The E-Beetle Bot Function Generator Lab Project E-Beetle Bot can be easily transformed into a *walking* function generator by using the circuits described in the previous section. The following laboratory project will investigate creating a Square-Triangle Generator, using the LEGO P-Bricks PWM signal to assist in creating the non-sinusoidal periodic waveform.

Part 1 of the laboratory project will focus on generating the square wave PWM signal with Part 2 outlining assembly and test procedures for the triangle wave generation.

The BOM is as follows:

E-Beetle Bot

One 10KΩ, 1/4 -W Resistor

BricxCC IDE

NQC programming language software

One Radio Shack sensor lab breadboard or equivalent

One single- or dual-trace oscilloscope

One modified LEGO electric wire

Part 1: Square-Wave Generation Test and Measurement Set Laboratory Project

The objective of this laboratory project is to validate the square-wave frequency generation from E-Beetle Bot's Output *B* port. The existing software code embedded in program slot 1 will be modified to generate a 125 Hz square wave signal that is visible on the oscilloscope screen.

Assembly and Test Procedure The procedure is as follows:

1. Build the E-Beetle Bot square-wave generator test setup, using Figure 3-11.
2. Open the BricxCC IDE software and modify the code used in Listing 1 of this chapter with the following NQC command: *SetPower(Motor_B, 3)*. Place this command in the code, as shown in Listing 2.
3. Compile and download the modified code into program slot 2.
4. Turn on the oscilloscope and set the measuring instrument to the following parameters: vertical setting - 2V/div., horizontal setting - 5mS/div.
5. Run the modified code located in program slot 2. E-Beetle Bot should move forward for 10 seconds and stop. On the screen of the oscilloscope, there should be a 125 Hz square wave from output B of the LEGO P-Brick. If not, stop the program and check the code for incorrect output port assignment. Also, check the settings of the oscilloscope for correct waveform viewing. Check the modified LEGO electric wire and connector for correct orientation. Finally, check the external 10KΩ resistor attached across the electric wire. Once the error has been corrected, repeat this step.
6. Run the code for final validation of the E-Beetle Bot square wave generator functionality.

7. Turn off the E-Beetle Bot. Prepare for final circuit assembly of the op-amp integrator outlined in Part 2 of the laboratory project.

Part 2: Triangle Wave Generation Test and Measurement Set Laboratory Project
The objective of this laboratory project is to validate the triangle-frequency generation from an op-amp E-Beetle Bot's Output *B* port. The existing software code embedded in program slot 1 will be modified to generate a 125 Hz triangle wave signal that is visible on oscilloscope screen.

The BOM is as follows:

E-Beetle Bot, with Modified LEGO electric wire and 10KΩ resistor

One 82 KΩ, 1/4 -W resistor

One 100 nF capacitor

One 100KΩ, 1/41/4 -W resistor

One TLS272 dual op-amp

Assembly and Test Procedure The steps for the procedure are as follows:

1. Wire the op-amp integrator, as shown in the circuit schematic on Figure 3-12, using the Radio Shack sensor lab breadboard or equivalent circuit-prototyping unit. For proper operation of the op-amp integrator, the LEGO connector should be oriented on output *B* port with the wire pointing down. See Figure 3-11.
2. Check your wiring, correcting any errors before applying power to the circuit.
3. Attach the oscilloscope to the output pin of the op-amp integrator circuit. Turn on the measuring instrument. For best measurement results, set the oscilloscope for a vertical scale (volts) of 500mV/div AC and set the horizontal scale (time) of 5mS/div.
4. Apply power to the op-amp integrator circuit.
5. Run the square-wave frequency generator program stored in slot 1.
6. E-Beetle Bot should move forward for 10 seconds and then stop.
7. The oscilloscope should display a 125 Hz, 1.24$V_{P\text{-}P}$ (peak to peak voltage) and then turn off.
8. Repeat step 5, observing steps 6 and 7 for proper robot and signal generation operation.

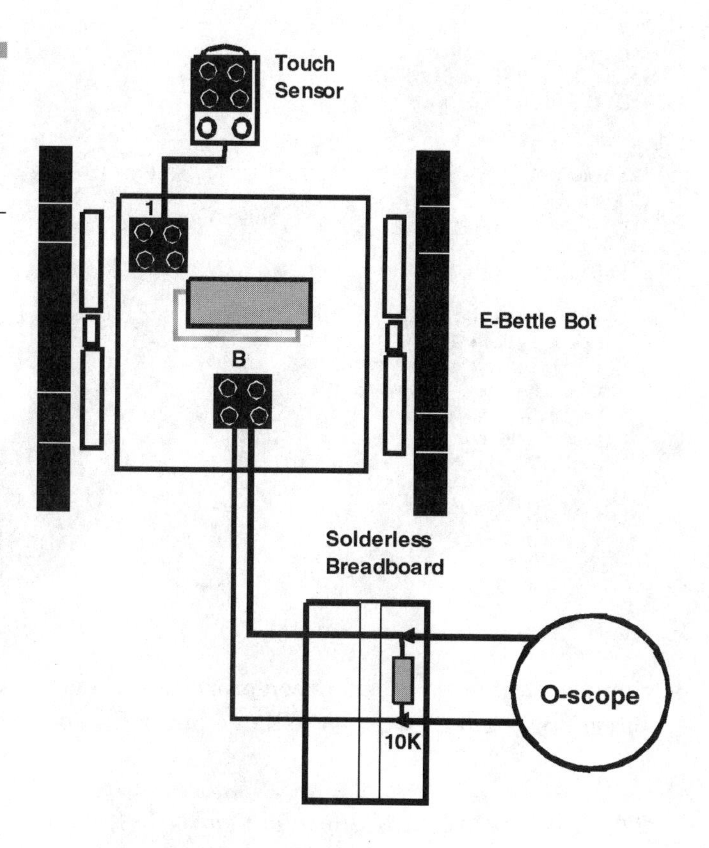

Figure 3-11
Test setup for the E-Beetle Bot square wave generator

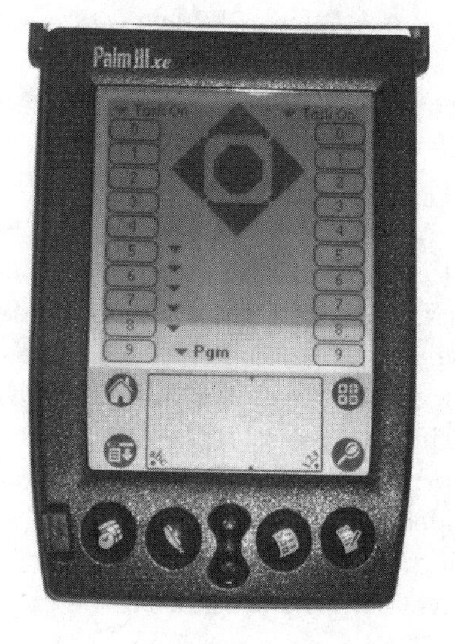

Figure 3-12
The brick remote software for the Palm handheld computer

Listing 3-1

```
#define Motor_A  OUT_A
#define Motor_B  OUT_B
#define Motor_C  OUT_C

task main ()
{
  SetSensor (SENSOR_1, SENSOR_TOUCH);
  SetPower(Motor_B, 3);
  while (true)
  {
    until (SENSOR_1==1);
    OnFwd (Motor_A + Motor_C);
    OnFor (Motor_A + Motor_C,  1000);
    Off (Motor_A + Motor_C);
    OnFwd (Motor_B);
    OnFor (Motor_B,  1000);
    Off (Motor_B);

  }

}
```

9. Stop the square-wave generation program and deactivate the op-amp integrator circuit power on the Radio Shack Sensor Lab.

TECH NOTE: *The Multimedia Control Lab, discussed in Chapter 1, is an excellent way to create a unique testing tool for electronic circuits. By using the BricxCC or the Java RCXTools IDE applications, the LEGO P-Brick becomes a versatile test instrument for validating audio circuits used as audible warning devices for Mindstorms mechatronics-based devices.*

Summary of E-Beetle Bot Function Generator Laboratory Project In Part 1 of the two-exercise laboratory project, the square wave generator was built using the LEGO P-Brick's internal PWM signal, controlled by the NQC command *SetPower(Motor_B, 3)*. The output signal was validated by visually viewing the non-sinusoidal waveform on an oscilloscope.

Part 2 consisted of wiring an op-amp integrator circuit, and attaching the inverting input pin to the modified LEGO electric-wire square wave output signal. After the embedded program of E-Beetle Bot was executed for 10 seconds of forward motion, a triangle waveform was observed on the oscilloscope's screen. The waveform was present on the measuring instrument for 10 seconds.

The idea behind this laboratory project was to illustrate the versatility of the LEGO P-Brick as a low-cost test instrument, and to demonstrate to the amateur roboticist a unique way to create two distinctive waveforms using P-Brick.

Palm Diagnostics and Robots

In Chapter 2, "Electronics Hardware Add-Ons," a discussion on the Palm handheld computer as a diagnostics controller was described, with a brief explanation of the concept represented in a pictorial diagram. The idea behind using the Palm computer in LEGO Mindstorms mechatronics-based work is to provide a pocket tool that will quickly validate a robot or intelligent device, without the overhead of a notebook computer or desktop machine.

Palm computing has found its way into various professions like medicine, law, engineering, and science, allowing practitioners to access information immediately, without the use of a notebook computer or desktop PC. As discussed in Chapter 2, the Palm handheld computer can access information about the embedded code residing inside the LEGO P-Brick, read sensor values, or directly control motor drives of a platform-based robot or intelligent machine.

In this final section, a Mindstorms software application for the Palm handheld computer will be explored through a hands-on laboratory project.

BrickRemote Software for Mechatronics Monitoring and Control of the LEGO P-Brick

This Palm handheld computer software application allows the amateur roboticist to directly monitor and control his or her LEGO Mindstorms robot or intelligent machine. The *BrickRemote* program was written by Michael Kory and made public (free software) to LEGO Mindstorms enthusiasts via the International LEGO Users Network (LUGNET, www.lugnet.com). Mike had a URL link listed on the Web site, but unfortunately the Web page doesn't exist anymore. Not to worry, the BrickRemote software can be found on the CD-ROM that accompanies this book.

The Palm software application is quite easy to use. It provides immediate control over the LEGO P-Brick's output ports using the infrared tower.

The infrared tower acts as the hardware transmitter and receiver host for sending and receiving data and motor drive commands to the P-Brick.

The following laboratory project will outline an assembly and test procedure for reading sensor data and controlling the motor drives of E-Beetle Bot.

The BOM is as follows:

E-Beetle Bot with attached touch sensor

BrickRemote software (found on CD-ROM)

One null modem adapter-9 pin, Radio Shack catalog number: 26-264

One gender changer-9 pin, Radio Shack catalog number 26-230

One LEGO Mindstorms serial infrared tower

One LEGO Mindstorms serial cable

One Palm handheld computer with cradle

Attaching the LEGO Mindstorms Serial Infrared Tower to the Palm Handheld Cradle Before proceeding with the lab project, the following assembly information for connecting the LEGO Mindstorms serial infrared tower to the Palm handheld cradle must be provided. The infrared tower serial cable and the Palms handheld cradle are pinned-out differently. The transmit-and-receive wire bundle in the infrared tower serial cable are different from the Palm wire harness.

Basically, the transmit and receive wires are crossed when directly connecting the tower serial cable to the Palm handheld cradle. A *null modem* adapter is required to correct this minor wiring problem. By placing the null modem adapter in between both cables, the transmit-and-receive pins will be connected directly to each other.

Another minor connectivity issue that must be resolved before the final cable assembly is to have the male pins plug into female terminals. A male-to-male gender changer will correct the problem by providing a final termination of the infrared tower, with a null modem adapter of male-to-female serial cable connection. Figure 3-13 shows the electrical component assembly for the serial communication connection of the infrared tower to the Palm handheld cradle. Both the null modem and gender changer adapters can be purchased from Radio Shack, at a cost of $6.59 each.

Assembly and Software Test Procedure The procedure steps are as follows:

1. Attach the Palm handheld cradle to serial port of the notebook computer or PC that has the Palm desktop on it.

2. Place the Palm handheld computer into the cradle in order to install the BrickRemote.prc file on it.
3. Install the BrickRemote.prc file onto the Palm Handheld computer, using the Palm Desktop software found on the CD-ROM that accompanies this book. Press the *HotSync* button on the cradle to start the Palm software download/install process.
4. Open the BrickRemote application installed on the Palm handheld computer. Figure 3-12 shows a screen shot of the Mindstorms software application.
5. Remove the Palm handheld cradle cable from the notebook or PC desktop serial port.
6. Attach the Palm handheld computer to the LEGO Mindstorms infrared tower, using the cradle as shown in Figure 3-13.
7. Place the infrared tower in front of E-Beetle Bot. Turn on the LEGO P-Brick.
8. In the center of the BrickRemote screen, touch your stylus to the *left* arrow. The left motor should activate, turning E-Beetle Bot slightly to the right.
9. With the stylus, touch the *right* arrow on the application screen. The left motor should activate, turning in the reverse direction. E-Beetle Bot should turn slightly to the left, to its original position.
10. Touch the *up* arrow on the application screen with the stylus. The right motor should activate, turning E-Beetle Bot slightly to the left.
11. Touch the *down* arrow on the application screen with the stylus. The right motor should activate, turning in the reverse direction. E-Beetle Bot should turn slightly to the right, to its original position.
12. At the top left-hand corner of the application screen, touch the stylus to the *Task On* down arrow. A dropdown box will appear. Touch *Task Off*, then back to *Task On*.

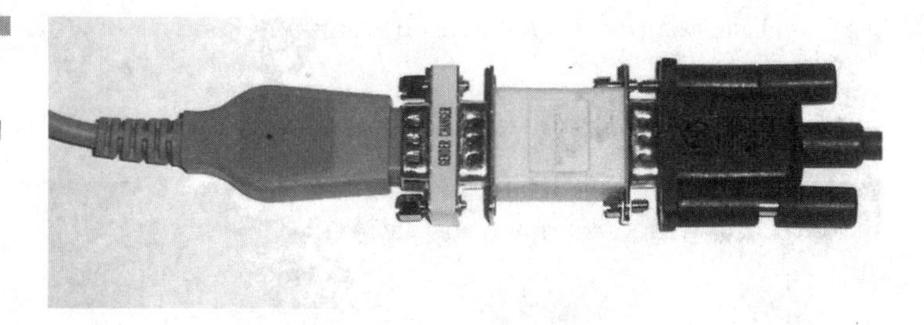

Figure 3-13 The electrical components used in attaching the Palm handheld computer to the serial infrared tower

13. Go to *Pgm* button arrow and a dropdown box will appear. Select *Pgm 1* with the stylus.
14. Touch the number *0* underneath the *Task On* button with the stylus. Program 1 should activate on the LEGO P-Brick, indicated by the little man running on the LCD screen.
15. Touch the stylus to change from *Task On* to *Task Off*.
16. Touch the number *0* underneath the *Task Off* button, with the stylus. Program 1 should have stopped running, indicated by the little man standing on the LCD screen.

The following procedure will outline the software assembly steps used in creating an E-Beetle Bot function generator.

1. Attach the op-amp integrator circuit, shown in Figure 3-10, to Output port B on the LEGO P-Brick.
2. Attach an oscilloscope to the output of the op-amp integrator circuit. For best measurement results, set the oscilloscope for a vertical scale (volts) of 500mV/div -AC and the horizontal scale (time) of 5mS/div.
3. Compile and download the square-wave generator NQC code, shown in Listing 3-1, to E-Beetle Bot. Accomplish this using the BricxCC IDE software. Store the program into slot 2 of the LEGO P-Brick.
4. Change to Program 2 with the stylus, using the *Pgm* dropdown box.
5. With the stylus, change the Task to *On* and touch the number *0*. A triangle wave should be displayed on the oscilloscope screen.
6. With the stylus, change the Task to *Off* and touch the number *0*. There should be no waveform displayed on the oscilloscope screen.
7. Touch the stylus to the dropdown box next to the number *5* on the BrickRemote application screen on the Palm handheld computer.
8. Select *Sensor 1* with the stylus. The binary value from the sensor should be *0*.
9. Press and hold the touch sensor attached to input port 1 on the LEGO P-Brick. A binary *1* should be displayed on the BrickRemote application screen.
10. Turn off the Palm Handheld computer and E-Beetle Bot.

This completes the Palm Diagnostics Controller laboratory project, the output of which is shown in Listing 3-2.

Listing 3-2

```
#define Motor_B  OUT_B
task main ()
{
    SetPower (Motor_B, 3);
    while (true)
   {
      OnFwd (Motor_B);
   }
}
```

Summary of BrickRemote Software Laboratory Project In the final laboratory project of this chapter, a Palm Diagnostics Monitoring Control tool was investigated. The BrickRemote software for the Palm handheld computer can be used with the portable tool to control LEGO Mindstorms mechatronics-based robots and intelligent machines quite easy and conveniently. The steps performed within the laboratory project demonstrated the capability of the BrickRemote software to read sensor data and control the movement of E-Beetle Bot's motor-driven legs.

The second part of the laboratory project explored creating a stand-alone function generator and controlling it with a Palm handheld computer. The square wave NCQ code, shown in Listing 3-2, allows E-Beetle Bot to generate a continuous stream of pulses that are converted into a triangle wave signal using the op-amp integrator circuit.

The intent of the presented material was to provide a foundation for creating RDA-based instruments that can perform a variety of functions. They can evaluate internal software controls. They can investigate external electromechanical interfaces and electronics-based interfaces for LEGO Mindstorms mechatronics-based robots and machines. The laboratory projects outlined in this chapter will be further explored (Chapters 5-10) of this book.

TECH NOTE: *On the CD-ROM that accompanies the book, L-Remote is another Palm handheld computer software application written by Mike Kory. This software application emulates the handheld Infrared Remote Control unit packaged with LEGO Mindstorms Ultimate Accessory kit. Features like turning on/off output ports A-C, beep control, forward/reverse motor direction, and selection of embedded programs are possible with the L-Remote software application.*

CHAPTER 4

Electronic Circuit Simulators and MLCAD

Using *computer-aided engineering* (CAE) tools, electronic circuits can be tested before the first wire is inserted into a solderless breadboard. Using industry standard design methods and practices, CAE-specialized software packages are used in the creation of mechanical and/or electrical/electronic products.

Industry trade publications such as *EDN Magazine*, *Electronic Engineering Times* (*EET*), and *Electronic Design* provide all the insider-news regarding *electronic design automation* (EDA), CAE tools, technology, business, and people. EDA is a design practice, whereby the product development process for creating electronic devices and systems is automated using software application tools. To an amateur roboticist, circuit-modeling and simulation tools can be invaluable resources. They can develop simple and complex electronics for *input/output* (I/O) interfaces for the Scout *programmable brick* (P-Brick) or the LEGO P-Brick (*robot command explorer,* or RCX). By creating a schematic wiring diagram on either a PC or notebook computer monitor and executing the electrical network model, the designer can quickly validate the function of the target circuit using the embedded simulation tools packaged with the circuit-simulator software. The results of the circuit simulator can be viewed on the screen after the simulation event or during the program execution of the target circuit.

Mikes LEGO Computer Aided Design (MLCAD) is a mechanical design software-package that enables the LEGO enthusiast the ability to build LEGO models on a PC or notebook computer screen. The software package is free, and a large support group of developers and users exist, in addition to an MLCAD Web page on the Internet.

By using MLCAD software, you have a virtual tool for capturing mechanical designs of robotic mobile-platforms, grippers (end-effectors), and body structures. MLCAD, along with circuit simulators, provides a complete set of design tools for developing mechatronics-based robots and intelligent machines. These tools allow you to go from concept to intelligent machine.

The chapter will explore circuit simulator tools and MLCAD with several laboratory projects. The following topics will demonstrate circuit simulation design-analysis and mechanical design techniques for LEGO Mindstorms mechatronics development:

- CircuitMaker 6 software
- PSpice-OrCAD
- MLCAD

CircuitMaker 6 Software

In Chapter 7 of *LEGO Mindstorms Interfacing*, the concept of *electronic computer-aided engineering* (ECAE) was explained on pages 236 through 240, using a *Power MOSFET* (PMOSFET) timer. In Chapter 8, pages 261 through 266, the concept was discussed using a monostable circuit. As illustrated in these discussions, CircuitMaker 6 software is a user-friendly design environment in which you can test and analyze electrical/electronic circuits for LEGO Mindstorms mechatronics-based robots and intelligent machines. The circuit schematics shown in *LEGO Mindstorms Interfacing*, as well as the schematics in this book, were done using CircuitMaker 6 software. The following paragraph describes the evolution of CircuitMaker. It is quoted from the Web site `www.microcode.com/`.

About CircuitMaker

"The CircuitMaker product family was originally developed by MicroCode Engineering, which was established in 1987 by three engineers from Signetics. During the first year, MicroCode primarily developed design and characterization software for Philips Semiconductors and Motorola on a consultant basis, then in 1988 released the first edition of CircuitMaker® for the Macintosh. The president and founder, Ozzie Boeshans, wrote the CircuitMaker product. Unable to find an easy-to-use, readily available product for simulating electronics, Ozzie decided to create one of his own, and CircuitMaker is the result.

"In 1998 MicroCode was acquired by Protel International (now Altium Limited), who shares a common interest in taking advanced technologies and making them easy to use.

"CircuitMaker products transform the designer's desktop into the complete virtual electronics lab, allowing users to quickly and easily develop designs, test the real-world performance of their designs through mixed-signal, SPICE-compliant simulation, and even generate prototype boards. Ideal for those interested in design exploration or rapid prototyping, CircuitMaker is characterized by exceptional ease-of-use and tight integration between schematic and simulator, and provides all the features of high-end software at a fraction of the cost."

The key to CircuitMaker is that it enables users to *quickly* and *easily* develop designs and test their performance. The following pages will illustrate this philosophy with detailed laboratory test procedures.

Setting up CircuitMaker 6 Software

Obtain CircuitMaker from the Web before running the laboratory test procedures. Their URL is `www.microcode.com/`. Figure 4-1 shows the homepage for the CircuitMaker software. The homepage contains a link to obtain the student version of the software. Click the link and follow the instructions to download a free version of the software, the CircuitMaker 6 package. Once the software is installed on your PC or notebook computer, the following laboratory projects can be carried out.

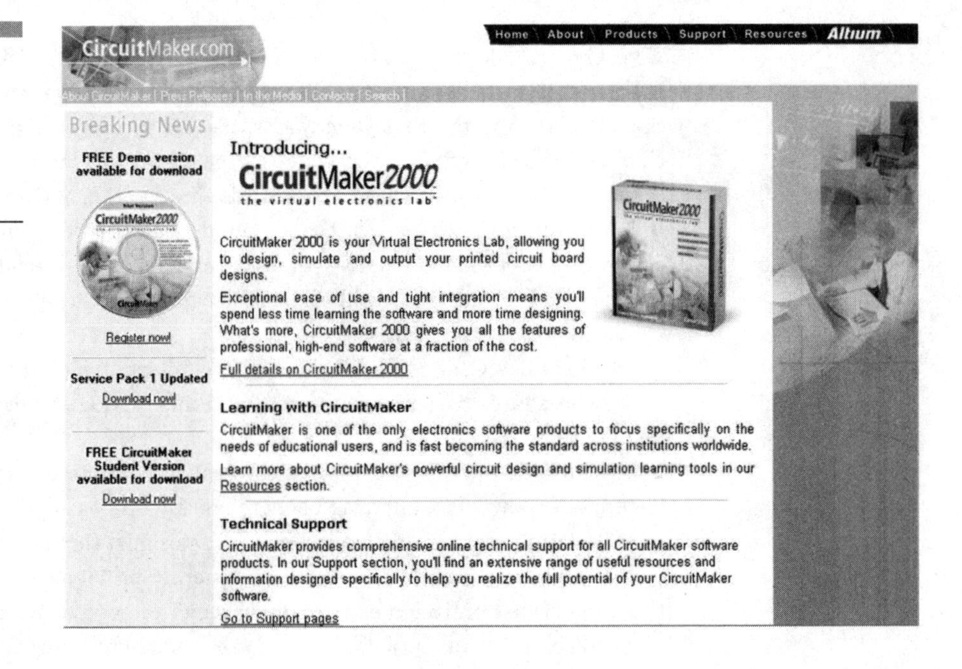

Figure 4-1 The Web homepage for obtaining CircuitMaker software

The Integrator Circuit Simulation Lab Project

The laboratory project test procedures outlined will illustrate how quick and easy CircuitMaker software is to use. This project will validate the development exercise discussed in Chapter 3, "Instrumentation and Robots," the exercise in turning E-Beetle Bot into a walking *robot digital assistant* (RDA) triangle-wave generator. Use following generic parts to build the basic integrator circuit discussed in Chapter 3.

The BOM for the CircuitMaker 6 software project consists of the following items:

Two resistors

One non-polarized capacitor

One ideal *operational amplifier* (op-amp)

One signal generator

The Circuit Model Assembly and Test Procedure The purpose of this laboratory project is to demonstrate the virtual function of the basic integrator circuit used in the triangle-wave experiment of Chapter 3. In addition, this simulation activity demonstrates the link between circuit software verification and physical prototypes, as related to electrical function and performance.

Creating a Virtual Square-Wave Generator Use the following steps to create a virtual square-wave generator:

1. Open the CircuitMaker 6 software, as shown in Figure 4-2.
2. Click *NEW* under the File drop-down menu task bar. A clean design sheet will be displayed on the PC or notebook screen.
3. Save this design onto your PC or notebook hard drive as *Integrator.ckt*.
4. Select the *transistor-capacitor* icon button from the Parts Library, located underneath the main menu task bar. Click the button and a Device-selection dialog box will appear on the screen. See Figure 4-3.

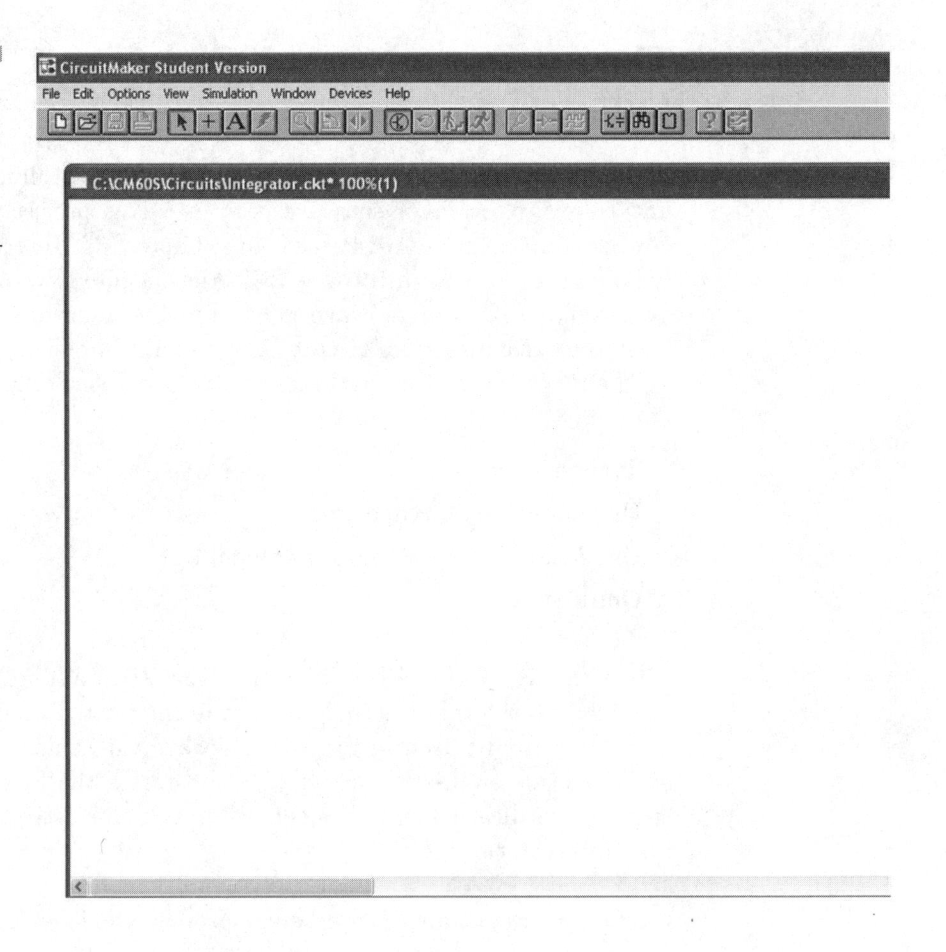

Figure 4-2 A clean design sheet upon opening CircuitMaker 6 software

5. Within the device selection dialog box, under the *Major Device Class* heading, scroll down until *Instruments* is found. Click *Instruments*. See Figure 4-3 for the instrument device selection choice.
6. Select *Analog* and *Signal Generator*, under *Minor Device Class* and *Device Symbol*. In the analog device box, a signal-generator icon will be present.

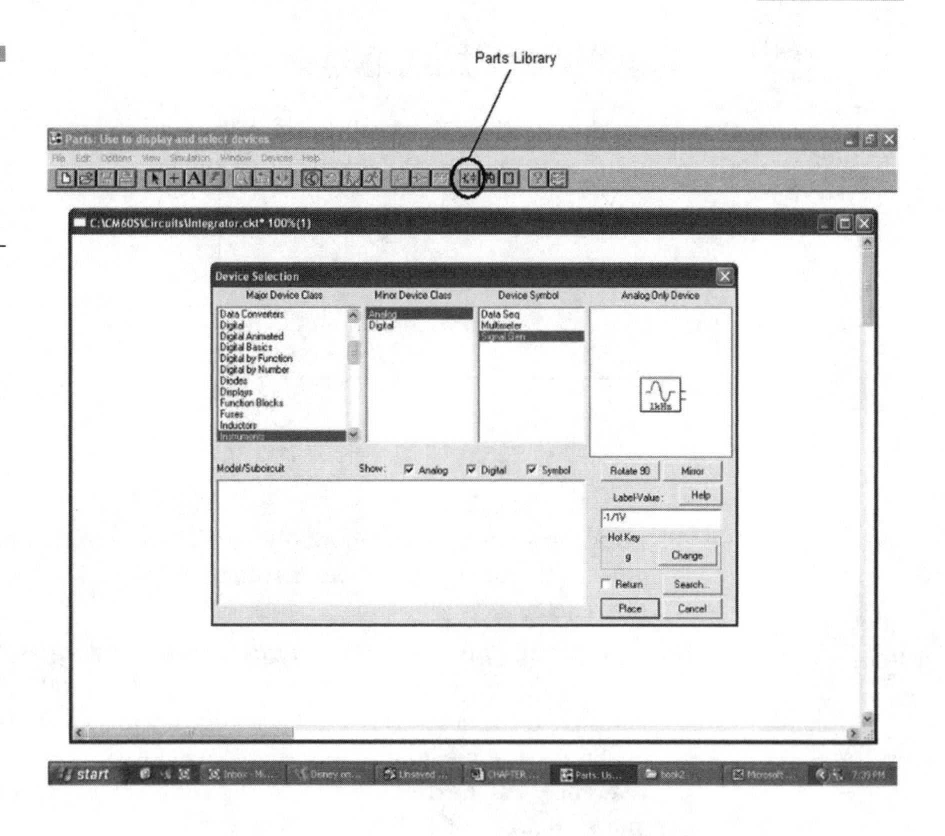

Figure 4-3 Preparation for adding the signal generator symbol to the design sheet

7. Click *Place*, at the bottom of the dialog box. The Device selection dialog box will be removed from the screen. The signal-generator icon will be on the screen. Move the cursor to right of the screen, dragging the icon. Click the left mouse button to place the icon on the screen. See Figure 4-4.
8. To change the signal pattern of the icon from a sine to square wave, right-click with the mouse. From the drop-down menu box, select *Edit Signal Generator*. The edit sine-wave data dialog box will be displayed on the screen.

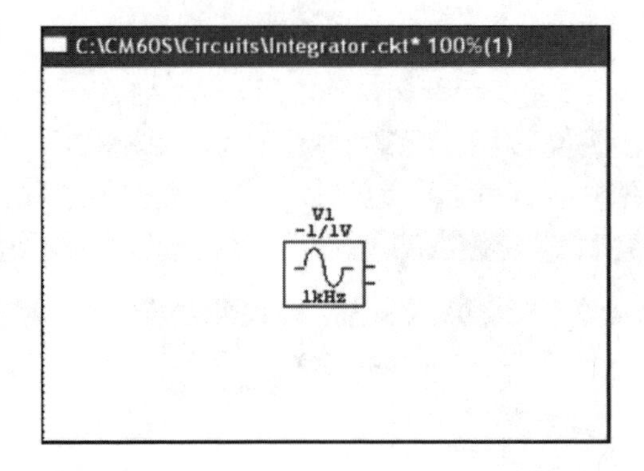

Figure 4-4 Final placement location for the Signal Generator instrument

9. Click the *wave* button, followed by the *Pulse* selection button. The edit pulse data dialog box will appear to the screen. See Figure 4-5 for reference.
10. Change the *Pulse Amplitude*, *Pulse Period* , and *Pulse Width* text edit boxes with the values shown in Figure 4-6. Click *OK* after entering the values. On the screen should be a square-wave signal with a frequency of 125 Hz and a signal amplitude value of 0/7V. See Figure 4-7 for reference. The next step will test the signal by using an oscilloscope and a virtual test probe.

Testing the 125 Hz Square-Wave Signal Using a Virtual Test Probe In order to test the 125 Hz square-wave signal or any circuit model using the software simulator, a ground must be provided. Use the following steps to test the 125 Hz square-wave signal, using a virtual test probe:

1. Using the Figures 4-8(a) and(b), wire the ground displayed on the PC or notebook screen to the square-wave signal generator, using the wiring tool located below the main menu task bar. After the final wire termination point is complete, press *Esc* to secure the wire to the target electrical/electronic icon on the screen.
2. Using Figures 4-9(a) through (c), wire the *Vin* terminal to the square-wave signal generator, using the wiring tool located below the main menu task bar. Click the *Rotate 90* button the appropriate number of times to orient it as shown in Figure 4-9(a).
3. Go to the *Simulation* main menu task bar. Click *Analyses Setup*. An analyses setup dialog box will appear on the screen.

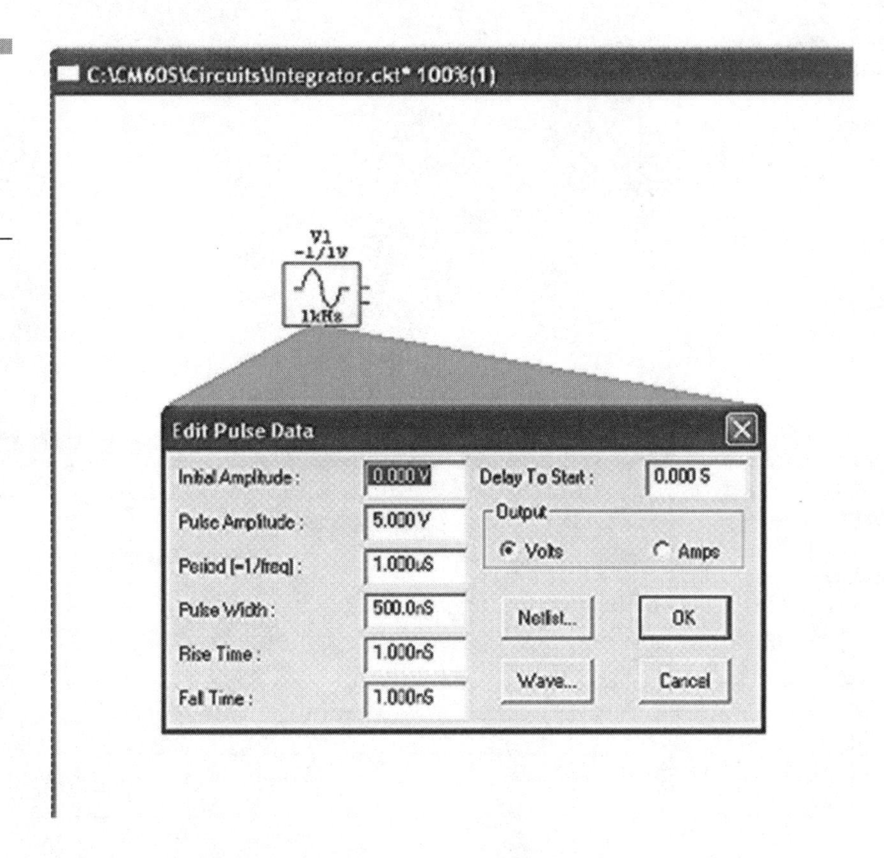

Figure 4-5 The pulse data dialog box, for creating a virtual square-wave signal generator

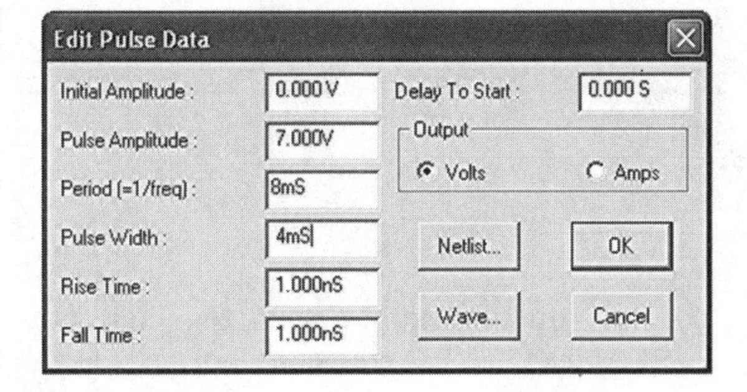

Figure 4-6 Creating a 125-Hz square-wave signal

4. Click the *Enable* check box next to the *Transient/Fourier* box. See Figure 4-10.
5. Click the *Transient/Fourier* box.

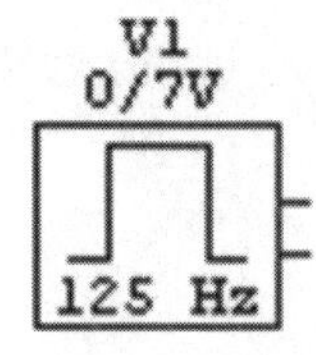

Figure 4-7 The finished 125-Hz square-wave signal generator

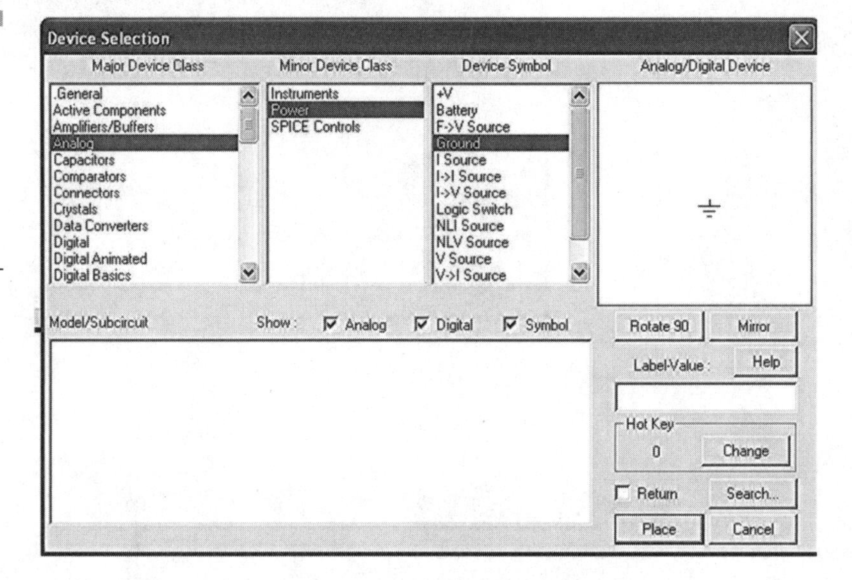

Figure 4-8(a) Adding a ground to the 125-Hz square-wave signal generator using the device selection dialog box

6. Enter the data into the text box, using the information as shown in Figure 4-11. Click OK . The transient/fourier-analyses dialog box will disappear from the screen.
7. Select the *Run Analyses* button on the *Analyses Setup* dialog box. An empty oscilloscope waveform display will appear at the bottom and a multimeter will be located at the left-corner screen.
8. Select the test probe located next to the stop sign underneath the main menu task bar.

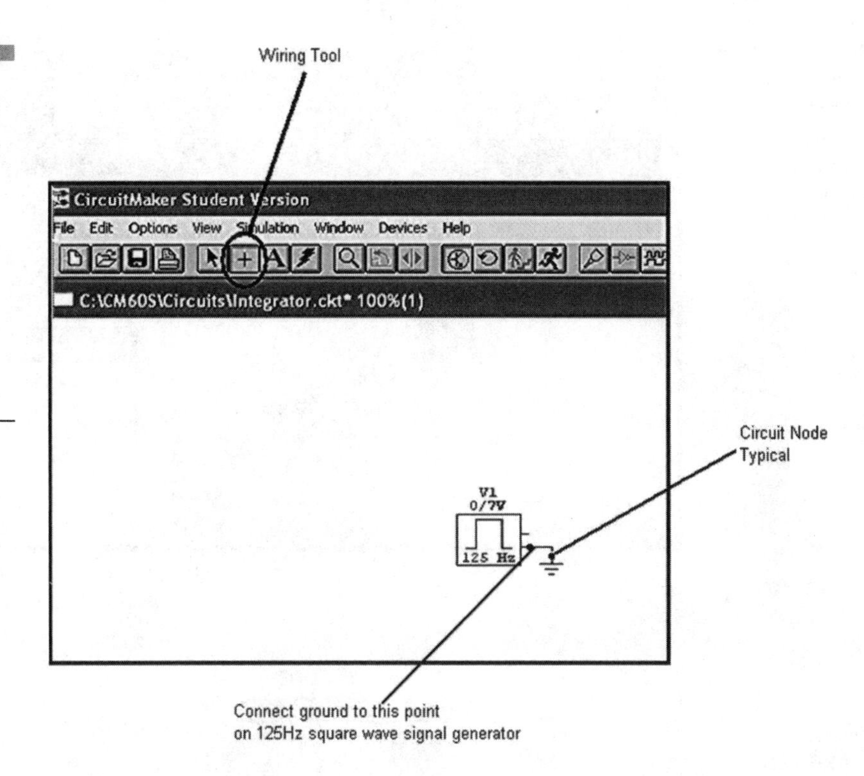

Figure 4-8(b) Use the wire tool to connect the ground to the 125-Hz square-wave signal generator. Note the circuit nodes on the ground and signal-generator instrument.

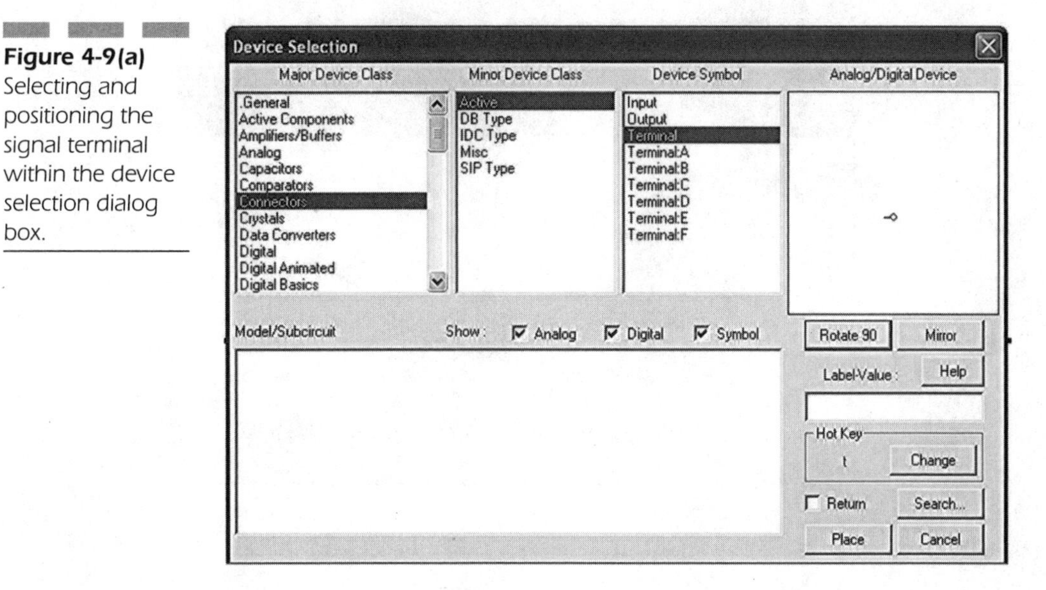

Figure 4-9(a) Selecting and positioning the signal terminal within the device selection dialog box.

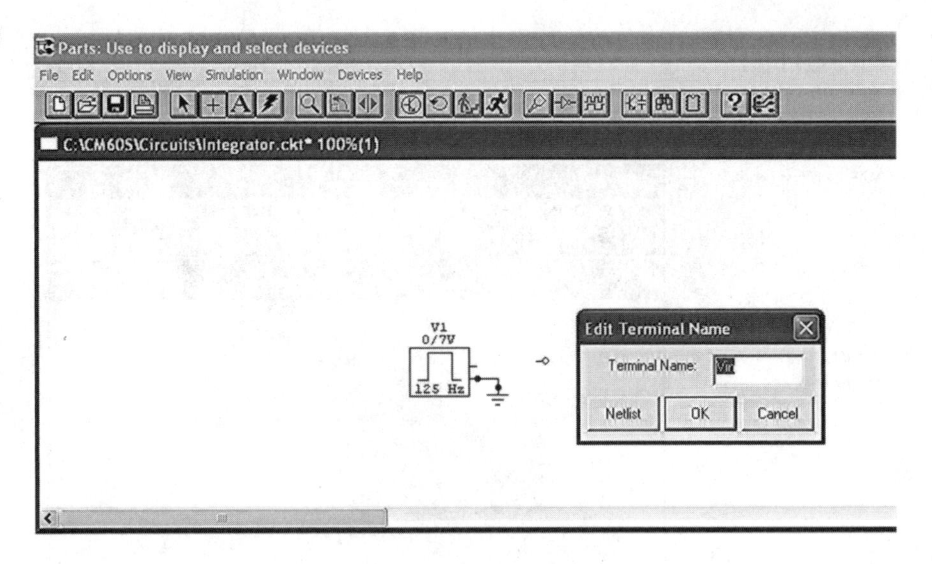

Figure 4-9(b) Naming the signal terminal with the edit terminal-name dialog box.

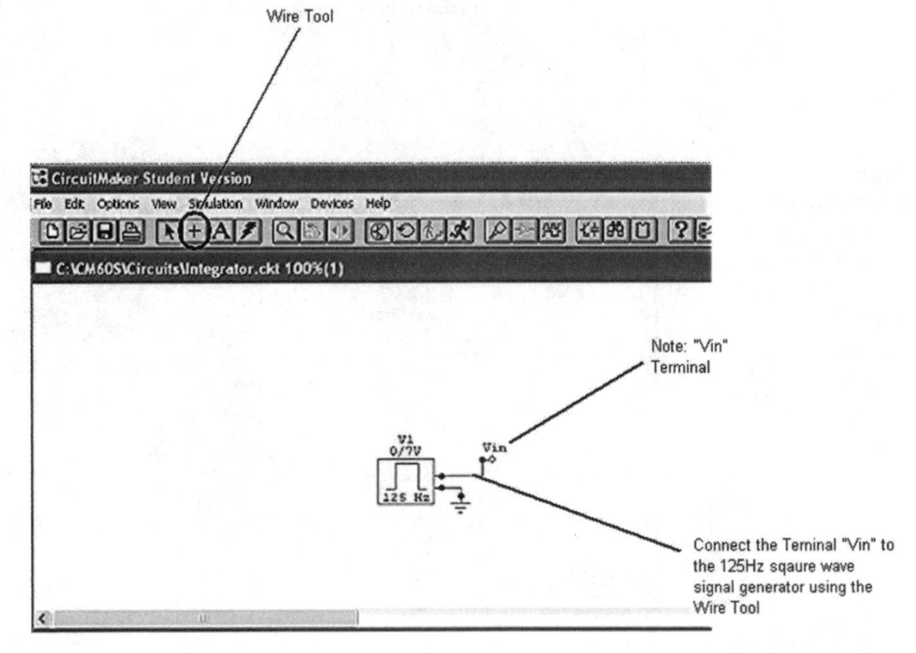

Figure 4-9(c) Using the wire tool to connect the Vin terminal to the 125-Hz square-wave signal generator.

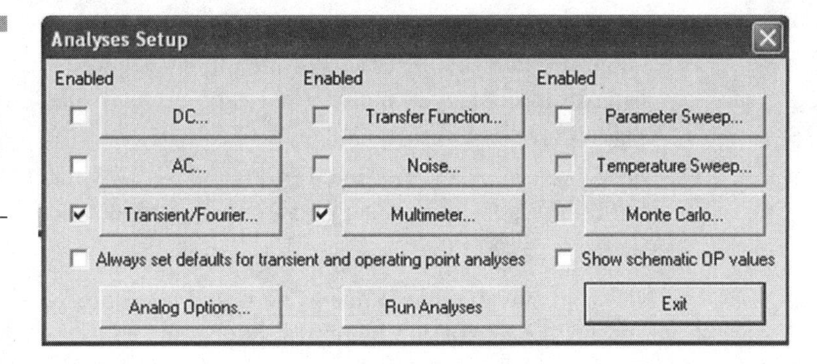

Figure 4-10 Analyses setup dialog box within the Simulation main menu task bar.

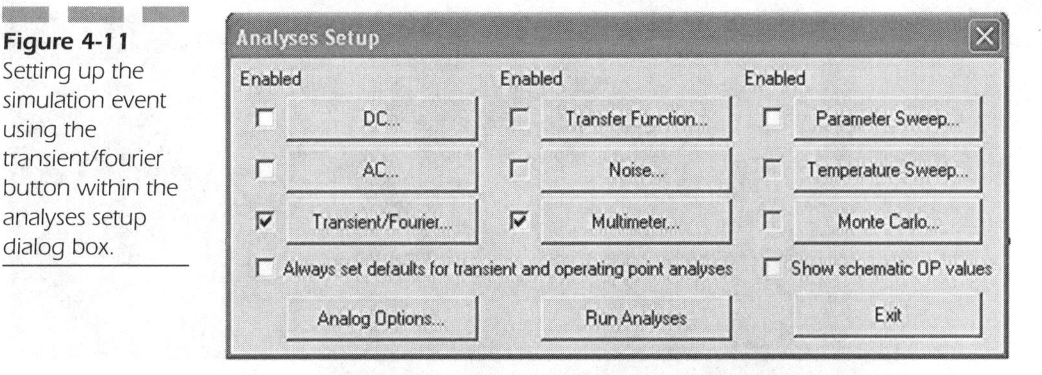

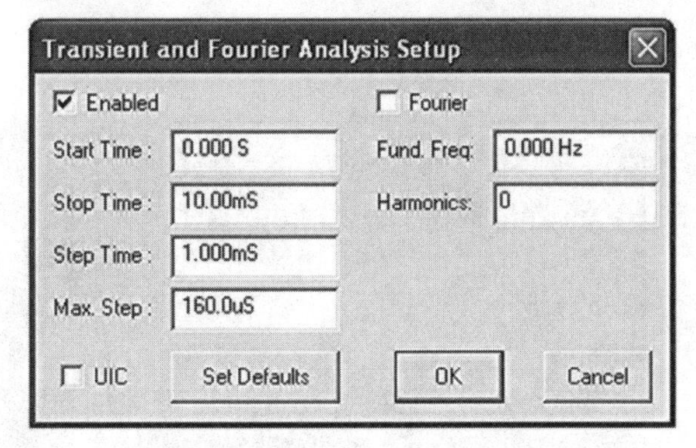

Figure 4-11 Setting up the simulation event using the transient/fourier button within the analyses setup dialog box.

9. Select the transient (oscilloscope) waveform display.
10. Touch the test probe at the *Vin* terminal node of the square-wave signal generator. A two-cycle square wave should be displayed on the oscilloscope. See Figure 4-12.
11. Move the *a* marker, as shown in Figure 4-13. Notice the frequency reading. What is the measured frequency displayed on the oscilloscope? $\text{Freq}_{\text{measured}}$__________.

Now that the virtual square-wave signal generator is operating correctly, the integrator circuit can be built.

Building the Basic Integrator Circuit The following assembly and test procedures will outline the technique of building a virtual basic integrator circuit using the electronic components listed in the software BOM. After

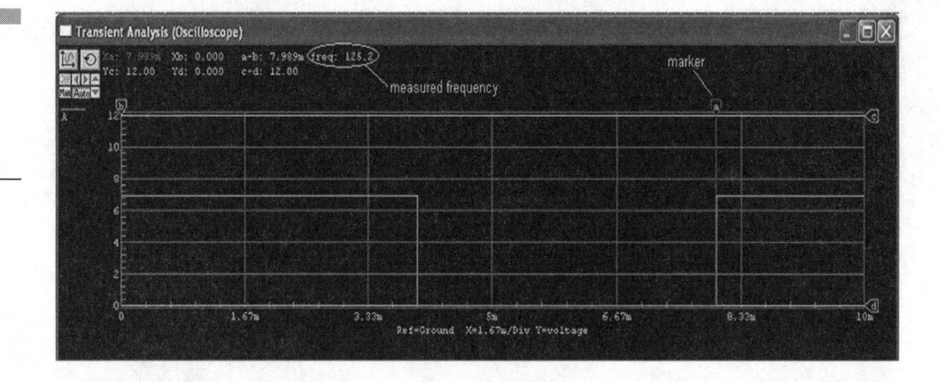

Figure 4-12 A two-cycle square wave measured with the test probe.

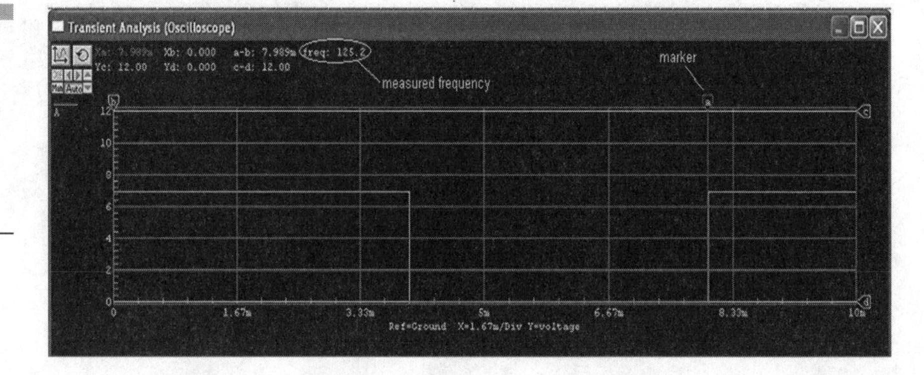

Figure 4-13 Measuring the frequency of the displayed square waveform, using the marker "A" tool.

wiring the circuit, the output signal will be a triangle waveform, as was the case with the actual breadboard device investigated in Chapter 3, using E-Beetle Bot.

In order to complete this procedure, gather the following electronic components from the device selection (parts list) dialog box:

- One ideal op-amp
- Two resistors
- One ground
- One capacitor

Use the following assembly and test procedure to build the basic integrator circuit:

1. Using Figure 4-14, wire the basic integrator circuit with aid of the wire tool, as discussed in the "Creating a Square-Wave Signal Generator" section.
2. Click the 1 KΩ resistor, and right-click the button to select *Edit Device Data*. An *Edit Device Data* dialog box will appear. See Figure 4-15.
3. Change the label value to *82k* and click OK. The new value should be displayed above the resistor. Repeat this procedure for the other resistor and the capacitor. The values for the other components are:

   R2 = 100 K

   C1 = 100 nF

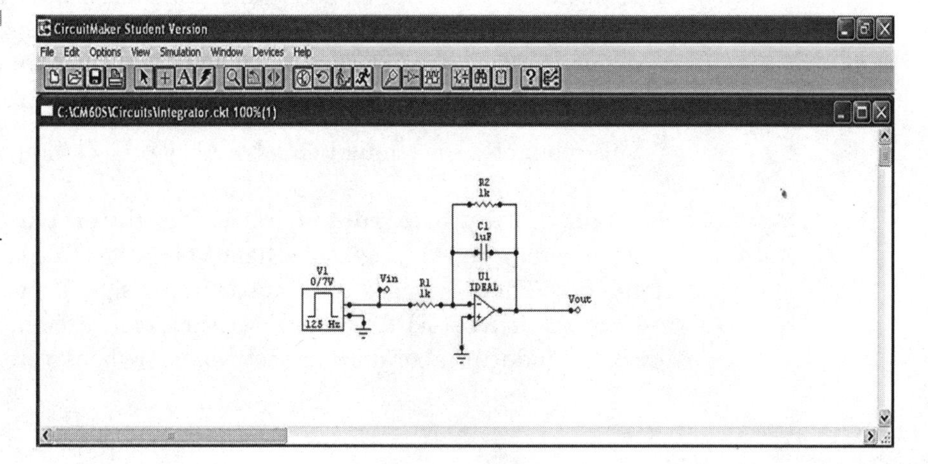

Figure 4-14 The electronic components from the device selection dialog box wired as basic integrator circuit.

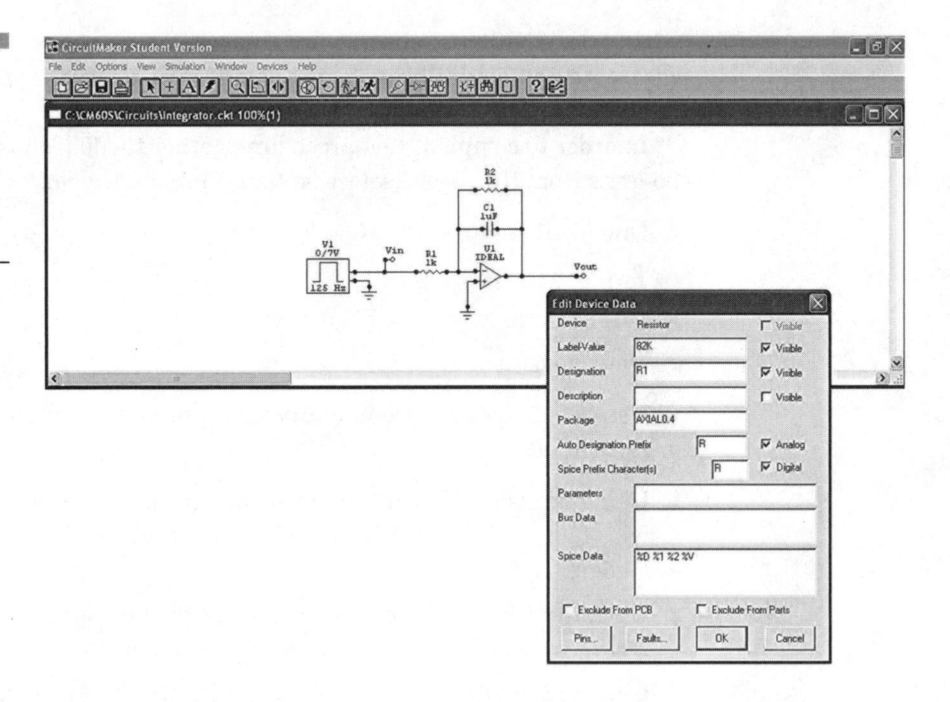

Figure 4-15 Changing R1 (1 KΩ) to an 82 KΩ value, using the Edit Device Data dialog box

4. Run the modified circuit model using the running man located before the test probe.
5. Select the transient (oscilloscope) window and touch the test probe at the *Vout* terminal. An elongated triangle waveform should be displayed on the oscilloscope window. See Figure 4-16.
6. To display more cycles on both the input square-wave and output triangle waveforms, edit the *Stop Time* using 30 milliseconds (mS) within the Transient/Fourier dialog box. Run the analysis again and note the number of cycles. See Figures 4-17(a) and (b).

This completes the simulation laboratory project. Stop the simulation.

Summary of the Integrator Circuit Simulator Laboratory Project

The laboratory project provided a hands-on look into circuit modeling and simulation. The steps used to create a virtual signal generator and circuit are common in most EDA and CAE software tools used in the industry. The intent behind this laboratory project was to demonstrate how the simula-

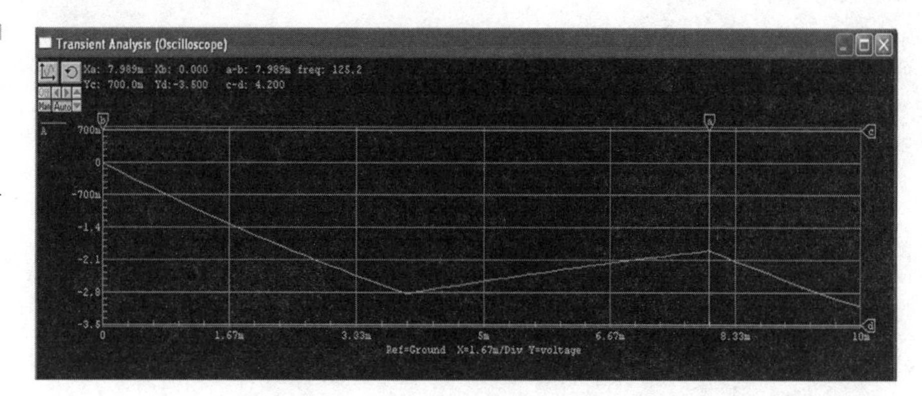

Figure 4-16 Output waveform of the basic integrator circuit simulation model.

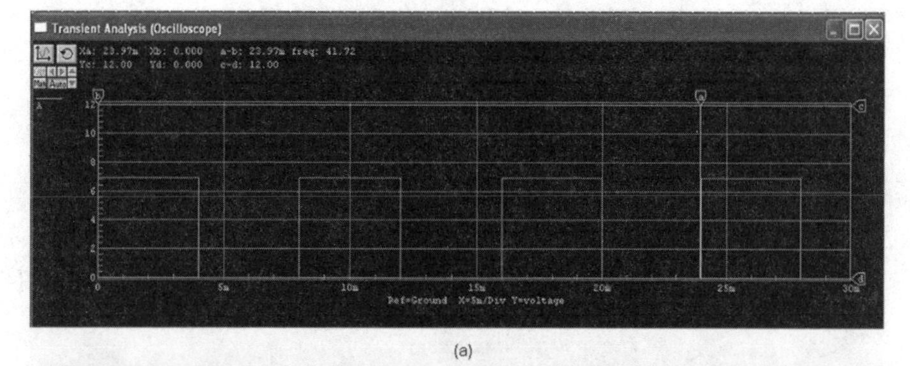

(a)

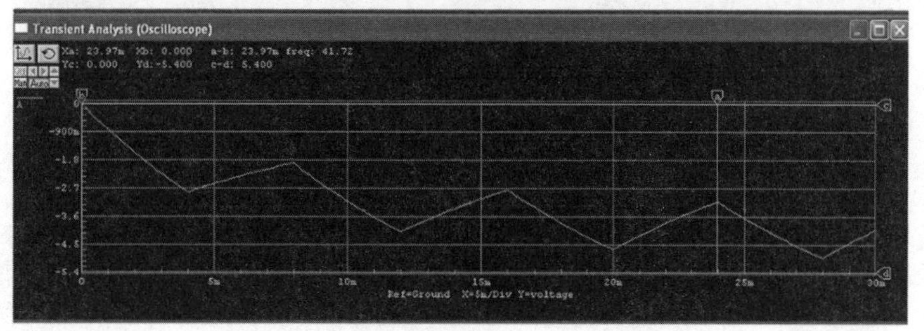

(b)

Figure 4-17 **(a)** Input waveform with four cycles. **(b)** Output waveform with three cycles.

tion of simple and complex circuits can aid in developing electronic circuit-interfaces prior to building prototypes using solderless breadboarding techniques and tools.

The basic integrator circuit built in Chapter 3 and the experimental virtual unit in this section should have similar input and output electrical characteristics. Therefore, by building a virtual circuit and running simulation tests, you will have a good understanding of how the prototype breadboard unit will work once it is running in the lab. Building virtual circuits and running simulations before building a prototype will be the initial step in designing electronic I/O interfaces for LEGO Mindstorms mechatronics-based robots and intelligent machines.

The next topic will look into using an industry-based CAE tool, known as *PSpice*-OrCAD™, for building virtual circuit models used in conducting electronics I/O interface design and analysis studies for LEGO Mindstorms mechatronics.

TECH NOTE: *Electronic circuit interfaces can be prevalidated using the simulation techniques discussed in the basic integrator circuit laboratory project. Use the steps outlined in the software assembly and test procedures to develop unique circuit blocks for I/O interfacing to the LEGO or Scout P-Brick.*

TECH NOTE: *The monostable circuit on page 262 in LEGO Mindstorms Interfacing shows little dots at every component and wiring connection. These dots are known as nodes. A node is an electrical junction that has two or more electrical elements connected together. During circuit analysis, these nodes assist in collecting data about the circuit when the test probe is used.*

PSpice-OrCAD

SPICE stands for *Simulation Program with Integrated Circuit Emphasis*. Spice was originally an analog circuit-simulator. The software was developed at the University of California at Berkley. PSpice is the commercial derivative of SPICE2 and was developed by MicroSim Corporation. SPICE2 belongs to the public domain because its development was based on U.S. public funds. SPICE2's robustness and power made it an industry standard. The major mathematical algorithms of SPICE2 are used by PSpice and conform to the input syntax of the modeling tool.

PSpice Circuit Design and LEGO Mindstorms Mechatronics

PSpice helps the designer simulate the circuit's design function graphically on the PC or notebook computer before building physical hardware. This is an appealing feature because the designer can make the necessary changes on the virtual prototype without modifying any circuit hardware. The benefit to you is that PSpice enables you to check the operability of the target circuit I/O interface to be validated using real-life simulation events. Because all the tests, designs, and modifications are made on the PC or notebook computer, you can save money. Buy only the appropriate electrical/electronic components required for circuit-block interfacing to LEGO Mindstorms mechatronics-based robot or intelligent machine.

Obtaining PSpice-OrCAD from the Internet

You can download the student version of industry software package free of charge. The Cadence Web site address is `http://www.cadencepcb.com/products/downloads/PSpicestudent/default.asp`. See Figure 4-18.

The version 9.1 software introduces the college or university student to circuit modeling and simulation techniques using a Windows-based *integrated design environment* (IDE). The methods used in the Circuit Maker 6 laboratory project, discussed in the previous section of this chapter, can be applied when using PSpice-OrCAD. To download the software under the *Products* heading, select the *PSpice 9.1 Student Version*. Download both PSpice 9.1 Student Version (28 MB) and the supporting documentation (13 MB) files to your PC or notebook computer hard drive. See Figure 4-19. Once the zipped file is stored on your hard drive, uncompress the folder to gain access to the *Setup* executable file. Run the executable file, following the software prompts for installing the program to your computer.

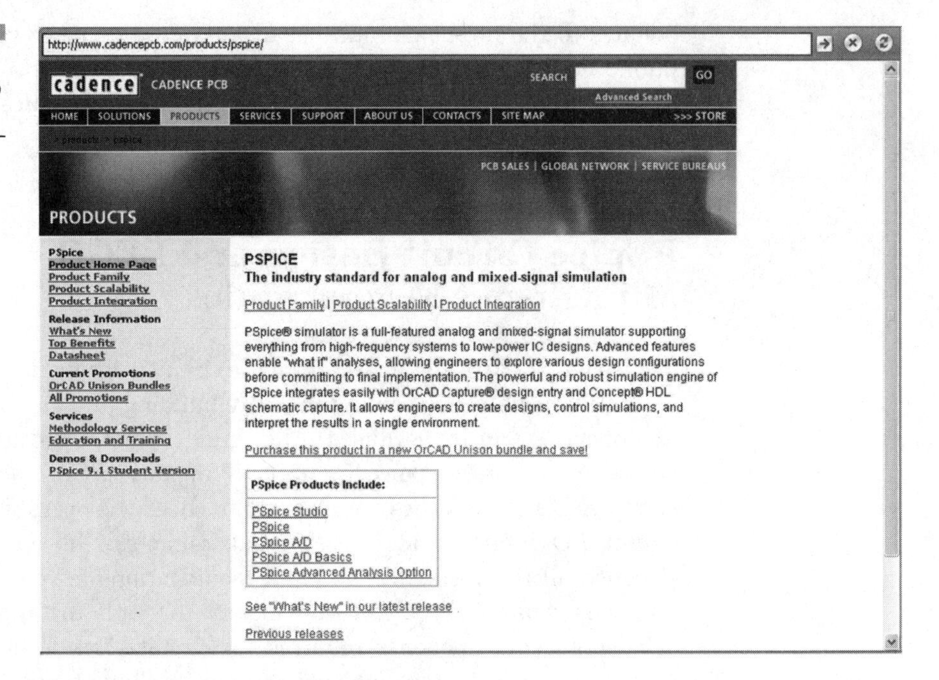

Figure 4-18
The Cadence web site

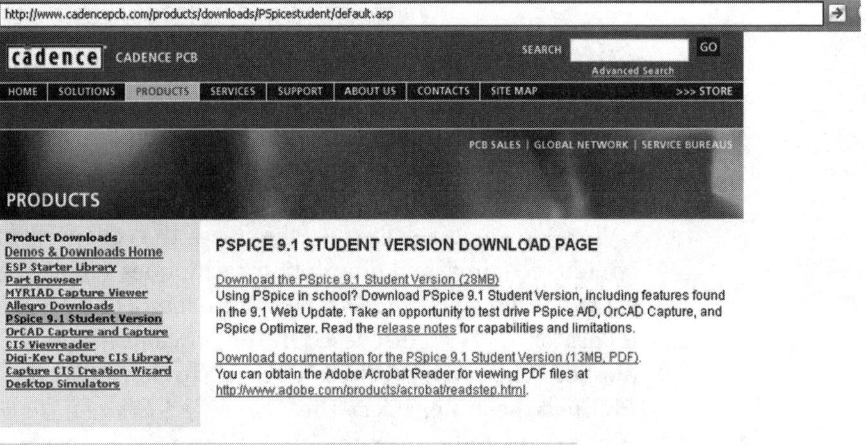

Figure 4-19
PSpice 9.1 Student Version download web page

Proof of Concept (POC) Simulation Study

The following POC will describe the functional objectives of an input circuit block for driving a LEGO CAM camera. A basic electronic switch using three resistors and a transistor will be described. Threshold conditions for turning the electronic switch on and off will be provided. This information will provide a baseline for validating the proper operation of the circuit using the PSpice simulation software. This circuit simulation will be further explored in Chapter 5, "Telerobotics."

A Rotary-Sensor Control Circuit Description This short telerobotics-control experiment requires a motorized DC drive for moving a LEGO CAM camera. A 50 KΩ potentiometer is wired as a rheostat to simulate a rotary sensor. A rheostat is a terminal resistor whose resistance can be varied using an attached rotating shaft. A voltage-divider bias transistor-circuit will take the threshold signal established by the rotary sensor to activate the electronic switch. Upon reaching an angular position from the rotary sensor, input port 1 of the LEGO P-Brick will read the output control signal from the control circuit. The LEGO CAM camera will rotate using the motorized DC drive. Figures 4.20(a) and (b) show the two conditions for turning on the LEGO CAM motorized DC drive.

Voltage-Divider Bias Basics Voltage-divider bias circuit topology is the basis for a single DC source to provide forward-reverse bias to the transistor. By using two series resistors, a voltage activation threshold can be established for switching the transistor on and off. Look at the following design analysis for applying a voltage-divider bias scheme to the rotary-sensor control circuit.

Using the voltage-divider rule, a 0.7V threshold is required for turning on the rotary-sensor control circuit:

$$V_A = \frac{R_A \times V_{CC}}{R_I + R_A}$$

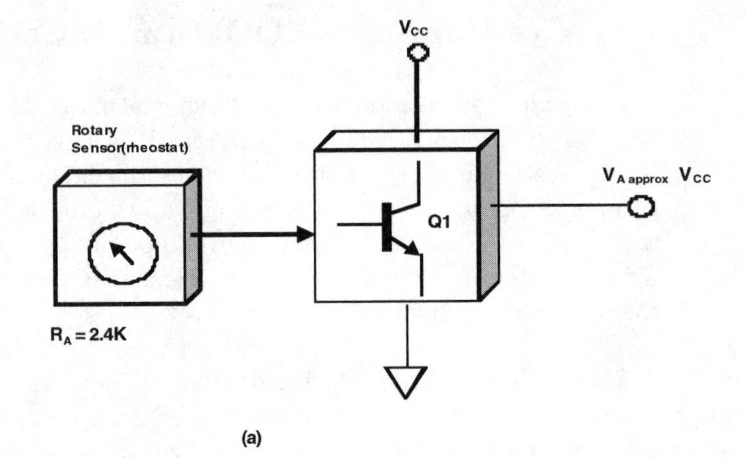

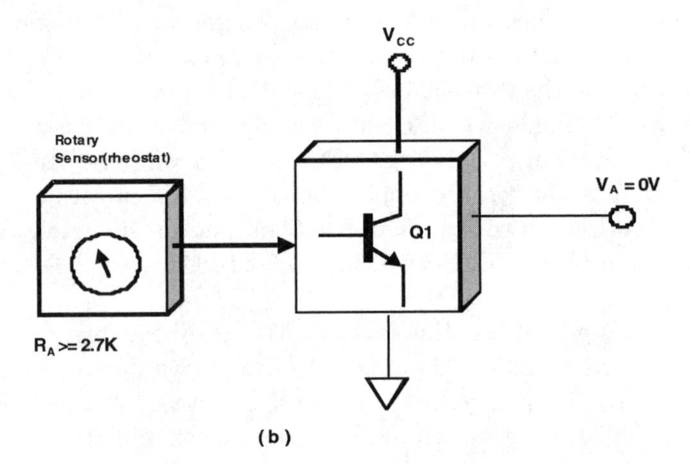

Figure 4-20 **(a)** Condition for turning Q1 transistor off, **(b)** Condition for turning Q1 transistor on

Solving for R_A:

$$V_B(R_I + R_A) = R_A \times V_{CC}$$

$$V_B \times R_I + V_B \times R_A = R_A \times V_{CC}$$

$$V_B \times R_I = R_A \times V_{CC} - V_B \times R_A$$

$$V_B \times R_I = R_A \times V_{CC} - V_B \times R_A$$

$$V_B \times R_I = R_A \times (V_{CC} - V_B)$$

$$\frac{V_B \times R_I}{V_{CC} - V_B} = R_A$$

With the equation below available and using R_1 as 33 KΩ, the appropriate resistance value to be dialed-in from the 50 KΩ potentiometer can be found. Substituting the correct values into the equation, R_A equals

$$R_A = \frac{0.6V \times 33K}{9V - 0.6V}$$

$$R_A = 2.357K$$

Therefore, adjusting the rotary sensor (rheostat) around 2.357 KΩ will turn off the transistor. The output voltage at pt. V_A will be approximately V_{CC}. If the rotary sensor is adjusted around 2.7 KΩ, the transistor will turn on, producing an output voltage of about 200 mV.

The following laboratory project will set up and execute a simulation session of the rotary-sensor control circuit, using PSpice-OrCAD Student Version 9.1 software.

A Rotary-Sensor Control Circuit Simulation Lab Project The following laboratory project will simulate the rotary-sensor control circuit operation. PSpice-OrCAD 9.1 version software will validate the conditions discussed in Figures 4-20(a) and (b).

The BOM for this project is as follows:

One 2N23904 NPN transistor

One 33 KΩ resistor

One 10 KΩ resistor

One 50 KΩ potentiometer

One bubble (signal terminal)

One analog ground

One $^{+}V_{DD}$ voltage source

Software Assembly and Test Procedures The steps for the procedure are as follows:

1. Open the PSpice software by selecting the *Design Manager*. The window in Figure 4-21 will appear on your PC or notebook screen.
2. Move your mouse cursor to the left, where the icons are displayed. Click the schematic symbol. Figure 4-22 shows the schematic icon location within the PSpice design manager. After clicking the icon, a

Figure 4-21 PSpice design manager displayed on the PC or notebook screen.

Figure 4-22 Locating the schematic icon on the PSpice design-manager window.

circuit schematic blank sheet and design environment window should be displayed onscreen.

3. Click *Draw* located at the main menu task bar. A drop-down box will appear. See Figure 4-23. Select *Get New Part*. A *Part Browser Basic* dialog box will appear on the screen. Scroll down the list to select a transistor, resistor, potentiometer, analog ground, bubble, and $^{+}V_{DD}$ voltage source.
4. Click the *Advanced* button to see the circuit symbols of each of these electrical components. Figure 4-24 shows the expanded window, with the visible circuit symbols displayed.
5. With the symbols placed on the schematic IDE window, the components can be wired using the *Draw Wire* tool. See Figure 4-25.
6. Using the *Draw Wire* tool, complete the circuit diagram, as shown in Figure 4-26.
7. Right-click each of the resistors to change their resistance values. A *Part Name* dialog box will appear on the screen, as shown in Figure 4.27, in order to edit the components' attributes. Click Save Attr and

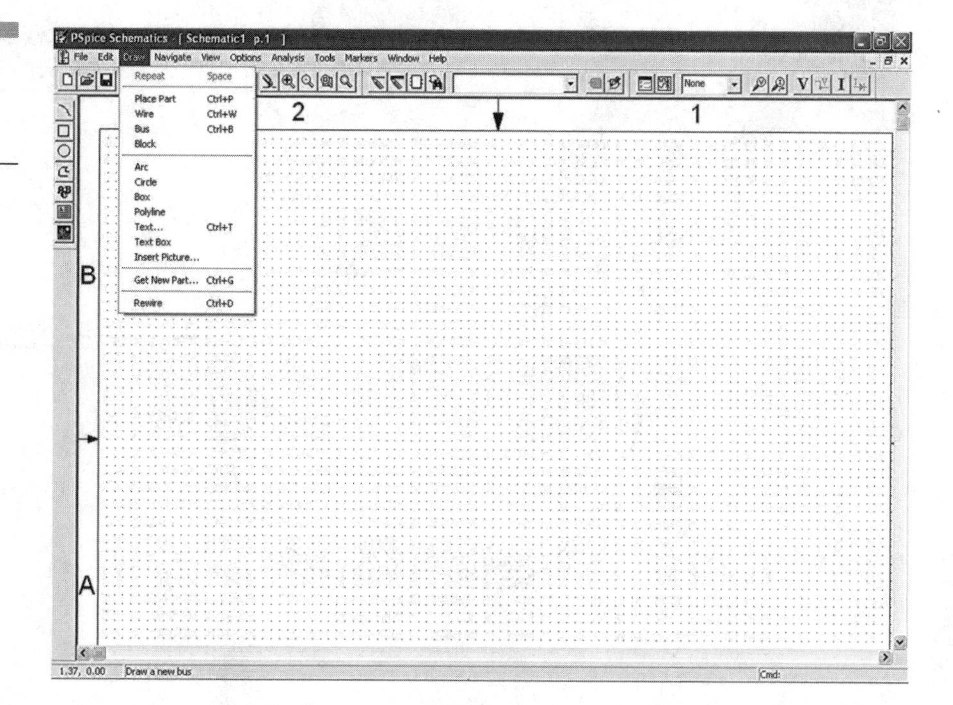

Figure 4-23 Getting to the parts library of PSpice

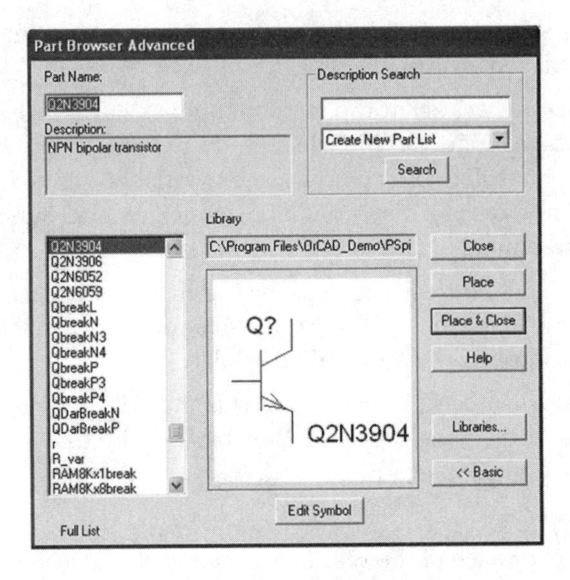

Figure 4-24 Advanced parts browser window within PSpice.

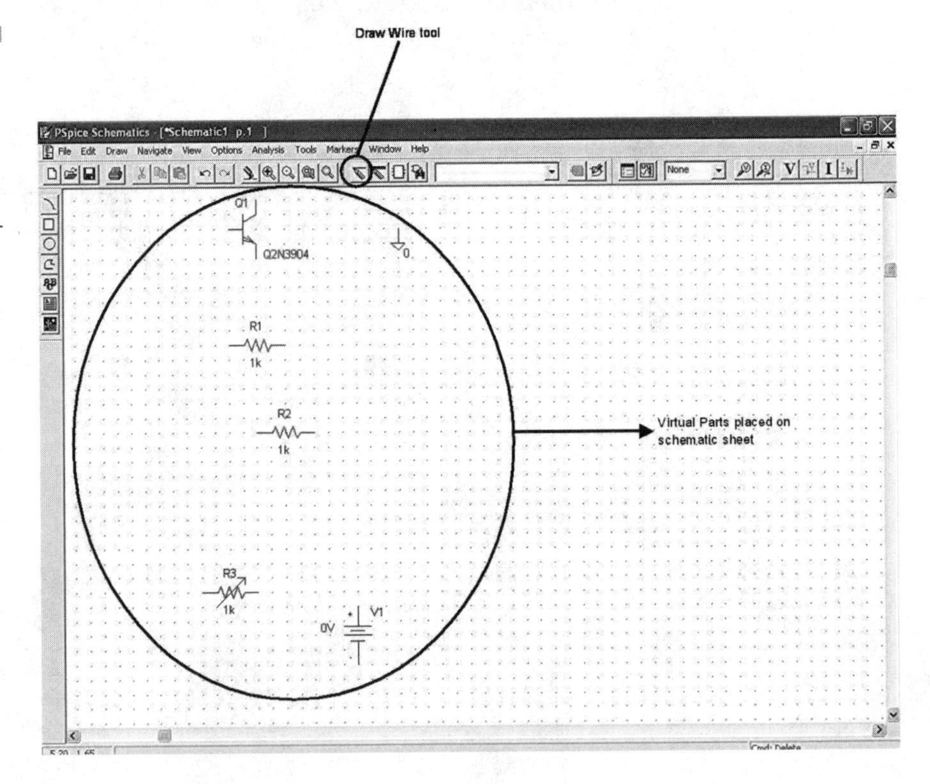

Figure 4-25 Parts placed on schematic sheet, ready to be wired with the draw-wire tool.

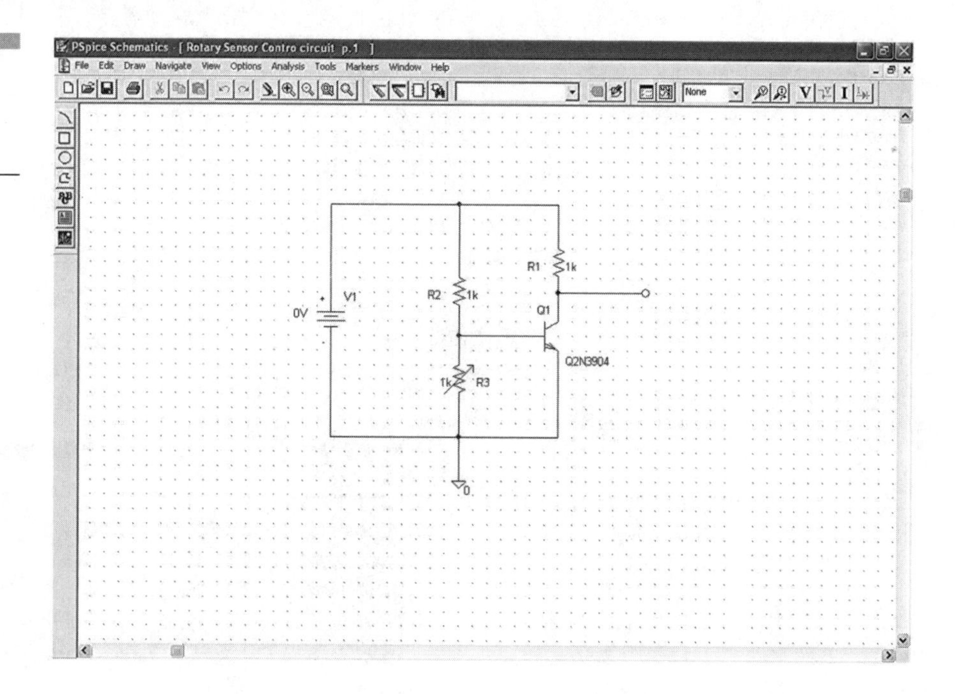

Figure 4-26 The finished rotary-sensor control circuit.

OK to lock-in the changed values. Note: For the R3 value, change the step value within the Part Name dialog box to 1, in addition to the resistance values.

8. Repeat step 7, changing the V1 to 9V.
9. Add text to the bubble by double-clicking the line. A label dialog box will appear on the screen. Type *Vout* into the text box and click OK. The terminal signal name will appear on the bubble.
10. The complete circuit schematic, with associated values, is shown in Figure 4-28.
11. Click the *Analysis* function from the main menu task bar to run the circuit simulation. Then select *Electrical Rule Check*, *Create Net List*, and *Simulate*.
12. Click the large boldface type ***V*** underneath the main-menu task bar. The schematic will display the voltages at the specified nodes denoted with dots. For the 50 KΩ setting, *Vout* equals 54.14 mV, meaning the transistor is turned on. See Figure 4-29(a).

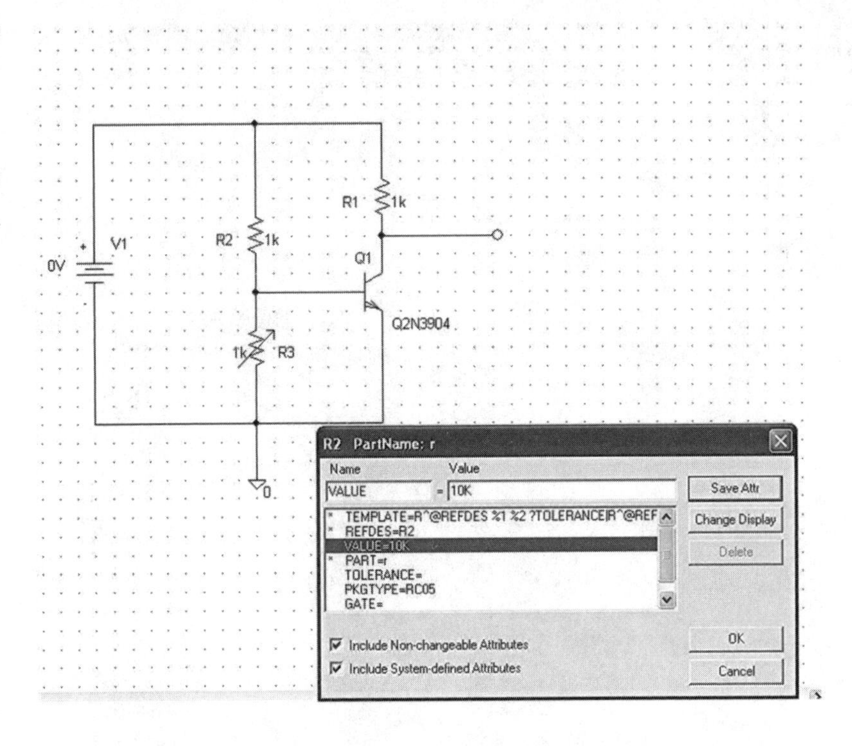

Figure 4-27 By right-clicking twice, the electrical component values can be changed.

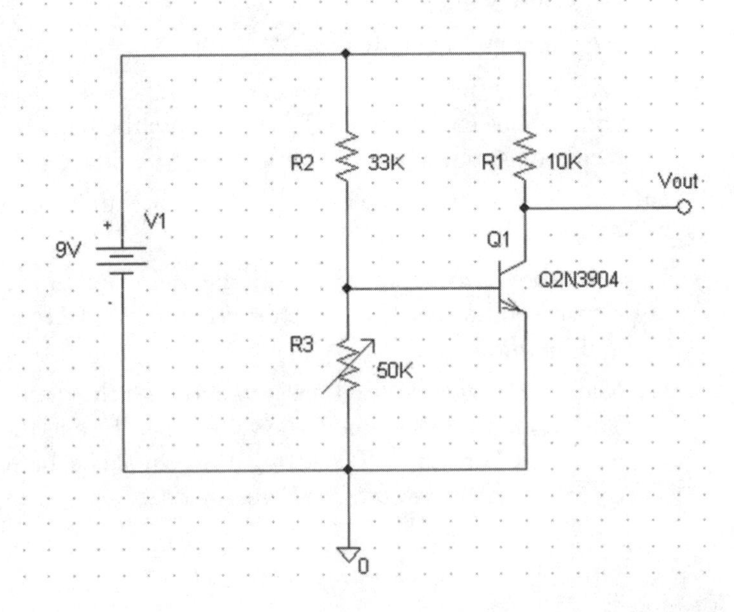

Figure 4-28 The complete circuit schematic with electrical values.

Figure 4-29(a)
With the rheostat set to 50 KΩ, Q1 transistor is turned on. Simulation results.

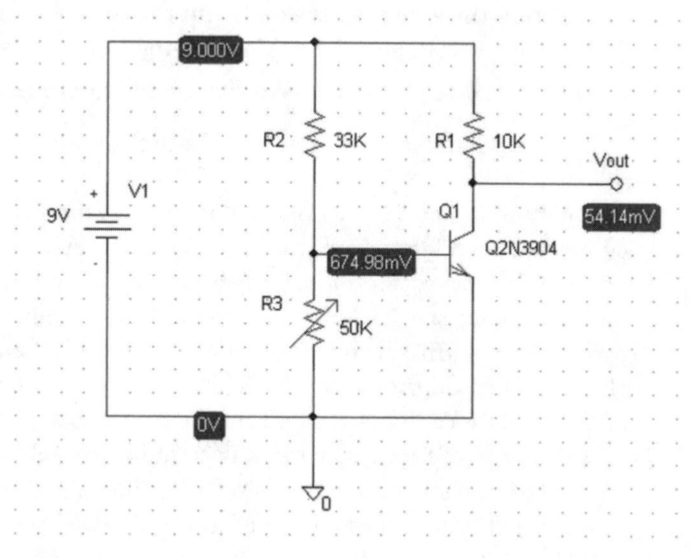

Figure 4-29(b)
With the rheostat set to 2.4 KΩ, Q1 transistor is turned off. Simulation results.

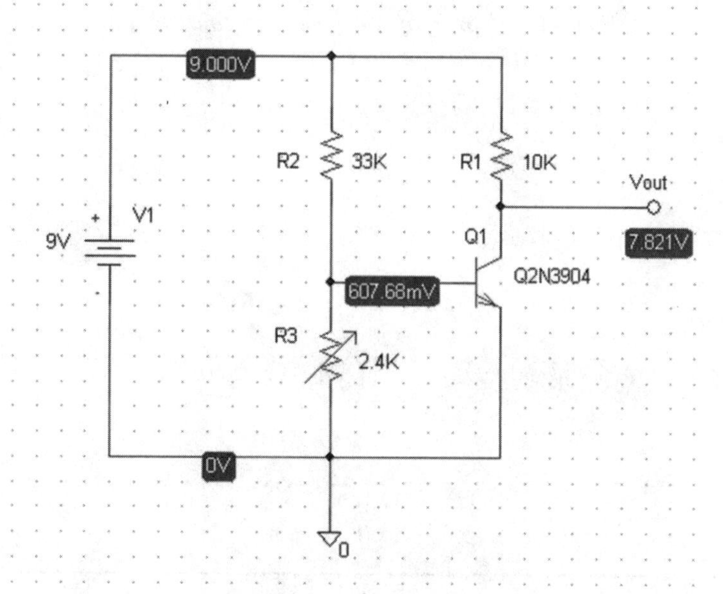

13. Change the rheostat value to 2.4 KΩ and run the simulation again, selecting the submenu tasks within the *Analysis* function in step 11. The output voltage is 7.821V, signifying the transistor is turned off.
14. Stop the simulation and save the file as *Rotary Sensor Control Circuit*.

Table 4-1 summarizes results of the simulation lab project.

Summary of Rotary-Sensor Control Circuit Simulation Lab Project In this circuit-simulation laboratory project, PSpice-OrCAD was explored. The operation of the rotary-sensor control circuit was validated using the rheostat and observing the NPN transistor's output voltage. A threshold voltage established by the series combination of the 33 KΩ and the varying 50 KΩ rheostat enabled the Q2N9304 NPN transistor to turn on or off. The data of the resistance value created by the angular movement of the rheostat and the state of the transistor was compiled in Table 4-1. This circuit will be further investigated in Chapter 5 for the validation of simulation function using a breadboard circuit and the LEGO CAM camera.

TECH NOTE: *In addition, besides validating LEGO Mindstorms mechatronics-based electronics I/O interfaces, both CircuitMaker 6 and PSpice-OrCAD software packages make great tools for documenting circuits and sharing them with others. By copying the circuits to the Windows clipboard and pasting them into either an MS Word document or PowerPoint presentation, the information can be shared through Internet downloads or email.*

Table 4-1

Results from the PSpice circuit simulation lab project

Rheostat Setting	Transistor Status
2.4 KΩ	OFF
2.7 KΩ	ON
50 KΩ	ON

MLCAD

Mechatronics is a multidisciplinary field involving mechanics, electronics, and software. In the last two sections of this chapter, simulation tools for designing and analyzing electrical/electronic circuits using two free software packages were introduced. These two simulation tools provide great flexibility in exploring simple or complex electrical/electronics I/O interfacing to Mindstorms mechatronics-based robots and intelligent machines.

The capability to create 3-D mechanical models on a PC or notebook computer is the next valuable tool for developing robots and smart machines. MLCAD is a free software mechanical-design package for modeling LEGO mechanics and structures. The following pages will explain the basic tools in the MLCAD software package used for building LEGO mechanical models and creating 3-D mechanical models on a PC or notebook computer for mechatronics-based applications.

MLCAD Basics

MLCAD is front-end-Windows-based *graphical user interface* (GUI) for the LDraw software developed by James Jessiman. The files needed for MLCAD are *complete.exe* and *ldraw027.exe*. Download these files from the Web site `www.ldraw.org/download/start/win/step1`. See Figure 4-30.

The *ldraw027.exe* file has all the LEGO constructs (building elements), parts, example brick models, and supporting software resources for running the application in MSDOS. The *complete.exe* file has the recent-parts library and constructs for building LEGO models. After running these two files, the MLCAD Windows-based GUI can then be executed. Obtain the MLCAD software as a zipped file from `www.lm-software.com/mlcad/`. See Figure 4-31.

Once the MLCAD software is downloaded to your PC or notebook computer, the design tool application will verify that the file path is correct for accessing the *LDraw* parts library. *C:\LDraw\LDraw* is the file-path structure that enables the software to run. With the MLCAD software running using the right file-path structure, the LEGO parts and constructs must be made available. Access them in the software. Figure 4-32 shows how to make these parts available to MLCAD.

The following lab project will explore the MLCAD design software for building a 3-D model.

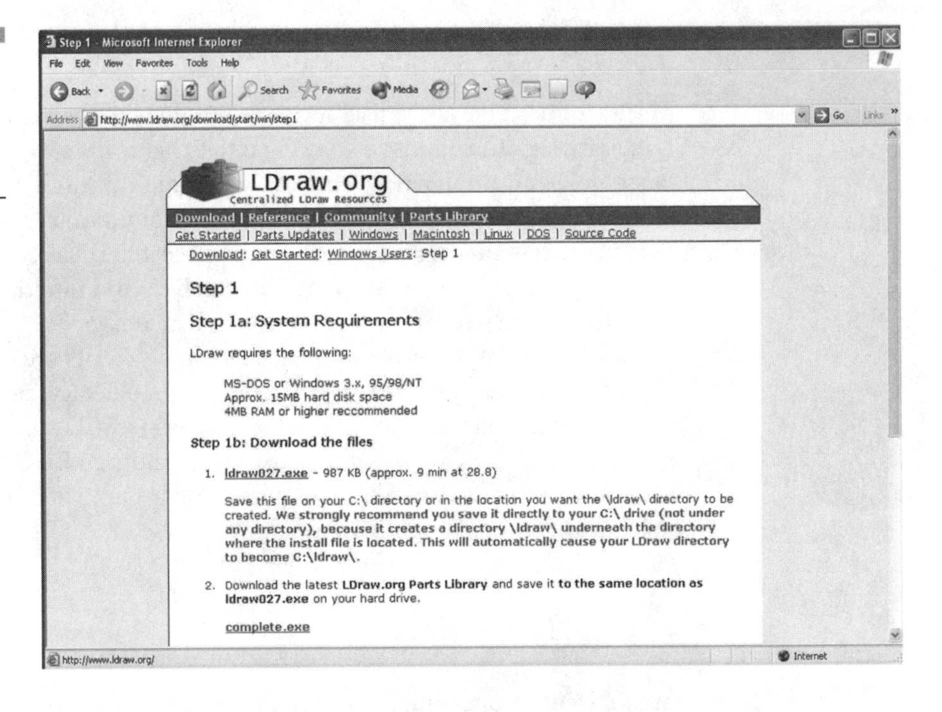

Figure 4-30 The LDraw files required for operating the MLCAD software.

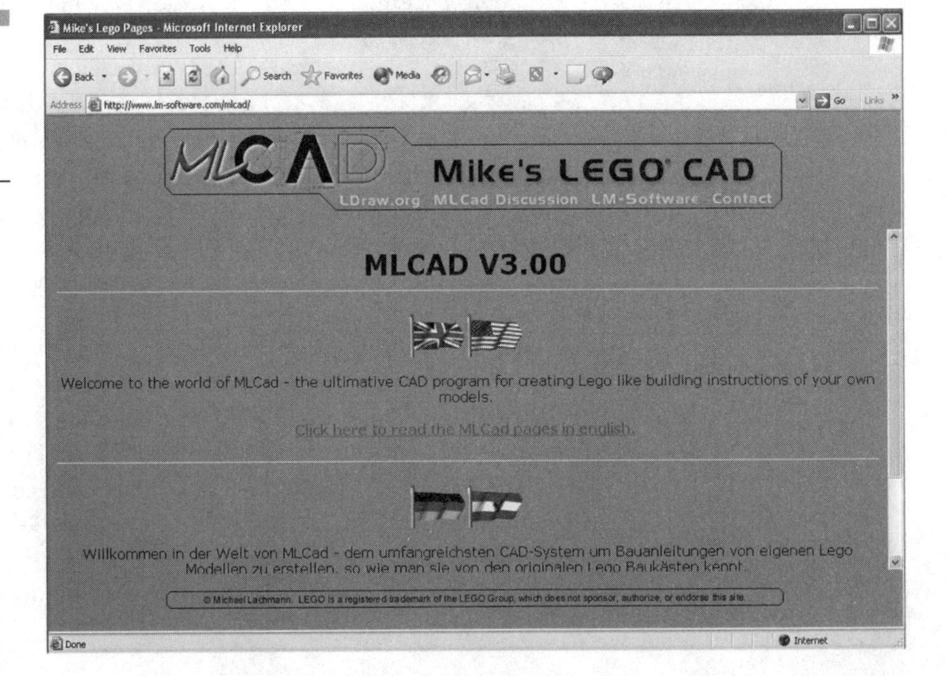

Figure 4-31 The download Web page for obtaining the MLCAD software.

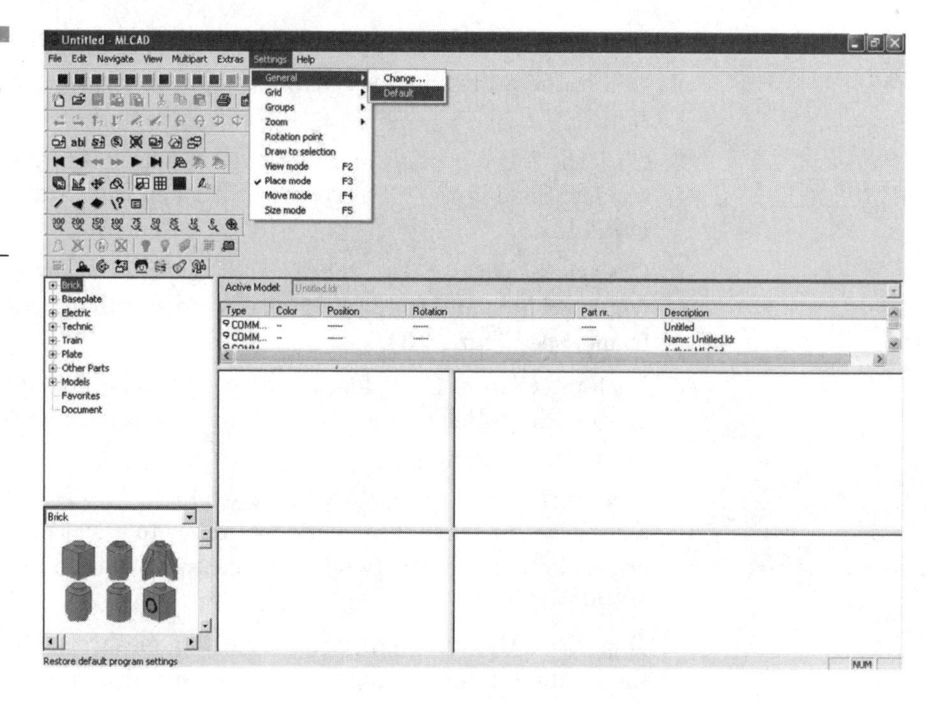

Figure 4-32 Making the LEGO parts and constructs available to the MLCAD design environment.

MLCAD Laboratory Project The following lab project will demonstrate the basics in building a 3-D model using MLCAD software. As a reference, the *Constructopedia Manual* version 1.5, pages 12–13, will be the mechanical design source for demonstrating MLCAD-basics.

Building the Drive-Base for E-Beetle Bot Using MLCAD This laboratory project will illustrate the basics tools within MLCAD that will allow you to build a LEGO 3-D model.

The BOM for this project is as follows:

Two 2×4 plates

Eight 1×2 plates

Eight 1×2 with door-rail plates

Two 2×10 plates

Eight 1×2 bricks

Four 2×2 plates

The following steps are only the fundamentals in building mechanical structures using LEGO bricks:

1. Open the MLCAD software package on a PC or notebook computer.
2. Within the parts window, expand the plates directory by clicking the + sign.
3. Click the 3-D model of the *Plate 2x10* and drag it to the lower-left window. See Figure 4-33. Save the model on the hard-drive as *DrvBase.dat*.
4. Change the color of the *Plate 2x10* by placing the cursor on the bottom window frame and right-clicking. A window will appear on the window frame. See Figure 4-34.
5. Click the *Change Color* feature. A selector color dialog box will appear on the screen. Select the *7 Light-Gray* color, then click OK.
6. Select *Plate 1x2* and drag it to the bottom frame. Using page 12 of the *Constructopedia Manual* version 1.5, place *Plate 1x2* on top of the 2x10 piece. Use the other three window frames to assist in the stacking-orientation task of the two LEGO plates. Save the file under the name *DrvBase.dat*.
7. To position the *Plate 1x2*, use the turning feature of the software. Use the rotate buttons located three rows down from the main menu task bar. See Figure 4-35 for the location of the rotate buttons.

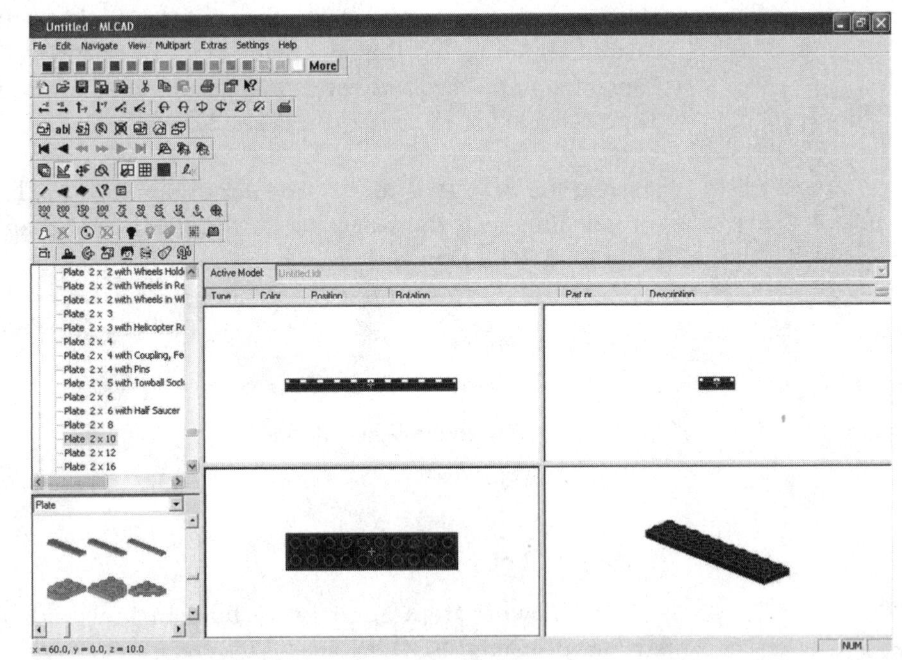

Figure 4-33 Drag and drop operation of the Plate 2x10 to the lower-left window of MLCAD

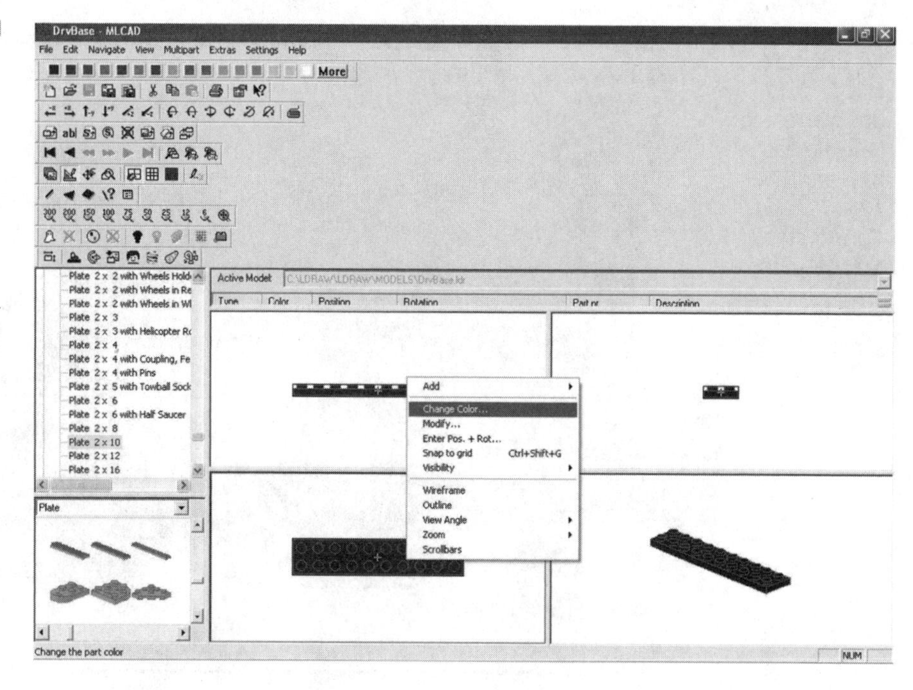

Figure 4-34 This is the dropdown menu for changing physical attributes of the LEGO parts. Color change is requested in this example.

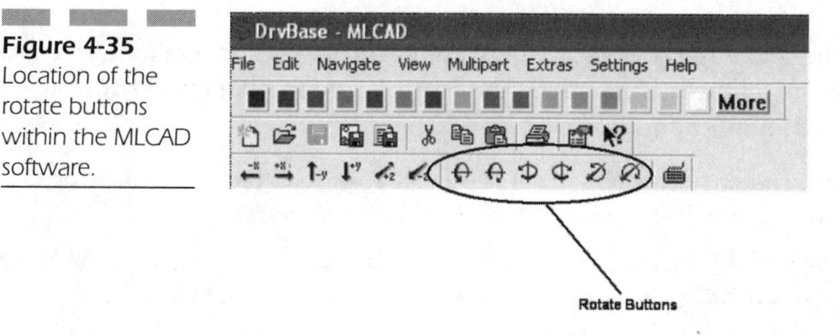

Figure 4-35 Location of the rotate buttons within the MLCAD software.

8. Continue using pages 12 through 13 of the *Constructopedia Manual* version 1.5 and steps 4 through 7 to complete the drive-base on page 13. Figure 4-36 shows the author's version of the MLCAD LEGO model.

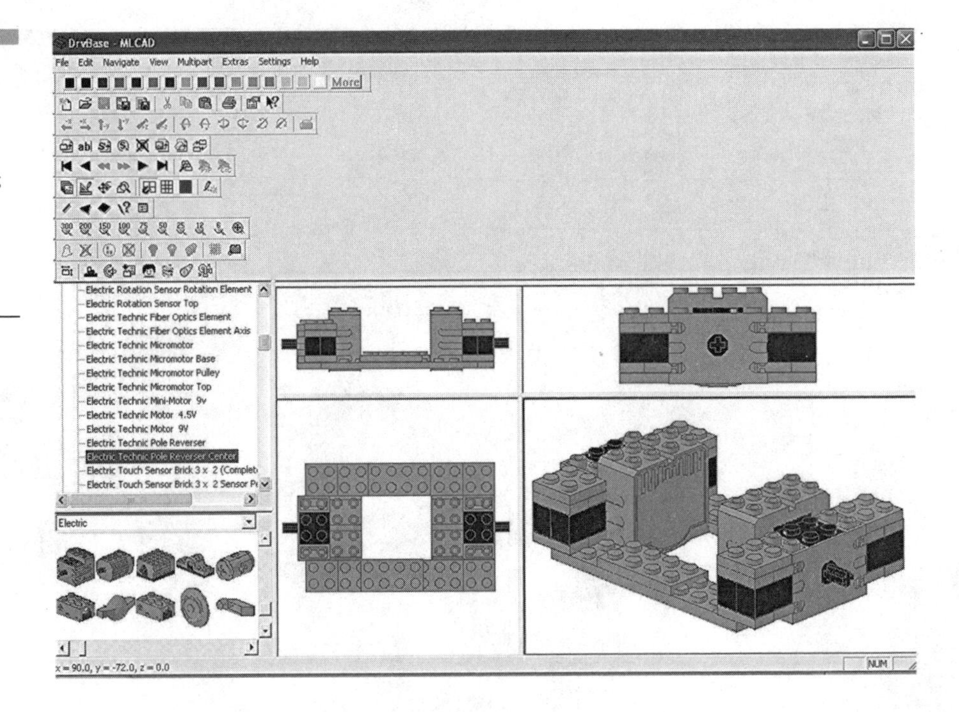

Figure 4-36 The finished driving-base for E-Beetle Bot, as shown on pages 12–13 of the Constructopedia Manual version 1.5

Copying and Pasting Objects This mini-exercise will investigate the copy and paste features of MLCAD. The Scout P-Brick will demonstrate how to use these duplication features.

1. From the parts window, click the *Electric* + to expand the directory. Scroll down and select *Electric Mindstorms Scout (complete)*.
2. Drag and drop the Scout P-Brick from the parts graphics window underneath the parts dialog box to the lower-left window.
3. Right-click the window to change the color of the P-Brick. Use *9-Light-Blue* as the base color for the Scout P-Brick.
4. Select the Scout P-Brick and press *Ctrl-C* . This two-keystroke event will copy the Scout P-Brick.
5. Press *Ctrl-V*. This two-keystroke event will paste a duplicate Scout P-Brick on top of the first one.
6. Drag the second Scout P-Brick and move it next to the first P-Brick, as shown in Figure 4-37.

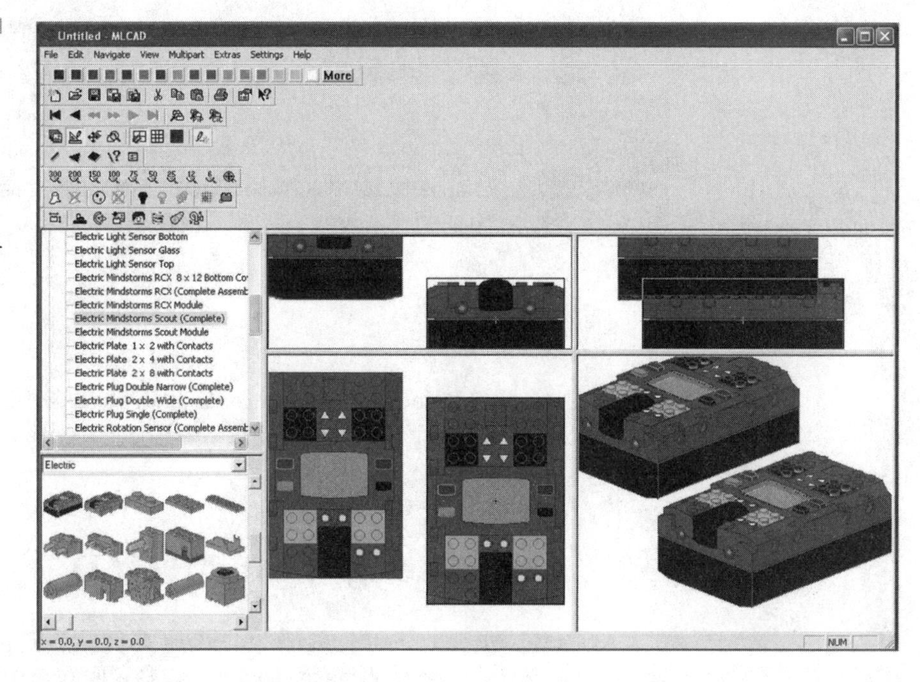

Figure 4-37 Creating a second Scout P-Brick using the copy and paste feature (Ctrl-C and Ctrl-V keys).

Copying and Pasting Objects 2 The following steps describe another way to copy and paste objects in MLCAD. Use the configuration window as well as *Copy* and *Paste* on the task bar.

1. Delete one of the Scout P-Bricks from the design window.
2. Go to the configuration window and select the *Electric Mindstorms Scout (complete)*.
3. On the main menu task bar, select *Edit*. From the drop down box, click *Copy*.
4. Go to the main menu task bar. Select *Edit*. From the drop-down box, click *Paste*. The configuration window will display a second *Electric Mindstorms Scout (complete)*. See Figure 4-38.

Summary of the MLCAD Laboratory Project This lab project allowed you to investigate a free mechanical-design software tool for LEGO building bricks. Some of the MLCAD features investigated in this lab project were copy and paste, changing colors, and rotating LEGO brick-objects. The user-friendly Windows GUI allowed for easy construction of 2-D and 3-D virtual LEGO models on a PC or notebook computer. The basic features outlined

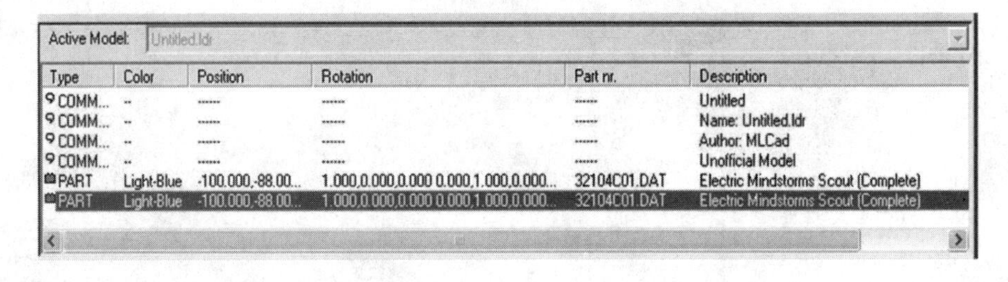

Figure 4-38 Creating a second Scout P-Brick, using the configuration window and the edit function from the main-menu task bar.

in the lab project allowed you to create a drive-base mechanism for E-Beetle Bot by dragging and dropping LEGO bricks onto a design Windows environment.

Additional Thoughts on MLCAD

MLCAD is the mechanical-design equivalent of Pro/E (Engineer) and Solidworks industry-based software packages. Pro/E (`www.ptc.com/products/proe`) and SolidWorks (`www.solidworks.com/`) allow mechanical designers to create 2-D and 3-D models using a Windows-based interactive environment. Objects created in both mechanical-design software packages allow the designer to manipulate their creations in either 2-D or 3-D space using the control keys, control buttons, and pull-down tasks from the main-menu bar.

MLCAD is based upon the same interactive design philosophy. You can manipulate LEGO Mindstorms mechanical models in 2-D and 3-D using the four-view display windows. As demonstrated in the hands-on laboratory exercises, MLCAD is a versatile design tool. It has the capability to create LEGO models using the same interactive controls in the industry-based software packages, SolidWorks and Pro/E. Although the basic software features of MLCAD that we discussed only scratch the surface of its design-modeling capabilities, the amateur roboticist might find certain sophisticated options invaluable.

Here is a short list, with a brief description of these advanced design-create features:

Subfiles MLCAD enables the grouping and manipulation of multiple elements. Subfiles allow these grouped LEGO-brick objects to be reused with target mechanical design. They can be moved and translated using the rotate buttons discussed earlier in this section.

Wireframes Wireframes break down the LEGO brick solid models into lines and circles. The use of wireframes during virtual construction speeds up the response time for the display refresh of complex LEGO models. If the response time seems slow while building the target LEGO model, consider using wireframes.

Changing the colors of multiple elements To change elements' colors, the configuration window, along with the selection of the target LEGO bricks, can be used within MLCAD. By selecting the parts you want in the configuration window and in the design frames, the chosen color will be applied to the entire group.

Final Thoughts

MLCAD is similar to most software packages. It requires you to experiment with its features to become a proficient user. The laboratory projects explored in this chapter provided the fundamentals of using MLCAD software to design LEGO models on the PC or notebook computer. Explore, experiment, and have fun with this unique and free design software tool.

Internet resources

Here is a list of Web-based resources referenced in this chapter:

CircuitMaker software:

`www.microcode.com/`

Cadence PSpice-OrCAD:

`www.cadencepcb.com/products/downloads/PSpicestudent/default.asp`

LDraw software:

`www.ldraw.org/download/start/win/step1.`

MLCAD software:

`www.lm-software.com/mlcad/`

Pro/E (Engineering) mechanical design software:

`www.ptc.com/products/proe`

SolidWorks mechanical design software:

`www.solidworks.com/`

CHAPTER 5

Telerobotics

Telerobotics is an interesting application of the mechatronics discipline. In addition to using electronics, mechanics and software vision become part of the intelligent system equation. By giving robots and intelligent machines the capability to see the world in which they live, these artificial automatons can provide a wealth of information to their human operators. The camera becomes the extension of the robot's panoramic view. The camera can remotely view the target environment of the robot, thereby eliminating the need for the human operator to be present with the mechatronics-based machine. For dangerous environments, telerobotics provides safety against hazardous biochemical materials.

In this chapter, a telerobotics system will be built to explore the basic building blocks of providing vision and control for the remote operation of a mechatronics-based machine. The rotary-sensor control circuit, discussed in Chapter 4, will be used to operate the robot with a rheostat-driven transistor switch.

The following topics to be investigated in this chapter are:

- What is a Telerobot?
- A Rotary-sensor Controlled Telerobot
- Build and Test Instructions for the Rotary-sensor Control Circuit
- Build and Test Instructions for the Telerobot Mechanical Drive
- Final Assembly and Test
- Vision Software

What Is a Telerobot?

A telerobot is a mechatronics-based system. A telerobot consists of a control unit that carries out the remote manipulation of the target robot, and a communication subsystem for wireless operation of the target robot. The control unit provides the input commands used by the operator to move the robot's mechanical drive for body movement and the vision camera. A button box, keyboard, or wireless controller may be used to provide *human-machine control* (HMC) input to the motion commands to that manipulate the mechanical movement of the robot. An analog controller, such as a joystick or trackball, provides another alternate means of HMC of a mechatronics-based unit.

Both control methods must provide a means of converting the motion input commands to a compatible wireless equivalent that the remote unit can use to manipulate the mechanical drives of the target robot. The remote unit is another term to describe the robot's mechanical drive control for hand, arm, and body movement.

The communication subsystem provides a wireless link to initiate the movements of the body and vision camera of the robot; using software, a microcontroller, and electronic circuit motor-drivers. Figure 5-1 shows a block diagram for a telerobotic system.

The following sections of this chapter will discuss each of the building blocks shown in Figure 5-1 with the construction of a small telerobot.

Rotary-Sensor Controlled Telerobot

In Chapter 4, a rheostat-based control circuit was simulated, using PSpice-OrCAD. The circuit works where the 10 KΩ potentiometer (wired as a rheostat) turns on the 2N3904 NPN transistor when the resistance value is greater than 5.8 KΩ.

Therefore, a certain rotary position enables the electrical load, wired on the low side of the transistor, to activate. In the circuit model, a 10 KΩ resistor provided the electrical load required for the sensor-based switch to work properly during the simulation session.

The actual circuit will use an electromechanical relay to provide the control signal for the LEGO *robot command explorer* (RCX) or Scout *programmable brick* (P-Brick) to process and drive the vision-camera mechanical drive unit. The electromechanical relay contacts, along with a series 1 KΩ resistor, will provide the necessary circuit impedance for proper input-interface detection and operation to the internal DC network of the LEGO RCX or Scout P-Brick. Figure 5-2 shows the wiring schematic for the rotary-sensor control circuit.

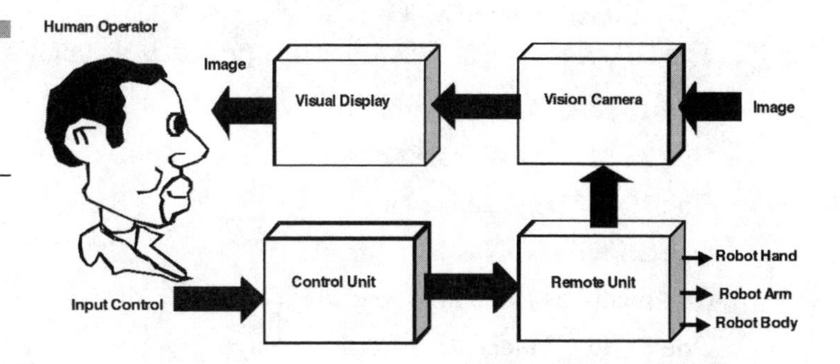

Figure 5-1
A block diagram for a telerobotic system.

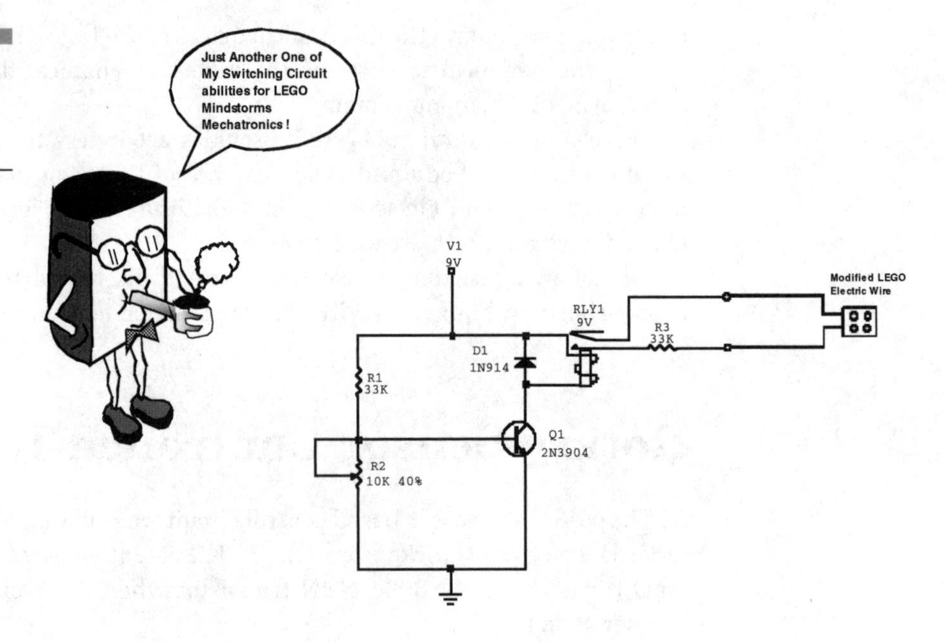

Figure 5-2 The rotary-sensor control circuit schematic.

Circuit Operation

The rotary-sensor control circuit will turn on the vision-camera mechanical drive unit, when a rheostat resistance value is greater than 5.8 KΩ. The mechanical drive is partitioned into two sections, an output B and an output A. Therefore, a boundary condition is required to activate the two output ports of the LEGO RCX or Scout P-Brick. A rheostat resistance-value of less than 3.5 KΩ will turn on output B, allowing the vision camera to move up. A rheostat resistance-value of greater than 5.8 KΩ will turn on output C, allowing the vision camera to turn right.

The input motion-control in Table 5-1 summarizes the rheostat to vision camera movement.

The input motion-control table will be used to validate the circuit's function for proper operation.

The *bill of materials* (BOM) for this project is as follows:

One 10 KΩ, 1/4-W resistor

One 33 KΩ, 1/4-W resistor

One 10 KΩ potentiometer

One 2N3904 NPN transistor

One modified LEGO electric wire

One 6V to 9V electromechanical relay

Table 5-1

Input motion-control table

Rheostat Value	Vision Camera Output Port
< 3.5 KΩ	Output B is ON/Output is C OFF
> 5.8 KΩ	Output C is ON/Output B is OFF

One Radio Shack Electronics Learning Lab breadboard or equivalent circuit prototyping station

One DC voltmeter or *digital multi-meter* (DMM)

Build and Test Instructions for the Rotary-sensor Control Circuit
This mini test procedure will validate the function of the rotary-sensor control circuit in operating the mechanical motor drives of a telerobot.

1. Wire the circuit shown in Figure 5-2, using the Radio Shack Electronics Learning Lab breadboard or equivalent prototyping station.
2. Attach an audible ohmmeter or continuity tester across the *Normally Open* (N.O.) contacts of the 9V electromechanical relay.
3. Turn on the power of the Radio Shack Electronics Learning Lab breadboard or equivalent circuit prototyping station.
4. Adjust the 10 KΩ rheostat slowly, until the audible ohmmeter or continuity tester produces a tone.
5. Turn off the power, and measure the resistance of the 10 KΩ rheostat using an ohmmeter. Record the resistance value here. _______KΩ
6. With a DC voltmeter or DMM, measure the voltage across the 10 KΩ rheostat.
7. Repeat step 2 of the test procedure.
8. Turn on the power of the Radio Shack Electronics Learning Lab breadboard or equivalent circuit prototyping station. Adjust the resistance of the 10 KΩ rheostat, using an ohmmeter. Record the resistance value here. ______KΩ
9. Turn off the power. Measure the 10 KΩ rheostat. Record the voltage value here. _____V
10. With a DC voltmeter or DMM, measure the voltage across the 10 KΩ rheostat. Record the voltage value here. ____V

11. Remove the DC voltmeter from the circuit. By rotating the 10 KΩ rheostat in both directions, the contacts of the electromechanical relay will switch on and off. A *click* sound will be heard due to the opening and closing of the electromechanical contacts.

12. The circuit is now ready to be connected to the LEGO RCX P-Brick. Attach the modified LEGO electric wire to the LEGO RCX P-Brick's input port 1.

The next section will describe the mechanical drive assembly of the telerobot.

Simple Machines and Telerobots

The mechanical drive for the telerobot is basically a power stand that moves the LEGO Cam up, down, and side-to-side. The four basic motions of the LEGO Cam movement rely on two motors. One LEGO electric motor provides up and down movement of the LEGO Cam. The other motor provides side-to-side motion. To accomplish these mechanical camera positional movements for the LEGO Cam, two simple machines are used: gears and belts-and-pulleys.

The belts-and-pulleys provide indirect drive to the gears, producing the four mechanical movements of the camera. The gear drive and the belt-and-pulley drive provide a convenient way to maximize the speed of the LEGO electric motor. They also maintain sufficient torque for the movement of the

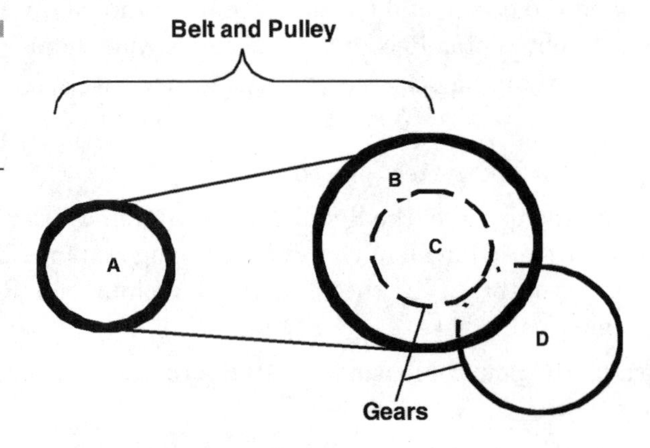

Figure 5-3 Basic gear to belt-and-pulley mechanical drive

camera and its rotating base or power stand. Therefore, the camera can be rotated and raised conveniently, and its speed managed sufficiently, through four mechanical movements. Figure 5-3 shows the basic gear to belt-and-pulley mechanical drive used in the telerobot project.

The gears provide the necessary torque for mechanical base and liftarm for the camera, while the belt-and-pulley drive maintains constant speed and positional control for the telerobot's LEGO Cam movement.

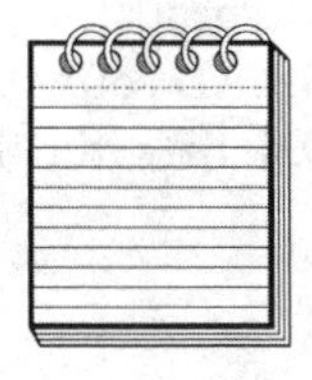

TECH NOTE: *Carnegie-Mellon University teaches a class on Rapid Design Through Virtual and Physical Prototyping.*

There is a six-page document produced by this course, entitled "Introduction to Mechanisms." This document explains simple machines and is a must-have document for the amateur roboticist involved in designing mechanical drives for LEGO Mindstorms mechatronics-based projects. The Web site is `www-2.cs.cmu.edu/People/rapidproto/mechanisms/ chpt2.html`.

Build and Test Instructions for the Telerobot Mechanical Drive

This section shows how to build the telerobot that will be used to move the Vision Command's LEGO Cam. The telerobot is based on the Power Stand design, shown on pages 34 through 37 of the *Vision Command Constructopedia*. As discussed in the previous section, the Power Stand uses two simple machines: gears and the belt-and-pulley. Two LEGO electric motors provide the electromechanical actuation for driving the simple machines. The motors drive the gears using the belts-and-pulleys. The positions of the two motors, gears, and belt-and-pulley drives provide the rotate and raise/lower motions for the LEGO Cam.

The simple-machine camera drives are perpendicular to each other, creating the necessary movements for panoramic viewing. The reduced speed of the gear drive provides for a smooth actuation of the LEGO Cam's rotate-and-raise mechanical movements. The Power Stand is built from four mechanical blocks, each repeated twice for the four basic movements of the camera.

The structure consists of the following mechanical blocks:

- One motor belt-and-pulley machine, driving two gears for the raise/lower motion of the LEGO Cam.
- One motor belt-and-pulley machine, driving two gears for the rotating motion of the LEGO Cam.

The assembly instructions for building the power stand are located on pages 34–37 of the *Vision Command Constructopedia*.

There are 82 parts required to build the Power Stand telerobot. The DAT numbers are provided for reviewing the parts, using MLCAD software discussed in Chapter 4. This Power Stand telerobot BOM lists the LEGO bricks required:

One black 3001.DAT brick 2x4

Two light gray 71427C01.DAT Electric Technic Mini-Motor (9V)

Six light gray 3023.DAT Plate 1x2

Eight light-gray 3710.DAT Plate 1x4

Seven light-gray 3666.DAT Plate 1x6

One light-gray 3034.DAT Plate 2x8

One light-gray 3832.DAT Plate 2x10

One blue 32034.DAT Technic Angle Connector #2

Two black 3704.DAT Technic Axle 2

One black 32073.DAT Technic Axle 5

One black 3706.DAT Technic Axle 6

One black 3707.DAT Technic Axle 8

One black 3737.DAT Technic Axle 10

Two light-gray 3749.DAT Technic Axle Pin

Four black 3701.DAT Technic Brick 1x4 (with holes)

Four black 3894.DAT Technic Brick 1x6 (with holes)

Four light-gray 3713.DAT Technic Bush

Ten light-gray 4265C.DAT Technic Bush (1/2 smooth)

Two red 3648.DAT Technic Gear (24 tooth)

One white 6588.DAT Technic Gearbox 2x4x3 1/3

Two light-gray 6632.DAT Technic Liftarm 1x3

Two black 152.DAT Technic Liftarm 1x9 (bent)

Four light-gray 3673.DAT Technic Pin

Four light-gray 4274.DAT Technic Pin ½

Three light-gray 3709B.DAT Technic Plate 2x4 (with holes)

One light-gray 32001.DAT Technic Plate 2x6 (with holes)

Two light-gray 4185.DAT Technic Wedge Belt Wheel

Two red 4716.DAT Technic Worm Screw

One black 2902.DAT Wheel 81.6x15 Motorcycle

One white 2903.DAT Wheel 81.6x15 Motorcycle

Instructions for Building the Power Stand Telerobot

The following instructions are used to build the Power Stand telerobot. Refer to Figures 5-4 through 5-13:

1. Construct the horizontal gear-axle drive bottom plate. (Figure 5-4)
2. Attach the reinforced front and rear walls to the horizontal gear-axle drive bottom plate. (Figure 5-5)

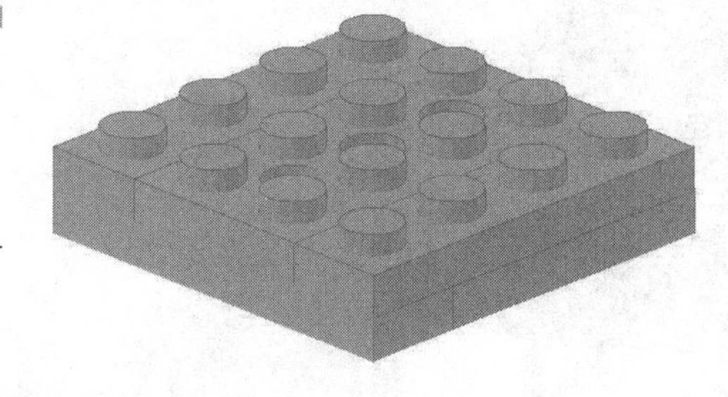

Figure 5-4 Construction of the horizontal gear-axle drive bottom plate

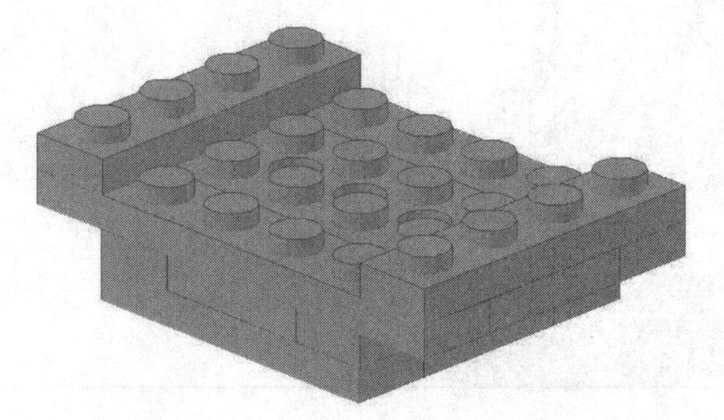

Figure 5-5 Horizontal gear-axle drive bottom plate with reinforced front and rear walls

3. Add the gear-axle to bottom plate. (Figure 5-6)
4. Mount the horizontal gear-axle drive onto the rubber tire stand. (Figure 5-7)
5. Add the worm gear to the horizontal gear-axle drive. (Figure 5-8)

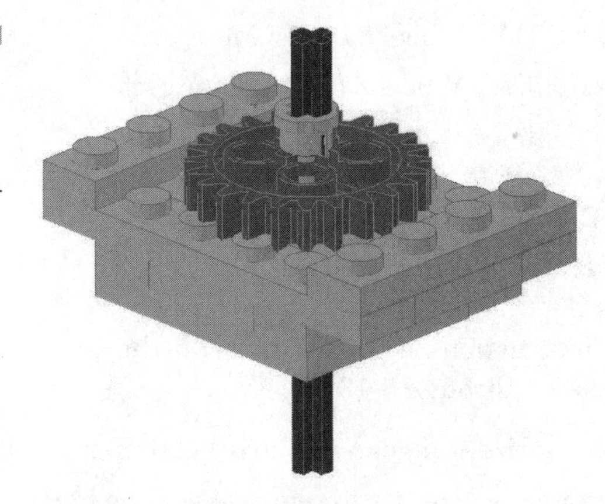

Figure 5-6 Adding the gear-axle to bottom plate

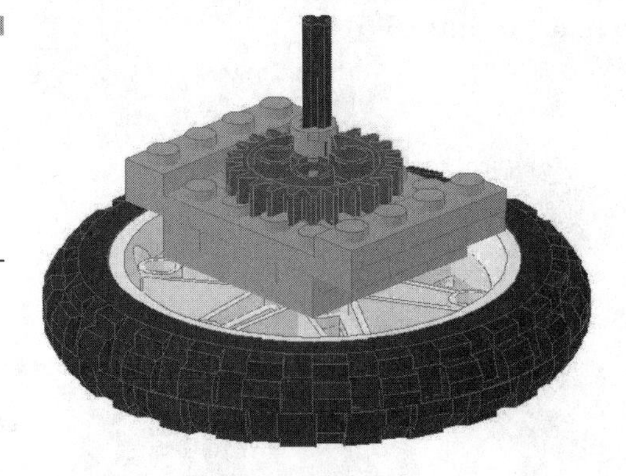

Figure 5-7 Mounting horizontal gear-axle drive onto the rubber tire stand

Figure 5-8 The worm gear added to horizontal gear-axle drive

6. Attach the T-angle axle connector to the front and rear walls. (Figure 5-9)
7. Add the sidewalls with liftarms to horizontal gear-axle drive. (Figure 5-10)
8. Mount the vertical motion gearbox plates on top of the horizontal gear-axle drive. (Figure 5-11)
9. Mount the vertical motion gearbox on top of the horizontal gear axle drive. Add two LEGO electric motors (with small and large pulleys) to the Power Stand telerobot. (Figure 5-12)

Figure 5-9
T-axle connector with front and rear walls

Figure 5-10
Sidewalls with liftarms added to horizontal gear-axle drive

Figure 5-11 Vertical motion gearbox plates mounted on top of the horizontal gear-axle drive.

Figure 5-12 Mount the vertical motion gearbox on top of the horizontal gear-axle drive. Two LEGO electric motors with small and large pulleys are added to Power Stand telerobot.

10. Add the bend liftarms with four pins to the Power Stand telerobot (Figure 5-13)

The final construction of the Power Stand telerobot consists of making the mechatronics device less machine-like. *Eyes* are added to the LEGO Cam mounted between the two bend liftarms. Figure 5-14 shows the completed Power Stand telerobot. Confirmation of the mechanics of the telerobot requires running a mini test procedure for quick validation.

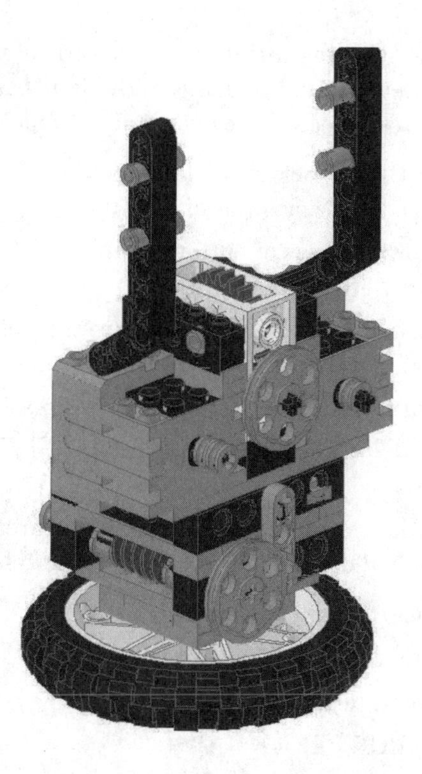

Figure 5-13
Adding the bend liftarms with 4 pins to the Power Stand telerobot.

Figure 5-14
The completed Power Stand telerobot.

Minitest Procedure Use the infrared remote control to test the motion of the Power Stand telerobot and to activate the four basic movements of the camera. The mini test procedure consists of the following:

1. Attach the LEGO RCX P-Brick to the telerobot by attaching the motorized drives to output ports B and C.
2. Press the up and down arrows of buttons *B* and *C* to observe the LEGO Cam rotary and raise/lower movements.
3. Try each camera motion individually, to assure the pulleys are moving freely with no binding of the small rubber bands.
4. Try combinations of LEGO Cam movements by pressing the *B* and *C* buttons in different patterns to create discrete mechanical motions for the camera.

Here is a systematic test procedure using the infrared remote to check the LEGO Cam motorized drivers of the Power Stand telerobot:

1. Attach the left LEGO motor to output port B.
2. Attach the right LEGO motor to output port A.
3. Have the Power Stand telerobot facing you.
4. Turn on the LEGO RCX P-Brick.
5. Press and hold the *B* up arrow button. The LEGO Cam should move to the left.
6. Press and hold the *B* down arrow button. The LEGO Cam should move to the right.
7. Press and hold the *C* down arrow button. The LEGO Cam should move down.
8. Press and hold the C up arrow button. The LEGO Cam should move up.
9. Press and hold the *B* up and the *C* down arrow buttons simultaneously. The LEGO Cam should move to the left and down, in one sweeping motion.
10. Press and hold the *B* down and the *C* up arrow buttons simultaneously. The LEGO Cam should move to the right and up, in one sweeping motion.

Quick Lab Project Summary

Up to this point, the telerobotics project consisted of building the rotary-sensor control circuit and mechanical Power Stand. The rotary-sensor control circuit has been modified slightly to provide the switching interface for activating the LEGO RCX P-Brick. This wiring task was accomplished by replacing the 10 KΩ resistor with an electromechanical relay.

The electronic control circuit was tested using an audible ohmmeter or continuity tester connected to its switching contacts. The two threshold resistor-values of the 10 KΩ rheostat for turning the NPN transistor on and off were measured with and recorded on the lab project test procedure page. The two bias voltage values created by the voltage divider circuit are applied to the base of the NPN transistor. The two values provide a switching threshold for turning the NPN transistor on and off.

Once the rotary-sensor control circuit was working properly, the Power Stand telerobot was built using either the *Vision command Constructopedia* or the MLCAD assembly diagrams. The left LEGO electric motor was attached to output port B while the right motor was attached to output port C of the LEGO RCX P-Brick, using two LEGO electric wires. The mechanical operation of the Power Stand telerobot was examined for gear and pulley binding, and effective lifting of the attached LEGO Cam. A simple test procedure used the infrared remote to validate the mechanical operation of the Power Stand telerobot.

The next phase of the project will consist of writing software for the directional control of the Power Stand telerobot. *Interactive C* (IC) programming language will be used to develop code for detecting the open and close contacts of the electromechanical relay.

A Robotics Vision Development System

A robotics-vision development system enables the amateur roboticist to build and experiment with mechatronics-based prototypes. You can validate designs, circuit block interfaces, software, and mechanics of the target robot. The solderless breadboard, LEGO Mindstorms *Robotic Invention System* (RIS), LEGO Cam, and electrical/electronic parts form the basis of a robotics-vision development system. The software programming tools, electrical/electronic circuit interfacing techniques, and CAD discussed in the earlier chapters are also some of the essential building blocks for developing robotics-vision systems.

This chapter is an essential link to robotics-vision development, because of the discussions of LEGO robots and integrating vision using a mini-camera. The remaining sections of this chapter will provide you with closure on building a robotics-vision development system.

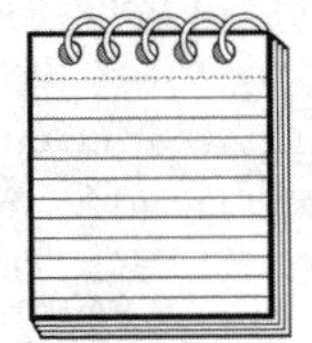

TECH NOTE: The Vision Command Constructopedia *serves as an excellent mechanical design aid in developing motorized drives for telerobots. The various motorized drives for* pan *(to turn) and* tilt *(to raise or lower) can be developed using the designs shown in the* Constructopedia.

Final Assembly and Test

Now that the rotary-sensor control circuit and the Power Stand telerobot are working properly, wire the two sub-components together. As stated in the mini test procedure, attach the left motor drive (pan) to output port B and the right motor drive (tilt) to output port C. The rotary-sensor control circuit is connected to input port 1 using a modified LEGO electric wire with a 1 KΩ resistor in series to the mini-wiring harness.

Switching Resistor Quick Test

To check for proper switching of the 1 KΩ series resistor use the following test procedure:

1. Attach the test leads of an ohmmeter to the modified LEGO electric wire connector, as shown in Figure 5-15.
2. Adjust the ohmmeter scale to 2 KΩ, for best measurement reading of the 1 KΩ series resistor.
3. Activate the Radio Shack Electronics Learning Lab, or an equivalent circuit prototyping station.

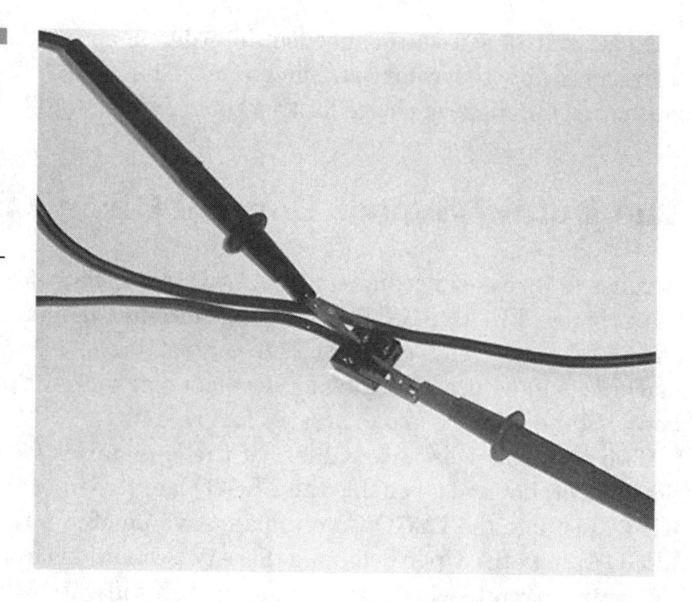

Figure 5-15 Attaching the test leads of an ohmmeter to the modified LEGO electric wire connector.

4. Turn the 10 KΩ rheostat until the contacts of the electromechanical relay click. The ohmmeter reading should be around 1 KΩ. Note: The author's resistance reading value using a DMM was 0.992 KΩ.
5. Turn the 10 KΩ rheostat in the opposite direction, to open the contacts of the electromechanical relay. The ohmmeter should be reading *OL* for *open load* or infinity.
6. If your circuit displayed results other than what was shown in the last two steps, deactivate the rotary-sensor control circuit and check your wiring. Verify the 1 KΩ resistor is properly wired in the circuit.
7. After finding and correcting any wiring errors, repeat the test procedure.

This minitest procedure is important to the telerobot project because the rotary-sensor control circuit enables the amateur roboticist to operate the robot using two resistance positions for pan and tilt operation of the mechatronics-based machine. Therefore, it is critical to perform the mini test in order to assure the pan and tilt functions of the telerobot will be operable when using the rotary-sensor control circuit.

The rotary-sensor control circuit can now be attached to input port 1 using the modified LEGO electric wire.

Next, control software is needed, in order to activate the Power Stand telerobot using the rotary-sensor circuit. The LEGO RCX P-Brick programming language of choice for this bot is *Interactive C version 4.0* (IC4).

The Rotary-Sensor Control Circuit Software

Because of its ease in coding and the LCD text display feature, IC4 will be the primary LEGO RCX P-Brick programming language for the Power Stand telerobot. The input motion control table, as well as the state machine, will be used as software development tools for programming the Power Stand telerobot's pan and tilt movements.

Two rheostat resistance values set the boundaries for when the Power Stand telerobot will be enable the LEGO Cam to pan or tilt. A value of ,3.5 KΩ will enable the LEGO Cam to pan. A value of .5.8 KΩ will enable the LEGO Cam to tilt. Upon detecting these two conditions from the rotary-sensor control circuit switching contacts, the IC4 software will initiate the pan and tilt commands to the respective motor drives. The state machine in Figure 5-16 graphically illustrates the pan and tilt functions of the Power Stand telerobot. The IC4 code will operate in the following sequence events as follows:

1. LEGO RCX P-Brick is off.
2. A RUN event will activate the LEGO RCX P-Brick, but the Motor B and C will be off.
3. With the rotary-sensor control circuit turned on, a <3.5 KΩ resistance value from the rheostat will activate Motor B for 5 seconds.
4. After 5 seconds have elapsed, Motor B will turn in the opposite direction for 5 seconds.
5. After 5 seconds have elapsed, Motor B deactivates. This turn on and off condition of Motor B will exist as long as the <3.5 KΩ rheostat resistance value event is true.
6. If the <3.5 KΩ rheostat resistance event is true resistance value exists.
7. Motor C will activate for 2 seconds.
8. After 2 seconds have elapsed, Motor C will turn in the opposite direction for 3 seconds. This turn on and off condition for Motor C will exist as long as the >5.8 KΩ resistance value event is true.
9. If this event is not true, then the operator has initiated the RUN event.
10. Motor B and C are off.

Writing the software using this sequence of events, along with the use of the state machine, helps in coding the IC4 instructions. The software listing for the rotary-sensor control circuit interface is shown in Listing 5-1.

The operation of the code performs the function of the state machine, shown in Figure 5-16.

After downloading the code to the LEGO RCX P-Brick and pushing the run button, the software is executed as follows:

```
while(1)
```

The program will always be active because of the *while (1)* function being true. The following function checks for a binary status of the rotary sensor's electromechanical relay contacts.

```
if (digital(1)==1
```

When the rotary-sensor control circuit contacts close, a binary value 1 event exists. The *if (digital(1)==1* statement is used to detect the electromechancial relay activating due to the 1 KΩ resistor being connected to input port 1 internal 10 KΩ resistor network circuit.

Listing 5-1

```
void main()
{          while(1){
        if (digital(1)==1){
            printf("C ON");
            fd(3);
            sleep(5.0);
            bk(3);
            sleep(5.0);
            brake(3);
        }
        else {
            printf("B ON");
            fd(2);
            sleep(2.0);
            bk(2);
            sleep(3.0);
            brake(2);


        }
    }

}
```

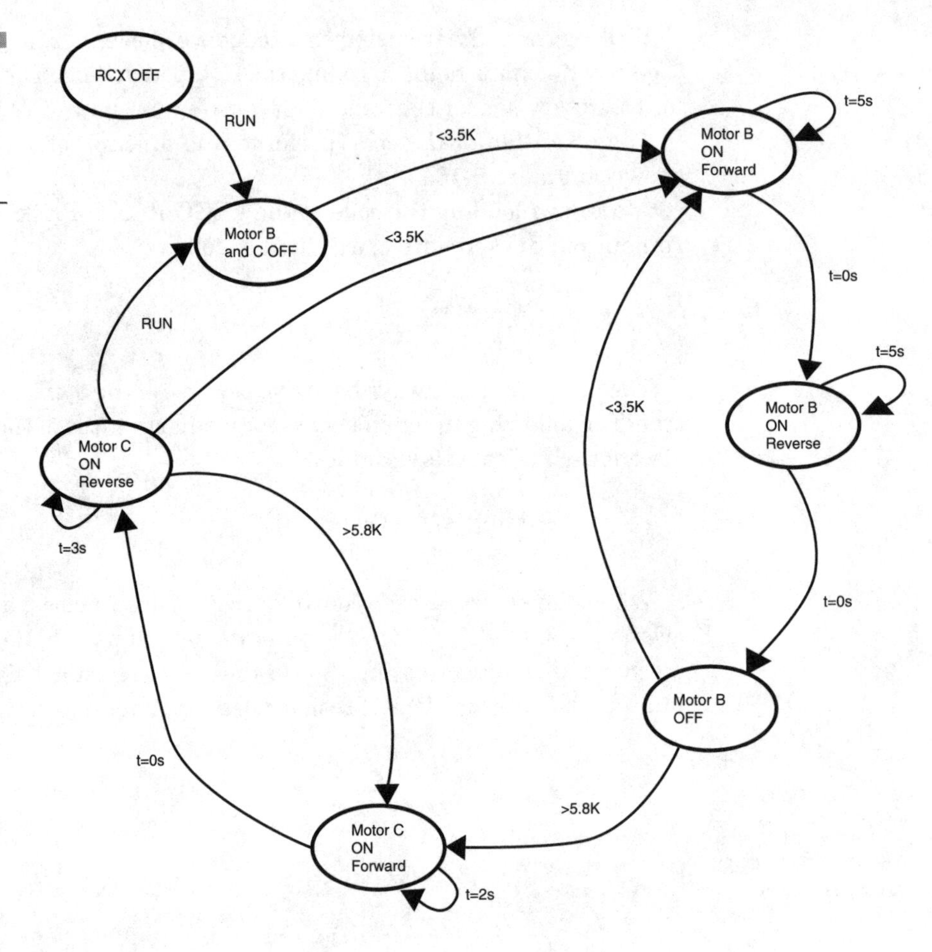

Figure 5-16 State machine for the Power Stand telerobot's pan and tilt function

```
printf("C ON");
```

The LEGO RCX P-Brick will display *C ON* on the mini-LCD screen using the *printf("C ON")* IC4 function. This mini text-message function enables you to see which motor is active, via this alpha-character information.

```
fd(3)
```

Output port C will be turned on using the *fd(3)* function within IC4. The LEGO Cam will then tilt the vision device upward, with the aid of the right motor drive attached to this output port.

```
sleep(5.0)
```

Before reversing the direction of the LEGO Cam, a 5-second delay is used to provide smooth motor directional-switching between the forward and backward movements of the camera.

```
bk(2)
```

The IC4 function will reverse the motor drive, lowering the LEGO Cam's viewing position.

```
sleep(5.0)
```

To provide a smooth transition to the next state the *sleep(5.0)* enables a 5-second delay to be initiated.

```
brake(3)
```

This function stops all motor drive operations for tilting the LEGO Cam, by stopping the small DC rotating machine immediately.

```
Else
```

If the rheostat value of the rotary-sensor control circuit is >5.8 KΩ, then the *else* function statements will be initiated.

```
printf("B ON");
```

The LEGO RCX P-Brick will display *B ON* on the mini-LCD screen using the *printf("B ON")* IC4 function. This mini text message function enables you to see which motor is active via this alpha-character information. Due to the limited number of segments for displaying alpha-characters on the LCD, the *B* will be in lowercase.

```
fd(2);
```

Output port B will be turned on using the *fd(2)* function within IC4 programming language. The LEGO Cam will then pan the vision device, allowing it move side to side with the aid of the left motor drive attached to this output port.

```
sleep(2.0);
```

Before reversing the direction of the LEGO Cam, a 2-second delay is used to provide smooth motor directional-switching between the forward and backward movements of the camera.

```
bk(2);
```

This IC4 function will reverse the motor drive, enabling the side to side viewing position of the LEGO Cam.

```
sleep(3.0);
```

To provide a smooth transition to the next state the "*sleep(3.0)* enables a 3-second delay to be initiated.

```
brake(2);
```

This function stops all motor drive operations for panning the LEGO Cam by stopping the small DC rotating machine immediately.

Upon running the program and switching between the pan and tilt functions using the 10 KΩ rheostat, the previous camera motion will complete one cycle before moving to the next LEGO Cam mechanical movement. Quickly changing between the 3.5 KΩ and 5.8 KΩ resistor limits, using the 10 KΩ rheostat, will not be detected by the IC4 software and the camera motion drives. The digital detection IC4 software processing is faster than the camera motion, due to slow response of the motor drives and the mechanical assembly of the Power Stand telerobot.

Writing the software in IC4 enables the code to be pre-tested by using the interactive compiler feature. Typing the commands onto the interactive editor window enables you to test each line of code for input function validation of the rotary-sensor control circuit. It also enables you to test each line of code for the output response of the two motor drives. Input detection and mechanical motion timing errors, as well as camera movement errors, can be corrected before committing the final code to the LEGO RCX P-Brick.

Operation of the Power Telerobot's Software

After the rotary-sensor control circuit software is entered into the IC4 programming window editor, the *integrated design environment* (IDE) will check for parentheses and compiler errors. The software is downloaded to the Power Stand telerobot's LEGO RCX P-Brick using the infrared tower. Once loaded into the P-Brick, the camera motion's software is ready for execution by pressing *RUN*. Setting the circuit's rheostat to a resistance value less than 3.5 KΩ enables the Power Stand pan function to be active. The motor drive will move the LEGO Cam side to side, reversing the direction of the viewing device every 5 seconds.

The LCD will display the *b ON* text message, alerting you to which motor drive is active. Changing the 10 KΩ rheostat's position to a resistance value greater than 5.8 KΩ will stop the telerobot from its pan function, switching to a camera tilt motion. The LEGO Cam will oscillate up and down with mechanical movement cycles of 2 seconds up and 3 seconds down. The mini-LCD screen will display the *C ON* text message, signifying that motor drive C is activated. The timing values used for the pan and tilt functions were experimented and validated by the author to give the optimum mechanical movement for the two camera position movements.

The Motion Control Software Test Procedure of the Power Stand Telerobot

This test procedure will allow you to validate how the software controls the two motor drives for the pan and tilt operation of the LEGO Cam. Once the code has been tested, the telerobot can then be attached to the *universal serial bus* (USB) of a desktop PC or notebook computer for image processing.

The following test procedure will validate the motion control software test procedure of the Power Stand telerobot:

1. With the LEGO RCX P-Brick attached to the Power Stand telerobot, turn on the P-Brick.
2. Using the IC4 IDE programming editor window, enter the rotary-sensor control circuit code shown in Listing 5-1.
3. Compile the code, correcting any errors displayed within the IDE programming editor window.
4. Download the code into the LEGO RCX P-Brick.

5. Using an ohmmeter, adjust the 10 KΩ rheostat for a resistance value less than 3.5 KΩ. Apply power to the rotary-sensor control circuit.
6. Execute the IC4 control circuit software by pressing the RUN button on the P-Brick.
7. The LEGO Cam should start to pan from left to right. Note the *b ON* text message on the LEGO RCX P-Brick's LCD.
8. Slowly increase the 10 KΩ rheostat until the LCD displays *the C ON* text message.
9. After completing one panning cycle, the LEGO Cam should start to tilt the vision device up and down. Note: The tilt function is initiated when the resistance of the 10 KΩ rheostat is >5.8 KΩ.

If the telerobot performs these steps correctly, the software is working correctly. If not, check for syntax errors in the code and verify the rheostat is wired correctly. Repeat the mini software test procedure, ensuring all rheostat boundary conditions are met with correct LEGO Cam pan and tilt position responses.

TECH NOTE: *Want to know how to improve the strength of your mechatronics-based robots? How about reducing or increasing the speed of a LEGO electric motor using gears, or changing its axis of rotation? There is an excellent mechanical design guide titled The Art of LEGO Design. Fred G. Martin, codeveloper of the LEGO RCX P-Brick, wrote the 19-page PDF document. Fred, a graduate of the MIT Media Lab, teaches Computer Architecture and Robotics at the University of Massachusetts, Lowell campus. The Art of LEGO Design is a treasure trove of mechanical construction techniques, from building stronger structures to reducing noise with pulley wheels. Download the guide from* `http://nike.wellesley.edu/~rds/rds00/handouts/ArtOfLEGODesign.pdf`.

Vision Software

Giving the machine computer vision is the final phase of the Power Stand telerobot. Computer vision means the use of a computer or a microcontroller-based device to analyze visual information. The vision hardware has some type of visual sensor to create an electronic or numerical analog of an object or visual scene. The sensor should have the ability to process images for the use of object identification and pattern visualization.

Industrial applications often use computer vision. Inspecting, locating, counting, measuring, and controlling industrial manipulators (telerobots) are all examples of industrial computer vision.

The basics of how a digital camera works will be explained. Then we will discuss vision experiments using off-the-shelf software and a home brew visual application program.

No-Film Cameras

A digital camera has a series of lenses that focus light to create an image of a scene. A conventional 35mm camera functions in the same way. Instead of focusing the captured light from the object or scenery onto a piece of film as the conventional 35mm camera does, the digital camera uses a semiconductor sensor that records the light intensity. A microcontroller breaks this recorded light intensity into digital data.

The semiconductor sensor used to convert light into electrical charges is known as *charged coupled device* (CCD). The CCD is a collection of tiny light sensitive diodes that convert light (photons) into negative electrical charges known as electrons. These light sensitive diodes are called *photosites*. Each photosite is sensitive to light. The brighter the intensity of light that hits a single photosite, the greater the number of electrons that will accumulate at that site. Figure 5-17 shows a picture of a CCD sensor.

The next step, after electron charge accumulation, is to read each value at the designated cell in the captured image. In the CCD sensor, the electron charge is transported across the imaging component and read at one corner of the array. An *analog-to-digital converter* (ADC) turns each array

Figure 5-17 A CCD image sensor, courtesy of www.howstuffworks.com

pixel value into its digital equivalent. The digital data is then read by a microcontroller for image processing and visual interpretation.

Vision software will then enhance the microcontroller's image-processed data, using visual algorithms (computer program procedures) that are displayed on a computer monitor. The LEGO Cam in the Steven Spielburg Movie Maker set (Cat.No.1349) and the Vision Command product (Cat No.9731) use this imaging technology for displaying objects and scenery on a desktop PC or notebook computer screen.

The next section will discuss the Power Stand telerobot and how you can use off-the-shelf vision software for the image processing of objects and scenery present in the robot's line of sight.

Robolab Vision Center

Besides allowing you to program digital behaviors into mechatronics-based robots and intelligent machines, the Robolab software has embedded vision center software for image processing experiments into prototyping telerobotic vision systems. The following discussion will explain how to use this software with the Power Stand telerobot.

Opening the Robolab Vision Center The vision center is an advanced tool within the Robolab programming environment. Therefore, an Inventor 4 programming session must be opened to gain access to the vision software. Before running the following vision center procedure, connect the LEGO Cam to the USB of the PC or notebook computer. Place the Power Stand telerobot in a good viewing location and mechanical movement area.

The following steps will guide you to obtaining the vision center software:

1. Run the Robolab software by double-clicking its icon located on the Windows desktop.
2. Click the *Programmer* button.
3. Double-click *Inventor 4* under *Inventor*.
4. Within the Robolab *Untitled 1.vi diagram* main tool bar, click *Tools*. A drop-down menu box will be displayed on the screen. See Figure 5-18.
5. Scroll down to the *Select Camera* menu choice and click it. A *Select Camera* box will appear on the screen. See Figure 5-19.
6. Click the scroll down arrow once. The Microsoft WDM Image Capture (Win 32) will appear within the *Select Camera* box.

7. Click the check mark with the mouse. The *Select Camera* box will disappear from the screen.
8. Click *Tools*.
9. Scroll down to select *Vision Center* and click it. The *Vision Center* window will appear on the screen. See Figure 5-20. Note the image within the *Vision Center* window.
10. Turn on the LEGO RCX P-Brick.
11. Turn on the power for the rotary-sensor control circuit.
12. Run the rotary-sensor control circuit software.

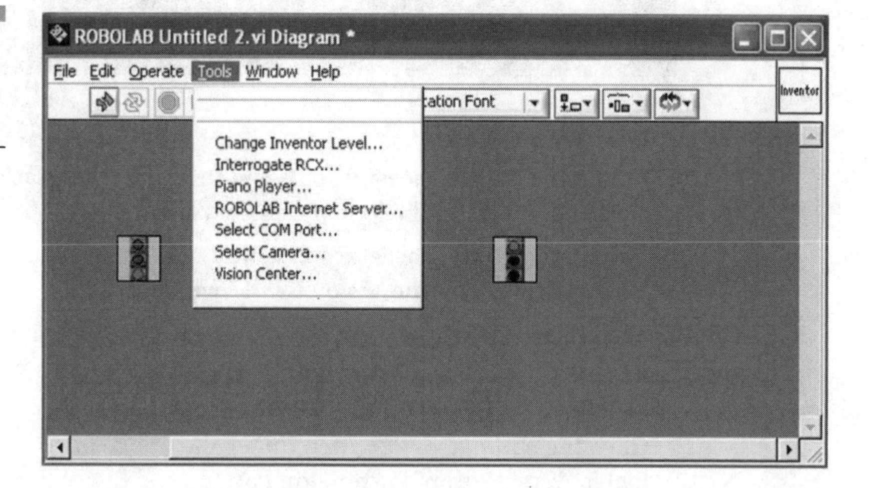

Figure 5-18 A Tools drop-down menu box.

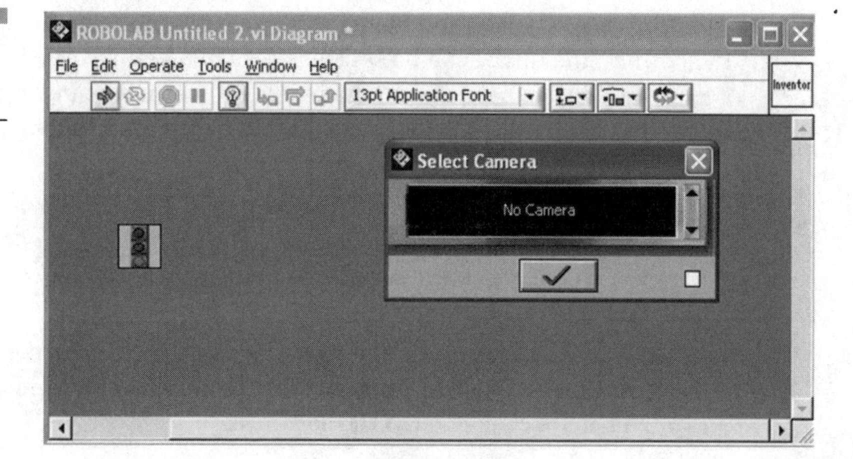

Figure 5-19 Select Camera box.

Figure 5-20 Vision Center window, with author's desktop table being displayed

13. Adjust the 10 KΩ rheostat for a pan function (rheostat value <3.5 KΩ). Note the image on the *Vision Center* window.
14. Adjust the 10 KΩ rheostat for a tilt function (rheostat value >5.8 KΩ). Note the image on the *Vision Center* window.
15. To save an image, click the *Define Sensor* icon with the mouse that is located next to the "Plane" color bar. The *Define Sensor* window will appear on the screen. See Figure 5-21.
16. Click the small arrow located under the image operation (designated by blank text) drop-down menu box.
17. When the *Enter the name of the file you wish to save* dialog box appears, type in the file name. By clicking the *Save as type* arrow to *All files*, the image can be saved as a .bmp (default) or a .jpg image. After the file has been saved, the image can be view using any paint or picture publishing software. Figure 5-22 shows the image the author saved using the *Define Sensor* feature.
18. To exit out of the *Vision Center* window, click the *Select Camera* icon located next to the *Define Sensor*. The *Select Camera* box will appear on the screen. Click the scroll down arrow until the *No camera* is displayed within the window.
19. Click the check mark and then the x in the right corner of the *Select* box. The screen will show a raster pattern on the window, signifying no camera is connected to the software.

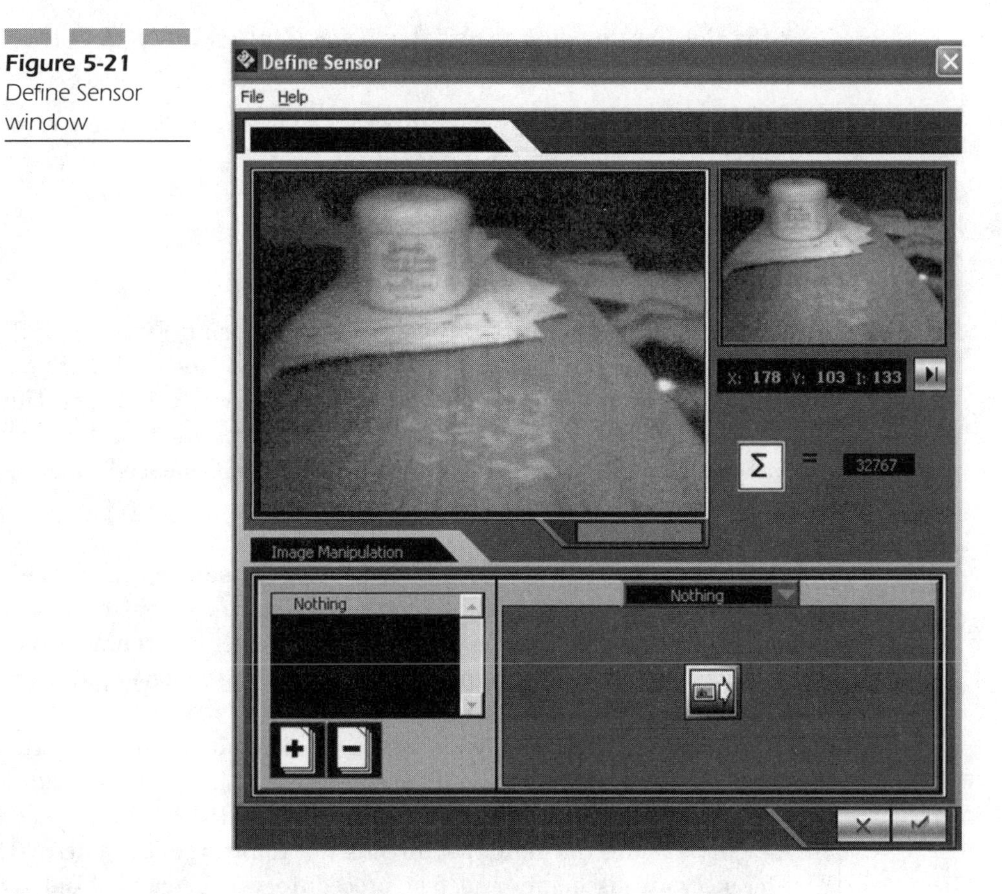

Figure 5-21 Define Sensor window

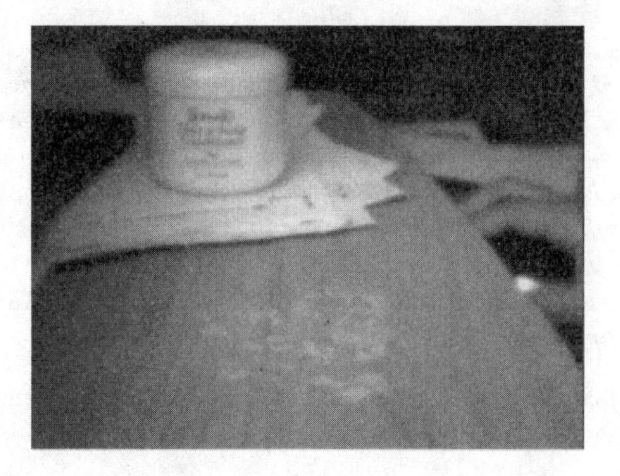

Figure 5-22 The author's saved image, taken by the Power Stand telerobot

20. Click the *Back* button with to exit the *Vision Center* window. Exit the Robolab software.

The next software lab project will investigate writing a *Visual Basic for Applications* (VBA) vision center for Excel.

A Home brew Vision Center

A home brew vision center can easily be developed using Excel VBA. By using a Logitech *software development kit* (SDK) and some basic Active X components in VBA, you can easily build a vision center within Excel. This telerobotic vision software application can open up a wealth of math-driven controls for operating a LEGO Mindstorms mechatronics-based robot, as well as provide a visual tool for the smart machine.

Excel VBA-based Vision Center Software Assembly and Test Procedure Logitech SDK is the important software element necessary for building a vision center within an Excel spreadsheet environment. The SDK has all of the camera drivers necessary for building vision applications for the image-processing device (the LEGO Cam).

The Logitech SDK is available on the CD-ROM that accompanies this book. To install the files, follow the instructions of the installation wizard. After completing the installation of the Logitech SDK, a folder titled *QCSDK1* will be under the main root directory of your target hard-drive (C or F). The software assembly and test procedure can be carried out for building the Excel VBA-based vision center.

TECH NOTE: *In the instructions on adding reference and controls for Active X components, the author assumes the reader has some knowledge or familiarity in doing this software enhancement within the Excel spreadsheet environment. For those readers who are not familiar with this software addition, step-by-step instructions can be found in LEGO Mindstorms Interfacing on pages 30–45 of Chapter 2, "Developing GUIs: Software Control Basics." Use the* vportal 21.0 type library *for the reference and the* videoportal class *as the control.*

The following lab project will outline a software assembly and test procedure for building an Excel VBA-based vision center:

1. Open the Excel spreadsheet software application.
2. Add a command button to the open spreadsheet using the controls toolbox in design mode.
3. Place the command buttons on the spreadsheet, as shown in Figure 5-23.
4. Right-click *View Code*. The *Visual Basic Editor* (VBE) window will appear on the screen.
5. Type the following code into the VBE:

```
Private Sub CommandButton1_Click()
UserForm1.Show
End Sub
```

6. Close the VBE by clicking the Excel *x* symbol located in the left-hand corner of the main menu bar. The spreadsheet, with command buttons, will be displayed on the screen.
7. Exit the design mode by clicking the *pencil and right triangle* icon on the menu bar.
8. Enter the VBE environment.

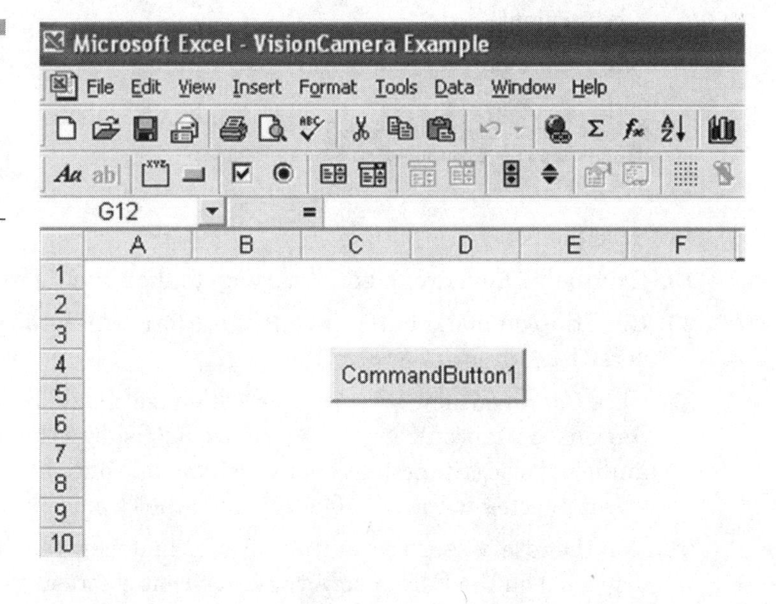

Figure 5-23 Placement of the Active X command button on the Excel spreadsheet.

9. Click *Insert* from the menu bar.
10. Click *Userform* from the pull-down menu selection items.
11. A userform will be displayed on the VBE IDE window. Adjust the userform's size to look similar to Figure 5-24 by dragging the lower-right corner of the Active X component.
12. Save the spreadsheet as *VisionCamera Example* on your desktop PC or notebook computer hard-drive.
13. Add the additional buttons as shown in Figure 5-24 from the toolbox. Add the names to these Active X controls using the *captions* feature of the property window.
14. Add the VBA code to the *Connect* button (Active X component) by right-clicking it and selecting the *View Code* function.
15. Enter the following code in the VBE window:

```
Private Sub CommandButton1_Click()
VideoPortal1.PrepareControl "QCSDK", "HKEY Local
  Machine\Software\QCSDK", 0
VideoPortal1.EnableUIElements UIELEMENT_STATUSBAR, 0, 1
VideoPortal1.ConnectCamera2
VideoPortal1.EnablePreview = 1
End Sub
```

16. Repeat step 14 of this test procedure for the *Disconnect* button (Active X component).
17. Enter the following code in the VBE window:

```
Private Sub CommandButton2_Click()
VideoPortal1.DisconnectCamera
End Sub
```

18. Save the file.
19. Exit the VBE environment, returning to the Excel Spreadsheet.
20. Click the command button and the userform will be displayed on the spreadsheet.
21. Click *Connect*. A few seconds later, an image should be visible within the videoportal control window. Figure 5-25 shows the image of the author's Power Stand telerobot viewing angle once the VBA software was connected to the LEGO Cam using the *Connect* button.
22. Run the rotary-sensory control software and view the images being displayed on the Excel userform vision center screen.
23. Click the *Disconnect* button on the userform and the image should be frozen.

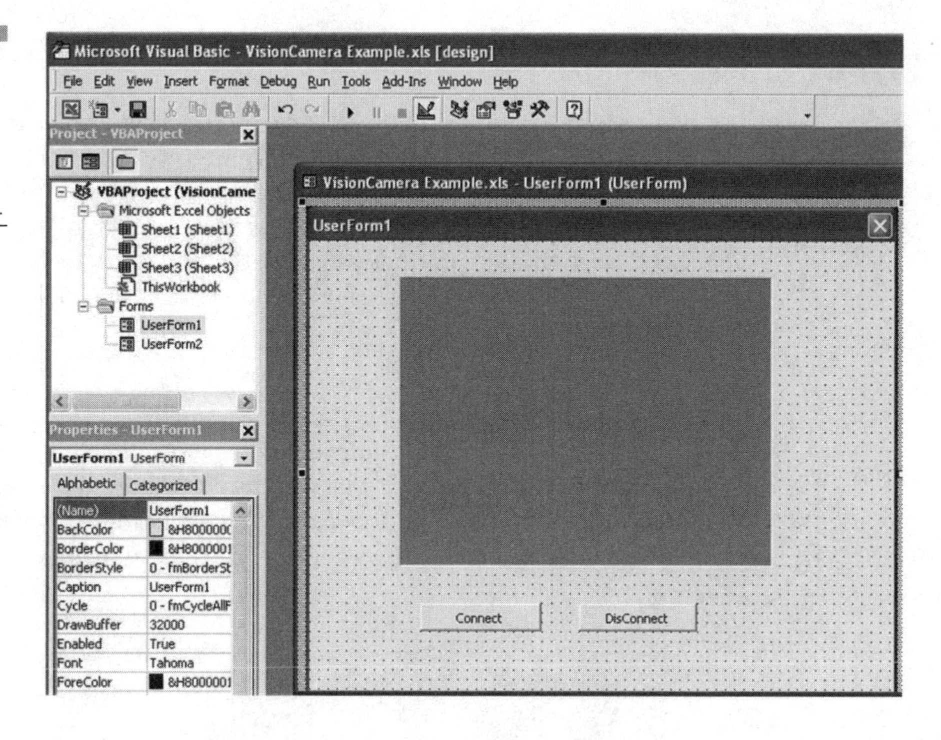

Figure 5-24 Size of userform for placement of the videoportal Active X control (camera window).

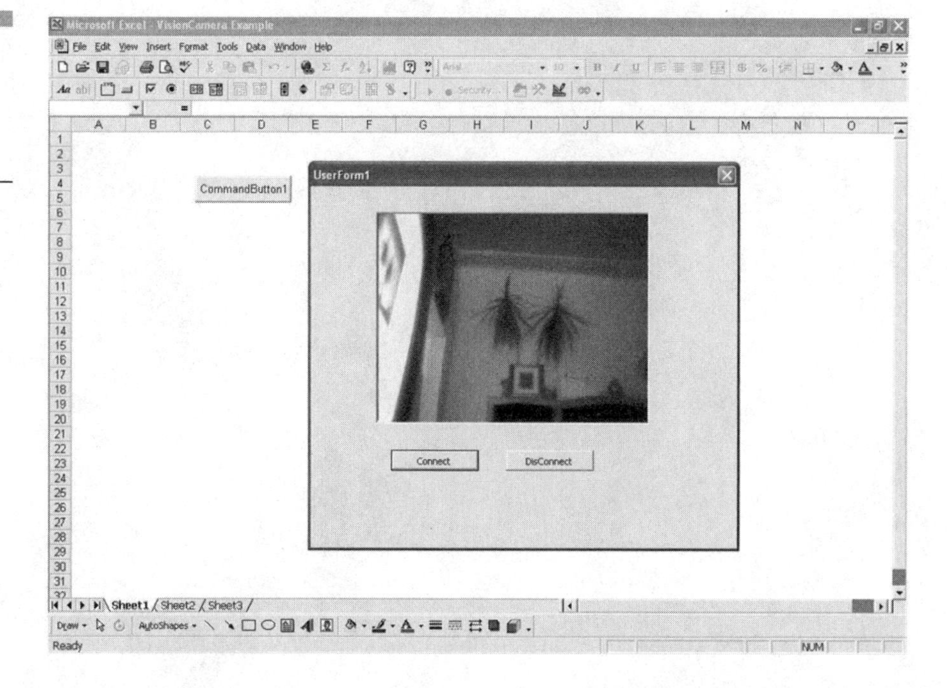

Figure 5.25 The image of the author's Power Stand telerobot viewing-angle.

Power Stand Telerobot Summary

This chapter demonstrated the basics of telerobotic using a hands-on LEGO Mindstorms mechatronics project. The electronics hardware, mechanics, and software tools used in this project allowed you to build a low-cost robotic-vision development system for the prototyping of a tabletop telerobot. I hope this chapter has provided the necessary tools for you to investigate complex telerobotics for robot contest or engineering education projects.

TECH NOTE: *For Windows XP users, a software package named Indeo 5.2 Codec must be installed on the PC or notebook hard-drive in order to use the Logitech SDK. Download the software from* `www.ligos.com/indeo/downloads`. *The software costs $14.95.*

CHAPTER 6

Smart Hand Controllers

The essential component in the telerobotics system is the hand controller. This input device is essential in telerobotics because it enables the operator to control the movement of the robot or smart machine using electromechanical or mechatronics-based input devices. Some hand controllers are as simple as a series of small toggle or slide switches mounted inside of a box. Other controllers are as sophisticated as small handheld computers with wireless interfaces to the target electronic device or machine. Passive components that can be used in hand controllers are electric switches, potentiometers, and pressure sensors.

Adding software to a microcontroller enables these switching elements to provide position and force data, thereby making the final product a smart hand controller. The smart hand controller can easily be connected to a robot using electronic interface circuits.

In this chapter, smart hand controllers will be explored using hands-on laboratory projects. Electronic circuits, the LEGO *robot command explorer* (RCX) *programmable brick* (P-Brick), and the Scout P-Brick will be used to experiment with *human machine interfaces* (HMI) to control robots and intelligent machines. This chapter will explore the following topics:

- A One-switch Hand Controller
- Diagnostics Software Validation of the One-switch Hand Controller
- The Burglar Alarm Tester-Demonstrator
- A Wireless One-switch Hand controller
- A Two-switch Hand controller
- The LEGO P-Brick and Smart Hand controllers.

A One-switch Hand Controller

A mechatronics-based hand controller can be built using one touch sensor, a P-Brick and several LEGO pieces. The one-switch hand controller can be programmed using a P-Brick to perform either a hard turn off, on or delay on and float off function. These switching functions allow the amateur roboticist to create some unique modes of operation for a robot or intelligent machine.

The one-switch hand controller enables the operator to control an electronic output indicator, an electric motor, or both. Figure 6-1 shows the components for building a one-switch controller.

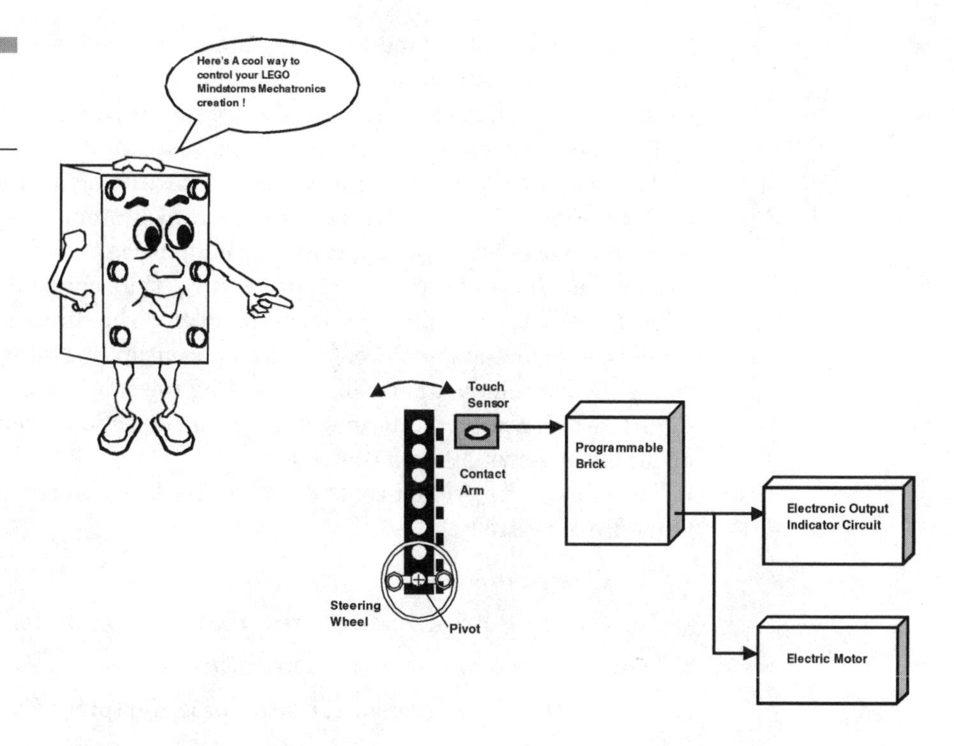

Figure 6-1 A one-switch hand controller

The touch sensor contact arm rotates around a pivot located at the far end of the long LEGO brick. The contact arm swings sideways, allowing it to depress the yellow button on top of the touch sensor. The internal series 1 KΩ and electric switch contacts are connected to the input of a P-Brick using a standard LEGO electric wire. As the contact arm touches the sensor, the P-Brick's +5V and 10 KΩ network circuits physically connect to the internal 1 KΩ resistor.

The software written by the amateur roboticist will turn the appropriate output port either on or off, controlling the function of the electronic indicator circuit or electric motor. Rotating the contact arm away from the button of the touch sensor disconnects the 1 KΩ resistor from the P-Brick's internal network circuit. The control software embedded inside of the P-Brick will provide the switching connectivity to the appropriate output port. The output port control signal will be applied to either the simple electronic indicator circuit or electric motor.

The flexibility behind the basic hand controller is the capability to program a unique output function for the external control of an electronic or electromechanical load. Therefore, the one-switch hand controller truly demonstrates the mechatronics design concept because of the programmability of the mechanical system using software. The mechanics of the one-switch hand controller are seamlessly interfaced to the controlled output using electronics or electromechanical components.

The one-switch hand controller can be thought of as a *Proof of Concept* (POC) study in developing a programmable switching device for controlling the operation of a mobile robot's drive-train system. The electric motor attached to one output of the P-Brick could be controlled to stop the small rotating machine, allowing the robot to turn, thus changing direction. The other motor would be allowed to spin, allowing the directional control feature of the robot to take place. An electronic output indicator would provide you with visual indications while operating the sensor-controlled machine. Figure 6-2 shows the state machine for the directional control of a robot, using the one-switch hand controller.

The robot directional-control state machine works with following sequence of events:

1. The P-Brick is off.
2. Turn on the P-Brick's power (Press the *Power On* button).
3. Output ports A and B are off (motors are off).
4. Execute the robot directional control program (press *Run* [*RUN* On]).

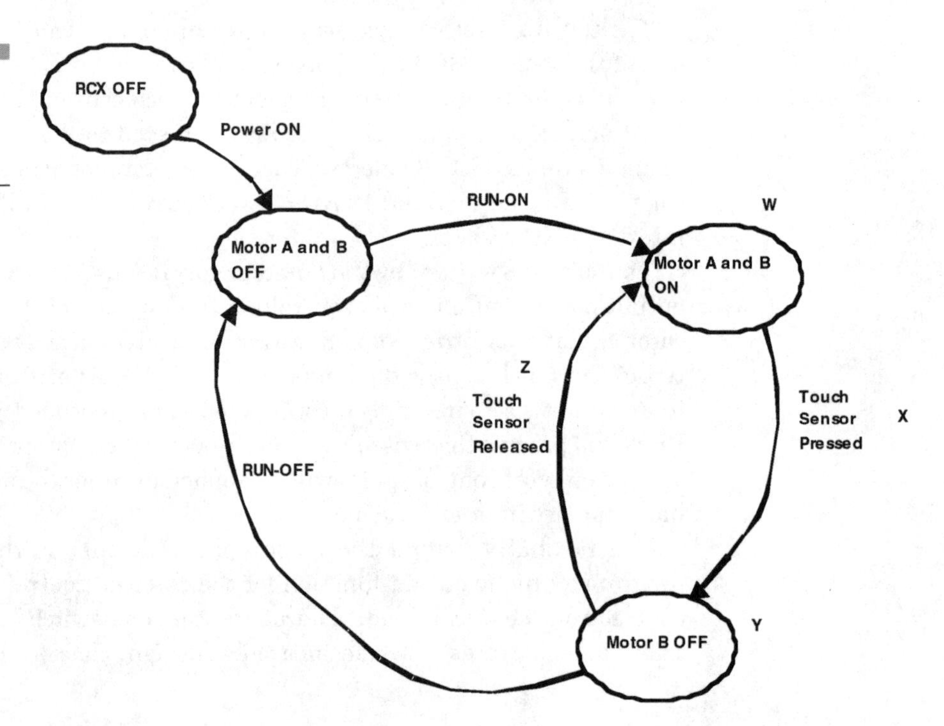

Figure 6-2 Robot directional-control state machine

5. Output ports A and B are on (motors are on) *w*.
6. Press the touch sensor with control arm (*Touch Sensor* pressed) *x*.
7. Output port B is off (motor B is Off) *y*.
8. Release the touch sensor with control arm away from the passive switch (*Touch Sensor* Released) *z*.
9. Output ports A and B are on (motors are on).
10. Stop execution of robot directional control program (press *RUN* [*RUN* off]).
11. Output ports A and B are off (motors are off).

The motion of the robot will be unaltered while in state *w*, as seen in Figure 6-2, until an event *x* is triggered. With this transition being true, the direction of the robot will change, as noted by state *y* in the state machine diagram. The robot will stay on this new directional path after the event *z* has occurred. Therefore, with output port B turned off while motor A rotates, the robot will turn, moving it in a new direction.

The following lab project will demonstrate a mechatronics-based hand controller using the assembly instructions and a test procedure.

TECH NOTE: *The Scout P-Brick can be programmed in a high-level language call* LEGO Assembler *(LASM). LASM enables the complex functions like logical* ANDing *two touch sensors. The Scout* integrated development environment *(IDE) tool enables you to write programs in LASM using its embedded text editor. The Scout IDE tool can be found on the CD-ROM packaged with the book.*

One-switch Hand Controller Laboratory Project

The following laboratory project will demonstrate the basic switching function of the hand controller. The Scout P-Brick will be used to test the one-switch hand controller.

The *bill of materials* (BOM) for this project is as follows:

One touch sensor

Two LEGO electric wires

One Scout P-Brick

One black Technic Brick 1x2

Two black Technic Bricks 2x4

One black Technic Brick 2x2x2/3 (with wire end)

One light-gray Electric Touch Sensor Brick 3x2 (complete assembly shortcut)

Two light-gray Plates 1x8

One light-gray Plate 2x2

One black Technic Axle 6

Two black Technic Bricks 1x2 (with holes)

Two green Technic Bricks 1x4 (with holes)

One black Technic Brick 1x8 (with holes)

One black Technic Brick 1x10 (with holes)

One light-gray Technic Bush

One light-gray Technic Liftarm 1x3

Three black Technic Pins

Two light-gray Technic Plates 2x4 (with holes)

One light-gray Technic Pulley (large)

Assembly and Test Procedure The following steps will allow you to assemble and test the one-switch hand controller:

1. Build the one-switch hand controller, using Figures 6-4 through 6-15. The following steps will test the electrical switching function of the hand controller:
   a. Attach an ohmmeter to the one-switch hand controller's LEGO electric wire, as shown in Figure 6-3.
   b. Adjust the scale of the ohmmeter to 2 KΩ, for an accurate reading of the touch sensor's internal resistor.
   c. Turn the steering wheel to the right, clockwise, locking the contact arm on top of the touch sensor's button.
   d. Record the ohmmeter reading _______KΩ
   e. Turn the steering wheel to the left, counterclockwise, unlocking the contact arm from the touch sensor's button.
   f. Record the ohmmeter reading ______KΩ.

2. Attach the hand controller to input port 1 of the Scout P-Brick.

3. Turn on the Scout P-Brick.

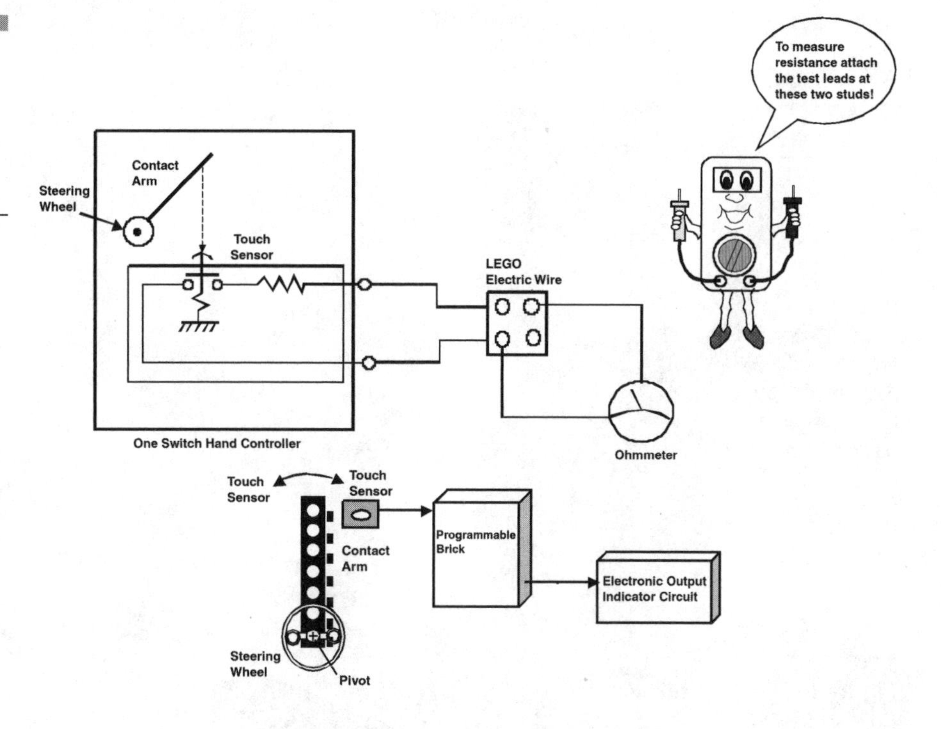

Figure 6-3 Attaching an ohmmeter to the one-switch hand controller's LEGO electric wire

4. Using the *Stand Alone Code* (SAC) feature of the Scout P-Brick, program the following function:

```
Motion     Touch   Light
Forward
           Wait
```

5. Press *RUN* on the Scout P-Brick.
6. Turn the steering wheel on the one-switch hand controller clockwise. The input port 1 LED and the A and B output port LEDs should be illuminated.
7. Turn the steering wheel on the one-switch hand controller counterclockwise. The input port 1 LED will be off while the A and B output port LEDs will be on.
8. Turn the steering wheel on the one-switch hand controller clockwise then counterclockwise. The input port 1' LED will toggle on and off and the A and B output port LEDs should be off.

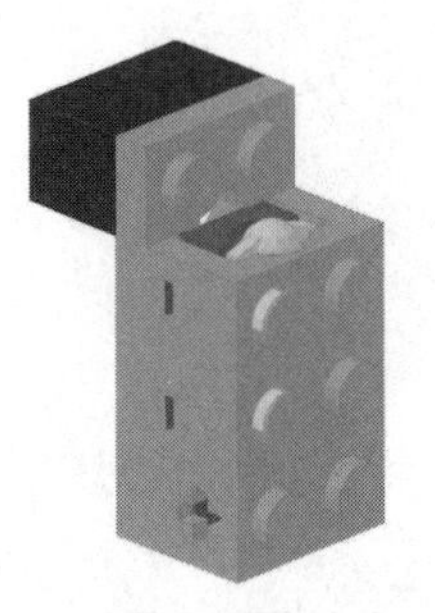

Figure 6-4
Build instructions for one-switch hand controller

Figure 6-5
Build instructions for one-switch hand controller

Figure 6-6
Build instructions for one-switch hand controller

Figure 6-7
Build instructions for one-switch hand controller

Figure 6-8
Build instructions for one-switch hand controller

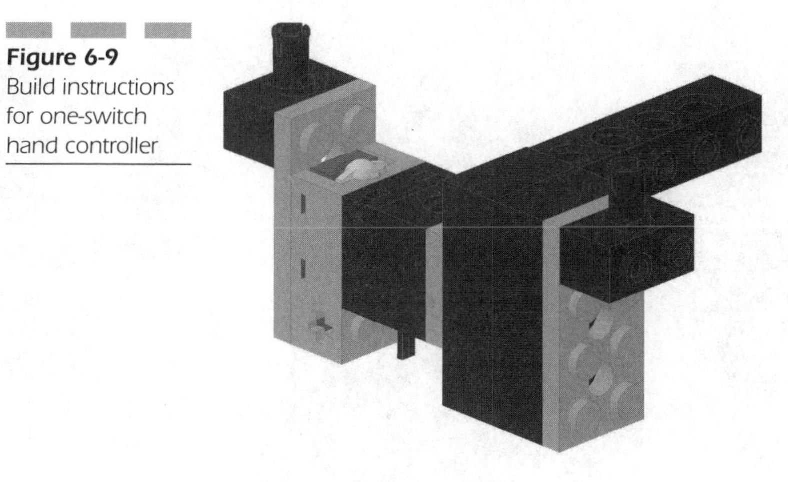

Figure 6-9
Build instructions for one-switch hand controller

Figure 6-10
Build instructions for one-switch hand controller

Figure 6-11 Build instructions for one-switch hand controller

Figure 6-11 Build instructions for one-switch hand controller

Figure 6-12 Build instructions for one-switch hand controller

Figure 6-13 Build instructions for one-switch hand controller

Figure 6-14 Build instructions for one-switch hand controller

Figure 6-15 Build instructions for one-switch hand controller

Summary of the One-switch Hand Controller Laboratory Project
This laboratory project demonstrated a procedure for building and validating a one-switch hand controller. The design of the switching device used a movable contact arm that pushes on the button of a touch sensor. By turning a small LEGO steering wheel that rotates on a pivot, the attached 1x7 brick with studs will make contact with the touch sensor. The hand controller has the capability to lock on to the touch sensor or provide a momentary switching function.

The internal resistor can be measured using an ohmmeter. When the button of the touch sensor is depressed, the ohmmeter is able to read the internal resistor value. When the contact arm is away from the touch sensor's button, the digital ohmmeter would display *OL* for open load.

After checking the switching function of the hand controller, the electromechanical switch was then connected to a Scout P-Brick. By attaching the LEGO electric wire of the one-switch hand controller to the input of the P-Brick, the final mechatronics product was tested using the Scout P-Brick. The microcontroller or core processor for the hand controller is emulated by the Scout P-Brick's two input and output ports.

The SAC mode of the Scout P-Brick was used to check the switching operation of the one-switch hand controller. The internal program of the Scout P-Brick selected among the nine preprogrammed functions embedded within the microcontroller's ROM. The *Wait For* instruction was used as the touch sensor's press/release event triggering signal for the Scout P-Brick. The embedded LEDs (input-1 and output A) provided visual indication of the event-transition operation of the one-switch hand controller's switching function.

Using the *Wait For* instruction with the one-switch hand controller created a toggle switching function. Upon turning the steering wheel clockwise or counterclockwise using the hand controller, the embedded LEDs at input 1 and output A would toggle the visual indicators. The one-switch hand controller in conjunction with the Scout P-Brick truly illustrates the concept of mechatronics, by creating smart products using electronics, mechanics, and software.

Diagnostics Software Validation of the One-switch Hand Controller

In addition to using an ohmmeter to analyze the one-switch hand controller, diagnostics software can be used to validate the operation of the electromechanical device. The Scout IDE can be used as a virtual instrument using the *advanced monitoring* feature of the tool. The following test procedure will outline the process for virtual validation of the one-switch hand controller.

Test Procedure

The following steps will validate the diagnostics software of the one-switch hand controller:

1. After installing the Scout IDE tool on your PC or notebook hard-drive, open the tool by clicking the icon located on the desktop or the listing of windows program files. The Scout IDE tool *graphical user interface* (GUI) will be displayed on the screen.
2. Place the infrared tower in front the Scout P-Brick.
3. Turn on the Scout P-Brick.
4. Click the *Advanced* button. The advanced monitoring GUI will be displayed on the screen.
5. Click the *enable-monitoring* button on the GUI. The light sensor bar should be active as indicated by the blue squares moving up and down as well as a series of beeps being emitted from the Scout P-Brick.
6. Rotate the steering wheel on the one-switch hand controller until the contact arm latches.
7. The touch sensor *one* button should be blue, indicating that the hand controller is switched on. The *255* value should have switched to *11*.
8. Rotate the steering wheel on the one-switch hand controller to unlatch the contact arm. The touch sensor *1* button should have returned its original gray color. The value of *255* should be displayed underneath the touch sensor *1* button.
9. Toggle the contact arm on the one-switch hand controller several times, noticing the numeric values switching between *11* and *255* with each latch and unlatch movement of the contact arm.

Summary of the Diagnostics Software Validation of the One-switch Hand Controller

The Scout IDE software tool is capable of examining the one-switch hand controller wirelessly, with the aid of the infrared tower. The advanced monitoring feature enables the one-switch hand controller's touch sensor to be validated when the contact arm depresses the button on the object detection component. When the one-switch hand controller's contact arm is latched to the touch sensor, the value read by the advanced monitoring GUI is 11. When the contact is unlatched, the numeric data value is 255.

An Inverting Logic Function

The one-switch hand controller can be programmed, with the aid of the Scout P-Brick, to produce an inverting logic function. The inverting logic function works by outputting the inversion or negation of the input control signal. If the input control signal is a logic *HI* or binary *1*, the output of the logic gate will be *0*. If the input control signal is a logic *LO* or binary *0*, the output of the logic gate will be 1. Table 6-1 summarizes the function of the inverting logic gate.

From a quick observation, the inverting logic gate may not seem practical as a core circuit element to use in a mechatronics application. However, the inverting logic gate is quite practical in providing an elegant design solution to a POC study developing a burglar alarm tester.

The basic concept behind a security system is to thwart the robber's attempt to enter a building by sounding an audible alarm. The audio tone of the security system is very high in *decibels* (dBs). A decibel is a unit of amplification (gain). The amount of gain can be very disruptive while testing the sensor switch and the control panel electronics at an engineering research and development center. Therefore, a tester must provide a visual indication to the operator that the output control signal is active. One solution to this design problem is to use a seven-segment LED display as the visual indicator for the burglar alarm tester. Figure 6-16 shows the system block diagram of the burglar alarm tester-demonstrator.

The one-switch hand controller will replace the sensor-switch component within the burglar alarm tester-demonstrator. The demonstrator will behave like an actual alarm, providing the negation function of an open-loop detection system. An open-loop detection system works when the

Table 6-1

Function of the inverting logic gate

Input Control Signal	Output
1	0
0	1

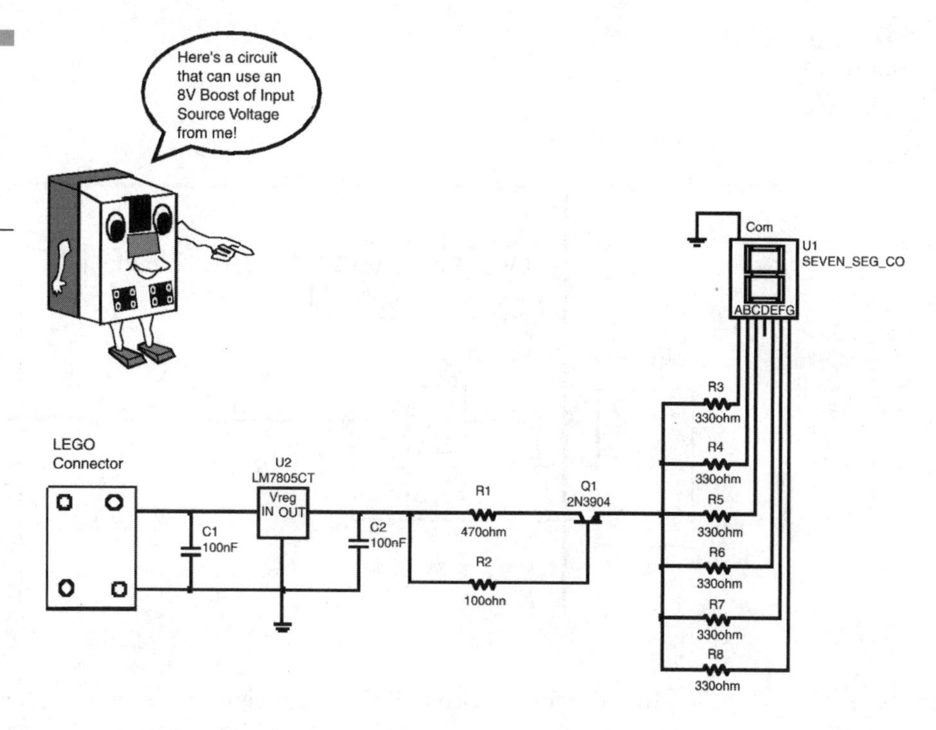

Figure 6-16 The system block diagram of the burglar alarm tester-demonstrator

Table 6-2 Open-loop detection system summary

One-switch hand controller	Seven-segment LED display
CLOSED	BLANK
OPEN	A

sensor-switch internal contacts are open, allowing the output control signal to drive an audible source. Table 6-2 summarizes the operation of the open-loop detection system using the one-switch hand controller, the Scout P-Brick, and a seven-segment LED display.

The seven-segment LED display is used because it is the simplest form of visual interface for an operator to use. The seven-segment LED display has enough resolution to display all numbers and some letters. The truth table provided shows the input possibilities and the corresponding outputs. When the one-switch hand controller is open, the seven-segment LED display will show the letter *A* for active output. Output port A will be turned on, allowing the letter A to be visible on the seven-segment LED display.

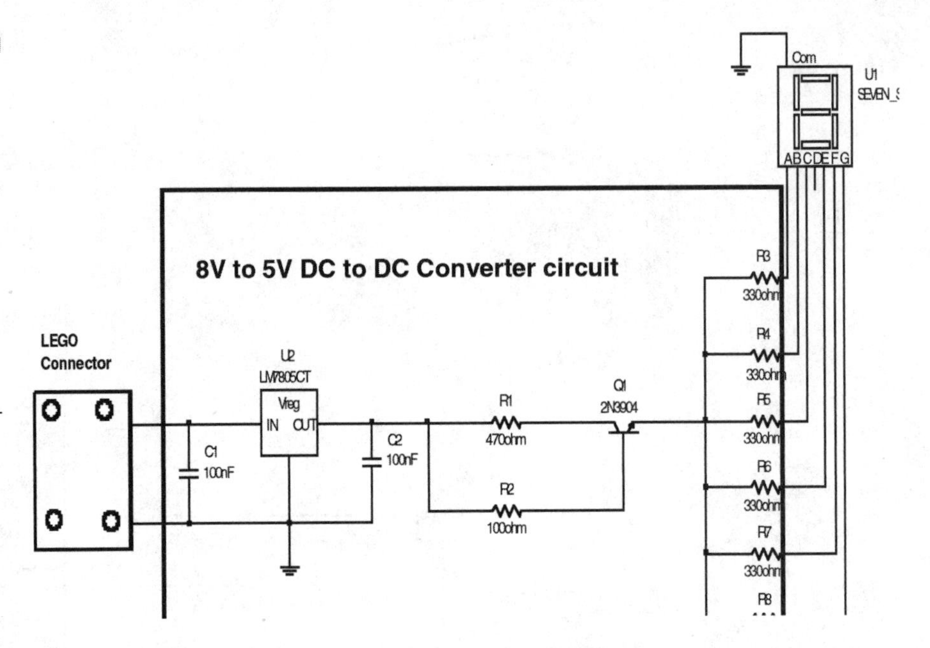

Figure 6-17 An electronic circuit capable of converting the 8V DC output voltage of port A and stepping it down to a constant 5V DC source using the 7805 Linear Voltage Regulator IC

In order for the Scout P-Brick to drive a 5V seven-segment LED display, an electronic driver circuit is required. The circuit in Figure 6-17 is capable of converting the 8V DC output voltage of port A and stepping it down to a constant 5V DC source using the 7805 Linear Voltage Regulator *integrated circuit* (IC).

The output voltage is applied to a high side driver capable of providing sufficient current gain, using the 2N3904 NPN transistor to drive the seven-segment LED display. Capacitors C1 and C2 provide filtering of transients (electrical noise) that may become coupled to the input and output pins of the 7805 linear voltage regulator IC from output port A of the Scout P-Brick. The series resistors R1 and R2 limit current flowing into the collector and base leads of the 2N3904 NPN transistor.

The last item to note about the high-side seven-segment LED display driver circuit is that the alphanumeric component is a common cathode optoelectronic device. It is called a common cathode LED display because it has only one ground or cathode pin, with the remaining seven segments being anodes. Therefore, like the ordinary silicon diode, in order for a segment to be turn on it must be *biased* (operating point) correctly. Therefore, the anode segment required to form the letter *A* will be wired to the output circuit of the driver using the emitter lead of the NPN transistor. The electrical node will be the +5V bus, while the common cathode pin will be connected to ground. The current will flow through each segment of the optoelectronic device to ground. This current path will allow each segment to emit a red light, thus displaying the letter *A*.

The Burglar Alarm Tester-Demonstrator

The following assembly instructions and test procedure will outline the steps for building a burglar alarm tester-demonstrator. A *demonstrator* is a black box with switches that simulate the production-intended parts used on the actual system. This specialized box can be used to test new functions embedded within the memory registers of the microcontroller. The one-switch hand controller along with the Scout P-Brick and the high side seven-segment LED display driver will form a programmable demonstrator for checking out the function of a burglar alarm using a silent output indicator.

The BOM for the burglar alarm tester-demonstrator is:

Two 100-*nanofarad* (nF) capacitors
One 7805 Linear Voltage Regulator IC
One 470 Ω, 1/4-W resistor
One 100 Ω, 1/4-W resistor
One 2N3904 NPN transistor
One 7-Segment red LED display, common cathode
Six 330 Ω, 1/4-W resistors
One modified LEGO electric wire
One Radio Shack Electronics Learning Lab kit or equivalent circuit prototyping station
One digital multi-meter (DMM)

Assembly and Test Procedure

The steps for the assembly and testing of the burglar alarm tester-demonstrator are as follows:

1. Breadboard the high side seven-segment LED display driver, shown in Figure 6-18, using the Radio Shack Electronics Learning Lab kit or equivalent circuit prototyping station.

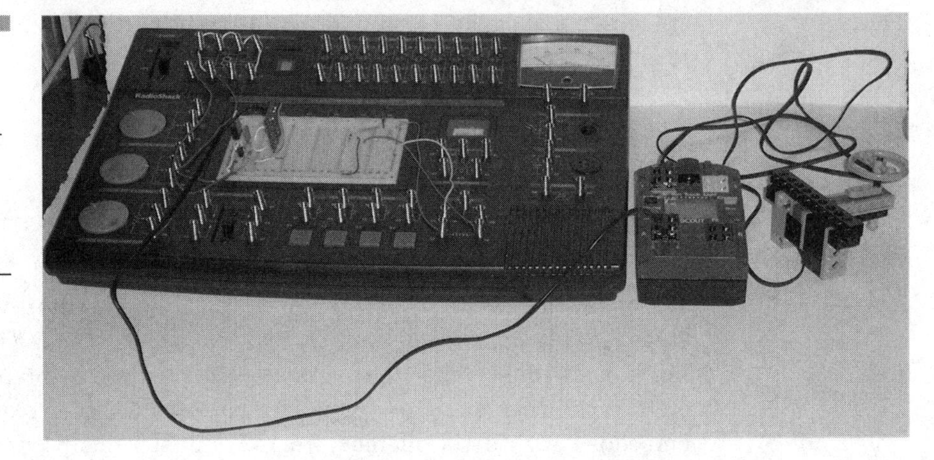

Figure 6-18
The high side seven-segment LED display driver is breadboarded using the Radio Shack Electronics Learning Lab kit.

2. Attach the driver circuit to the Scout P-Brick's output port A, using the modified LEGO electric wire. The wire attached to the connector should be pointing downward for correct polarity interface to the driver circuit.
3. Program the Scout P-Brick to operate as an inverting logic gate using the following SAC program.

```
Motion              Touch
Forward

       Brake
```

4. Run the code in SAC mode.
5. Using the steering wheel, rotate the contact arm of the one-switch hand controller clockwise, locking it into place. The seven-segment LED display should be blank or off.
6. Using the steering wheel, rotate the contact arm of the one-switch hand controller counterclockwise from the locking position. The seven-segment LED display should show the letter *A*. If not, turn the Scout P-Brick off and check for wiring errors, wrong resistor values, and misplaced transistor and IC leads to bread holes. After the errors are found and corrected, repeat steps 5 and 6 to obtain the proper results.

TECH NOTE: *The Scout IDE is a wonderful tool for checking the mechatronics-based input devices. By using the advanced monitoring feature, the mechatronics-based input device can be validated easily and quickly.*

The Burglar Alarm Tester-Demonstrator Summary

This laboratory project illustrated how the one-switch hand controller can be used as an input control signal for testing the inverting logic used in a burglar alarm system. The Scout P-Brick provided the negation function of the tester-demonstrator, by reading the input port 1 signal provided by the one-switch hand controller. A high side seven-segment LED display driver circuit was wired to output port A, using a modified LEGO electric wire. This simple display device will show the letter *A* for active output. This optoelectronic device will only be visible when the touch sensor of the one-switch hand controller is open. The display also provides a substitute silent alarm when testing the burglar alarm's detection sense function of the hand controller.

Exploring the Output Ports of a Scout P-Brick Using the One-switch Hand Controller

Now that the one-switch hand controller is operational, output port A can be explored using the Scout IDE tool.

Lab Project Procedure

1. The following lab project will investigate output ports A and B of the Scout P-Brick in relation to the hand controllers input control signal:
   1. Open the Scout IDE tool by clicking the icon located on the desktop or program files menu.
2. Turn on the Scout P-Brick, by pressing the *ON-OFF* button.
3. Click the *Init Ghost* button on the IDE GUI.
4. Click the *Stand Alone* button.
5. Click the *Advanced* button. The advanced monitoring GUI will appear on the screen.
6. Click the *Enable Monitoring* button. The power levels of both output ports should be at 7, provided the batteries are at sufficient charge.
7. Under the P-Brick Control section of the Scout IDE tool, click the *Run* button. The seven-segment LED display should be activated, as well as the two green triangle LED indicators. The LEDs (rectangle and

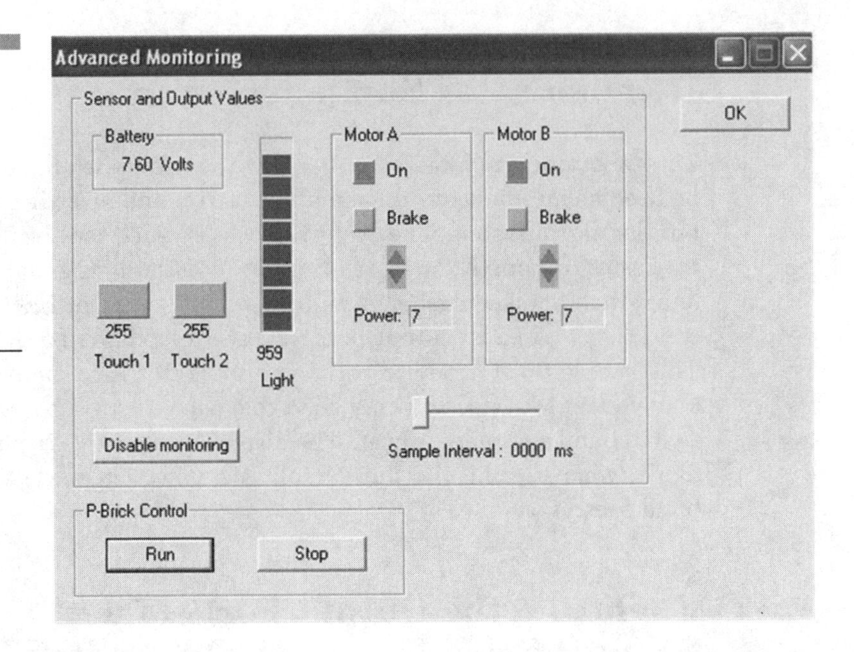

Figure 6-19 The LEDs (rectangle and triangle) for the output ports A and B on the advanced monitoring GUI will be illuminated.

triangle) for output ports A and B, on the advanced monitoring GUI, will be also be illuminated. See Figure 6-19.

8. Move the contact arm to the latch position by rotating the steering wheel clockwise. The seven-segment LED display, as well as the green triangle LED output port A, should be turned off. On the Scout IDE, the touch 1 should be blue with the number *11* or *12* displayed. Motor A should display the *Brake* red triangle LED indicator. Motor B will still be active as indicated by a green triangle LED indicator.
9. Move the contact arm to the unlatch position by rotating the steering wheel counterclockwise. The seven-segment LED display, green triangle LED, and Scout IDE GUI LED will be activated.
10. Stop the SAC program of the Scout P-Brick. Exit the IDE tool.

Summary of Exploring the Output Ports of a Scout P-Brick Using the One-switch Hand Controller The output ports of a Scout P-Brick can be remotely monitored, using the infrared tower and the IDE tool. By using the advanced monitoring feature of the Scout IDE software tool, output ports A and B of the P-Brick can be checked while running the Forward-Brake SAC program. As the seven-segment display is switching between blank to A using the one-switch hand controller, the output ports as well as the input port can be monitored with the advanced monitoring GUI.

TECH NOTE: *Turning a moving object requires one brake being applied to a rotating wheel while the other one spins freely. The* Forward-Brake *program used in the previous lab project can perform this turn function quite easily. With the one-switch hand controller latched on, output port A turns off while output B stays on. Therefore, an attached mobile platform will turn based on the one wheel spinning and the other one being stopped.*

A Wireless One-switch Hand Controller

A salvaged transmitter-receiver pair of a radio-controlled car can be added to the Scout P-Brick to create a wireless one-switch hand controller. By adding a simple transistor or IC relay driver circuit to the output port of the Scout P-Brick, the touch sensor of the one-switch hand controller will be able to activate the semiconductor-driven contacts of the electromechanical relay, thereby controlling the transmitter.

The receiver of the RC car will pick up a radio frequency signal. The receiver converts this wireless signal to a DC voltage capable of driving the hardwired electrical motors attached to the *printed circuit board* (PCB). The electrical motor of the RC car can easily be replaced with the LEGO electromechanical equivalent, by making the necessary wiring connections to the small DC actuators.

TECH NOTE: *A wireless one-switch hand controller can be built quite easily by following the build instructions in Chapter 1 of LEGO Mindstorms Interfacing. The "Wireless Basics" section of the book explains how to modify the transmitter-receiver controls of an RC car. Detail block diagrams, circuit schematics, and photos provide the technical information needed to make this wireless control modification.*

Human Machine Interface and Smart Hand Controllers

The objective of the HMI is to help the operator run a machine and manage its external and internal processes. A good HMI will improve the productivity of the operator. In addition, it will improve the productivity of the machines that affect the overall quality of the manufactured product. The function of the HMI varies based upon the type and complexity of the product produced, the machines used, the skills of the operator, and the automation used with the machinery. The types of HMI functionality usually include

- Graphic displays that provide machine operational information and status to the operator.
- User input that coordinates inputs from the operator for adjusting the machine operation.
- Data logging and storage that obtain machine history related to performance and operation of mechanical and electrical subsystems and interfaces.
- Trending or the use of graphs and charts to visually analyze data on current or past machine operation.
- Alarming, which alerts the operator of abnormal operation of the machine, using visual and audible warning indicators and systems.

The physical approaches to using HMI include the following:

- **Hardwired interface devices** Pilot lights (incandescent and LED), numeric displays (seven-segment LED displays and LCDs), and electrical mechanical switches (push-button, rocker, and toggle).
- **Proprietary hardware displays** Intelligent terminals ranging in size from small units, measuring 2x20 alphanumeric character displays with attached keypads, to color flat panel touch screens with full 15" color *Cathode Ray Tubes* (CRTs) with waterproof keyboards. These smart terminals usually run a proprietary operating system and are built using software provided by the hardware manufacturer.
- **PC and Palm-based displays** These HMIs consist of an industrially-hardened PC, display, and keyboard running a common commercially available software operating system. Off-the-shelf software, rather than the packaged program provided by the hardware manufacturer, is typical with these HMIs. Palm OS handhelds are becoming common in manufacturer facilities, providing monitoring and data acquisition capabilities using either a hardwired tether or wireless connectivity. Small *programmable logic controllers* (PLCs) with sealed keyboards and flat panel color screens allow mechatronics developers to create HMI systems using digital input cards for

hardwiring electromechanical switches or other input pointing devices.

The one-switch hand controller, as demonstrated in the previous lab projects, can be programmed to perform a variety of basic control and monitoring functions with the aid of the Scout P-Brick. As discussed in the section concerning available approaches for providing HMI functionality, the one-switch hand controller falls under the category of Hardwired Interface Devices because of its elementary electromechanical switching element, the touch sensor. For basic switching functions like negation, as demonstrated in the Burglar Alarm Tester demonstrator, the Scout P-Brick has the programming capability to perform this simple task. For sophisticated mechatronics-based HMI applications, the LEGO RCX P-Brick has the programming capability to read the one-switch hand controller and provide text data on the mini-LCD screen concerning the position of the contact arm.

The following lab project will explain how to enhance the switching capability of the LEGO mechatronics-based one-switch hand controller.

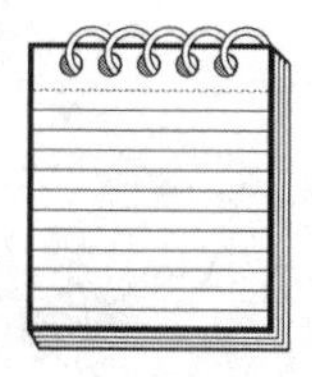

TECH NOTE: High Tech Services *(HTS) has a wealth of technical information on their Web site. Information on inputs/outputs, motion control, .NET, robots, and companies specializing in HMI and operator interfaces can be found on this well organized site. To learn more about operator interfaces and HMI systems, go to* `www.htservices.com/Tools/OperatorInterfaces/index.htm`.

A Two-switch Hand Controller

Adding an additional input control signal of the one-switch hand controller can be accomplished quite easily. By adding another touch sensor to the existing hand controller, an extra input control signal for a P-Brick is available. The modification for this added input control signal requires replacing a few LEGO bricks with a touch sensor, and four standard bricks and small plates. The additional touch sensor is mounted to the hand controller using the 2×4 Technic Plate (with holes), the 1×2 brick, and two 1×2 Technic Plates. They provide a mechanical structure and support for the touch sensor, as well as an attaching point to the top 1x11 horizontal brick.

With the additional sensor present, the contact arm now has additional travel, as well as another mechanical-electrical interface for input control to the P-Brick. Turning the steering wheel to the right or left will make or break a connection to the left and right touch sensors. The mechanically-selected sensor will then provide the corresponding control signal to the input port of the P-Brick.

Figure 6-20 shows the system block diagram of a two-switch hand controller. Figure 6-21 shows the MLCAD 3-D model of the mechatronics-based hand controller. The software embedded into the P-Brick will give the smart hand controller custom input detection and control functions for robots and intelligent machines.

Validating the Two-Switch Hand Controller with the Scout IDE Tool

The Scout IDE tool is capable of not only reading the input value of one touch sensor, but it can read a second detection device as well. The

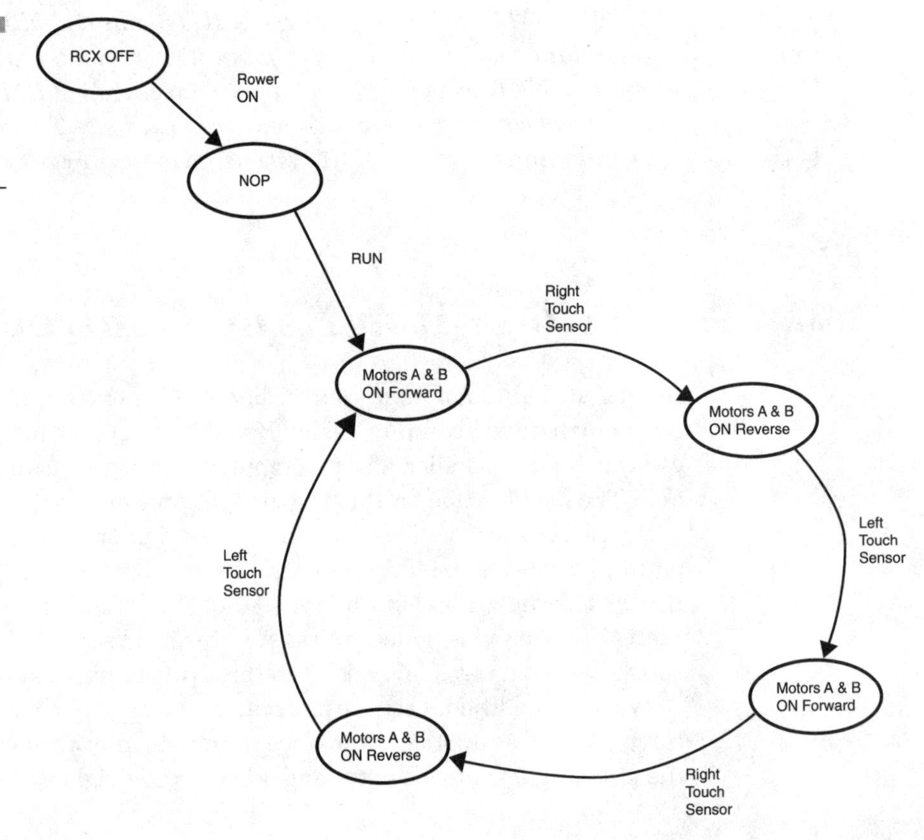

Figure 6-20 The system block diagram of a two-switch hand controller.

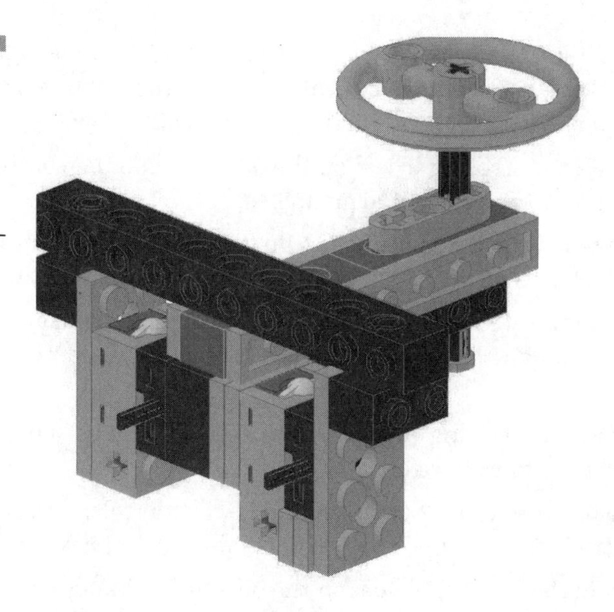

Figure 6-21 The MLCAD 3-D model of the mechatronics-based hand controller.

advanced monitoring GUI can easily read two touch sensors attached to the inputs of the Scout P-Brick.

The Two-Switch Hand Controller Minitest Procedure The following minitest procedure will outline how to validate the two-switch hand controller, using the Scout IDE tool:

1. Build the two-switch hand controller by modifying the one touch sensor device. Use Figure 6-21 as a modification guide to build the two-switch hand controller.
2. Attach the two standard LEGO electric wires to input ports 1 and 2 of the Scout P-Brick. The right touch sensor will attach to input port 2 and the left sensor to input port 1.
3. Place the infrared tower in front of the Scout P-Brick.
4. Activate the Scout P-Brick.
5. Open the Scout IDE tool by clicking the icon located on the desktop or program files menu.
6. Click the *Init Ghost* button on the IDE GUI.
7. Click the *Stand Alone* button.
8. Click the *Advanced* button. The advanced monitoring GUI will appear on the screen.

9. Click the *Enable Monitoring* button on the GUI. The light sensor bar should be activated, as indicated by the blue squares moving up and down as well as a series of beeps being emitted from the Scout P-Brick.
10. Rotate the steering wheel on the one-switch hand controller to the right, until the contact arm latches.
11. The touch sensor 2 button should be blue in color, indicating the hand controller is switched on. The 255 value should have switched to 11.
12. Rotate the steering wheel on the one-switch hand controller to unlatch the contact arm. The touch sensor 2 button should have resumed the value of 255.
13. Rotate the steering wheel on the one-switch hand controller left, until the contact arm latches.
14. The touch sensor 1 button should be blue in color, indicating the hand controller is switched on. The 255 value should have switched to 11.
15. Rotate the steering wheel on the one-switch hand controller to unlatch the contact arm. The touch sensor 1 button should have resumed the value of 255.
16. Exit the Scout IDE tool.
17. Turn off the Scout P-Brick.

Two-Switch Hand Controller Test Summary This mini test procedure allowed the amateur roboticist to quickly validate the two-switch hand controller. The Scout P-Brick provided the hardware interface and communication media to transfer the switch data of the hand controller to the IDE tool wirelessly, using the infrared tower. The Scout IDE displayed this numeric data through the advanced monitoring feature of the software tool. The *analog to digital* (A/D) value that corresponds to an open sensor is 255. The value for closed detection is 11. An A/D value is the conversion of a varying discrete analog signal to its equivalent digital representation. The numeric value is usually shown as a decimal number.

Testing the Two-Switch Hand Controller (Scout P-Brick SAC Mode)

The Scout IDE tool enables you to test the two-switch hand controller wirelessly, using the software's advanced monitoring feature. How can you validate the two-switch hand controller using the Scout P-Brick as a stand-alone tool? The Scout P-Brick has two embedded LEDs, one for each input port. By using these optoelectronic lights as visual indicators, the status of the touch sensors, using the position of the contact arm, can be monitored by the Scout P-Brick. Therefore, a portable tester can be realized using the Scout's external input ports.

Scout SAC Test Procedure The following laboratory project outlines the test procedure for validating the two-switch hand controller, using the Scout's external I/O ports:

1. Attach the left touch sensor to input port 1, and the right touch sensor to input port 2, as shown in Figure 6-22.
2. Activate the Scout P-Brick.
3. Using the steering wheel, rotate the contact arm to the left touch sensor until it latches. Did the LED turn on? Yes___ No__________
4. Using the steering wheel, rotate the contact arm to the right touch sensor until it latches. Did the LED turn on? Yes___ No__________

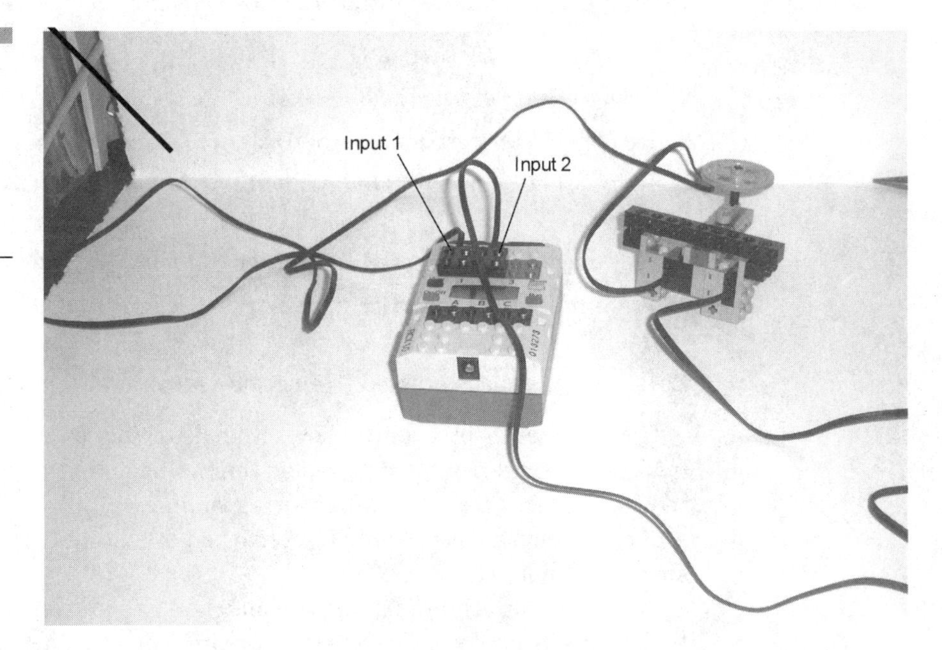

Figure 6-22 Attaching the left touch sensor to input port 1 and the right touch sensor to input port 2

***TECH NOTE:** The latching feature of the contact arm is accomplished by internal spring of the touch sensor. When the rotating arm makes contact with the yellow button on the touch sensor, the internal spring pushes up on the mini-electrical switch. The small yellow button makes contact with the rotating arm, providing a frictional force that minimizes movement from the rotating contact arm.*

An HMI Application: Robot Motion Planner

The Scout P-Brick has nine built-in motion patterns that, when combined together, will form various movements for the LEGO Mindstorms Robotics Discovery Set. An HMI application for the two-switch hand controller is the digital robot motion planner. By selecting a motion (state) with a touch (event), a defined movement pattern can be programmed on the Scout P-Brick. The programmed movement pattern is executed by pushing the *RUN* button on the Scout P-Brick. The output state of the behavioral movement pattern is initiated by the two-switch hand controller executing the motion via the appropriate output port(s) turning on or off. To illustrate this HMI motion planning, look at the following state machine diagram, shown in Figure 6-23, and the state-transition sequence of events:

1. RCX Off: RCX is turned OFF (state).
2. Power ON (transition): NOP (state).
3. *RUN* button pressed (transition): *Motors A&B ON Forward* (state).
4. Right touch sensor pressed (transition): *Motors A&B ON Reverse* (state).
5. Left touch sensor pressed (transition): *Motors A&B ON Forward* (state).
6. Right touch sensor pressed (transition): *Motors A&B ON Reverse* (state).
7. Left touch sensor pressed (transition): *Motors A&B ON Forward*(state).

This sequence of events shows the detection motion scheme of the *Reverse* movement pattern, based on the two-switch hand controller's input request (pressed) event and the corresponding output motion (state). This sequence of events can easily be converted to a state-transition matrix as shown in Table 6-3.

As shown in the diagram and the matrix, a closed-loop function exists for the *Reverse* behavioral movement program. The state machine and matrix are typical results for a reactive-based controller. A reactive-based con-

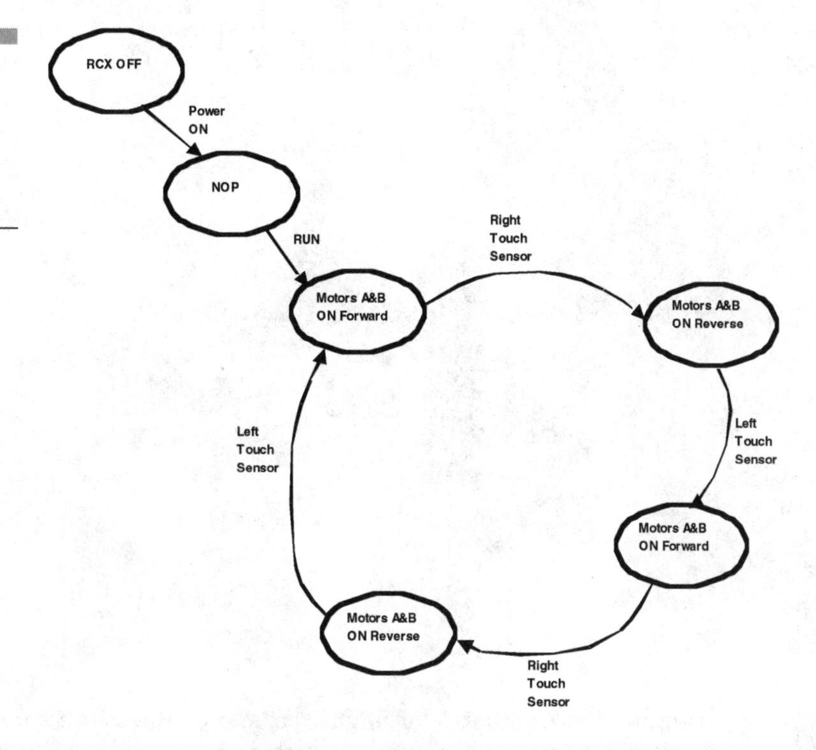

Figure 6-23 State machine diagram for Reverse movement pattern program

Table 6-3 State-transition matrix

Initial State	Transition	Next State
SCOUT P-Brick OFF	Power ON	NOP (No Operation)
NOP	RUN	Motors A&B ON Forward
Motors A&B ON Forward	Right Touch Sensor	Motors A&B ON Reverse
Motors A&B ON Reverse	Left Touch Sensor	Motors A&B ON Forward
Motors A&B ON Forward	Right Touch Sensor	Motors A&B ON Reverse
Motors A&B ON Reverse	Left Touch Sensor	Motors A&B ON Forward

troller is an embedded unit that reads either digital or varying discrete analog inputs, processes the signals with a core processor, and provides the appropriate outputs. In designing such a unit, the electronics hardware designer will create a state machine or matrix showing response of the behavioral-based control to known input events with an expectant outcome. To validate the function of the reactive-based controller, the state machine or matrix is used as a road map. It guides the designer through the various

Figure 6-24 The completed smart hand controller.

inputs that the controller might see, and produces the appropriate output responses.

Validation of the Robot Motion Planner Tool Laboratory Project
The following laboratory project will illustrate how the state machine can validate the *Reverse* movement pattern program, with the aid of the Scout P-Brick and the two-switch hand controller shown in Figure 6-24.

Validation Test Procedure Use the following steps to validate the *Reverse* movement pattern program:

1. Activate the Scout P-Brick and program the following SAC code:

```
Motion      Touch
Forward     Reverse
```

2. Turn off the Scout P-Brick. This step ensures the P-Brick will be in the NOP state.
3. Activate the Scout P-Brick and press the *RUN* button on the programmable device. The up green triangle LED indicators for both A and B output ports will be ON. This event puts the Scout P-Brick in the *Motors A&B ON Forward* state.

4. Using the steering wheel, rotate the contact arm to the right touch sensor. The down green triangle LED indicators for both A and B output ports will be ON. This event puts the Scout P-Brick in the *Motors A&B ON Reverse* state.
5. Using the steering wheel, rotate the contact arm to the left touch sensor. The up green triangle LED indicators for both A and B output ports will be ON. This event puts the Scout P-Brick in the *Motors A&B ON Forward* state.
6. Using the steering wheel, rotate the contact arm to the right touch sensor. The down green triangle LED indicators for both A and B output ports will be ON. This event puts the Scout P-Brick in the *Motors A&B ON Reverse* state.
7. Using the steering wheel, rotate the contact arm to the left touch sensor. The up green triangle LED indicators for both A and B output ports will be ON. This event puts the Scout P-Brick in the *Motors A&B ON Forward* state.

Validation of the Robot Motion Planner Tool Laboratory Project Summary In this laboratory project, the Scout P-Brick, along with the two-switch hand controller, was used to validate the state machine diagram shown in Figure 6-23. The seven-step test procedure outlined how to use the diagram to verify each state and transition by viewing the up/down position of the triangle LED indicators relative to the touch sensors input request as active or nonactive. An input request is the physical act of turning on an electrical switch by a press, slide, or toggle motion to its internal contacts, thereby providing an activate or nonactive control signal. This technique can be used to validate all of the Scout P-Brick's preprogrammed movements, with the aid of the state machine and the two-switch hand controller.

Thoughts on the Robot Motion Planner and Robotics Up to this point, seven-segment LED displays and discrete LEDs were used to provide motion behavior, based on a set of known input event conditions. The laboratory projects discussed can easily be adapted to mobile platforms by adding motorized drives to the target P-Brick for forward or reverse propulsion of the mechatronics-based robot or intelligent machine. The external (seven-segment LED displays) or internal (round and triangle-shaped LEDs) displays attached to the Scout P-Brick provide for a quick go-no-go check of the capability of the smart hand controller. Using these displays, you are able to validate that the smart hand controller is able to provide a stable input control signal to the P-Brick.

Simple machine gearing systems can easily be used in connection with the optoelectronic visual indicators to quickly provide the status of the mechanized motions' mode of operation of the robotics system to perform useful work. As a design challenge, take an idea for an existing robotics system and build a motorized drive that will give the mechatronics-based unit the required motion. The robot motion planner previously illustrated provides a user-friendly method to create unique mechatronics-based robots and intelligent machines using the Scout P-Brick and the two-switch hand controller as a physical design aid.

The LEGO P-Brick and Smart Hand Controllers

The Scout P-Brick is a useful programmable tool that enables quick and easy control of mechatronics-based robots and intelligent machines. The SAC mode of the Scout P-Brick makes it convenient for programming LEGO robots using the nine built-in motion patterns.

If more sophisticated machines are required, use the LEGO P-Brick. It has greater programming flexibility than the Scout P-Brick. Therefore, the LEGO P-Brick will be used along with the two-switch hand controller to demonstrate the switching capability and control of the mechatronics unit.

A LEGO P-Brick-Based Smart Hand Controller Demonstrator

The two-switch hand controller has the flexibility of controlling two electromechanical loads independently by switching them on or off with the programming aid of the LEGO P-Brick. By writing some software code, the two-switch hand controller will be able to activate an electrical load using one output port while keeping the other off until the hand controller provides the input control signal. The capability to independently switch between electrical loads is based on the following action. The LEGO P-Brick reads the input request of each touch sensor built into the hand controller and processes the input control signal to activate one output port while deactivating the other. The reversal input control signal is possible as well, using the embedded hand controller software.

The electromechanical actuation of each electrical load can be programmed to stay on for a predetermined amount of time or run continuously until the hand controller turns it off. Therefore, the hand controller has the intelligence to provide a continuous or time delay switching function to the attached electrical load initiated by the embedded software program.

The two-switch hand controller truly exemplifies a mechatronics design. The interaction between the mechanical actuation of the contact arm to the electrical touch sensors provides an input control signal to the LEGO P-Brick. The LEGO P-Brick processes the signal and makes a decision to turn on/off the appropriate electrical loads using embedded software.

Why Use the LEGO P-Brick as the Core Processor for the Smart Hand Controller?

The key to the design of a smart hand controller is flexible programmability. The robotic systems designer must be able to change the behavior of a robot using hardware and software components through programming customization. Embedded hardware must enable programmability because of the needs of the customer. Therefore, programmability is an essential element in software design because the code must be able to accommodate and process input data received by an external port or peripheral. In practicing the design philosophy of flexible programmability, the LEGO P-Brick is the logical choice in meeting this customization methodology for building smart hand controllers.

The Smart Hand Controller Demonstrator Laboratory Project

The following laboratory project will demonstrate the flexible programmability design philosophy used in building the smart hand controller. The smart hand controller is capable of controlling two electrical loads wired to output ports 1 and 2 of the LEGO P-Brick. The high side seven-segment LED display driver circuit, discussed in the "Burglar Alarm Tester-Demonstrator" section of this chapter, will provide a silent output indicator for the switching control function of the smart hand controller. The *Interactive C Version 4* (IC4) computer programming language will be used in this laboratory project because of its ease in software design and the capability to display small text messages on the mini-LCD screen of the LEGO P-Brick.

The BOM for this lab project is as follows:

One two-switch hand controller unit

One high-side seven-segment LED display driver circuit

IC4 computer programming language

Assembly and Test Procedure The following steps are used in the smart hand controller demonstrator lab project:

1. Assemble the smart hand controller, using Figure 6-23. Make sure the modified LEGO electric wire of the high side seven-segment LED display driver is pointing downward for proper circuit operation.
2. Open the IC4 software. Go to the program window editor, and type the following code:

```
void main()
{           while(1){
        if (digital(1)==1){
            printf("1 ON");
            sleep(2.0);
            fd(1);

        }
        else {
            printf("1 OFF");
            sleep(2.0);
            brake(1);
        }
        if (digital(2)==1){
            printf("2 ON");
            sleep(2.0);
            fd(2);

        }
        else {
            printf("2 OFF");
            sleep(2.0);
            brake(2);

        }


    }

}
```

3. Place the infrared tower in front of the LEGO P-Brick.
4. Turn on the LEGO P-Brick. Download the software into the P-Brick.

5. Place the contact arm of the two-switch hand controller in the center position of the LEGO electromechanical switch.
6. Press the *RUN* button to execute the code. The mini-LCD screen should toggle *1 OFF* and *2 OFF* messages every 2 seconds.
7. Rotate the contact arm to the right touch sensor of the two-switch hand controller. The mini-LCD screen should toggle *1 ON* and *2 OFF* messages every 2 seconds. The character *A* on the seven-segment LED display should present. If not, turn off the LEGO P-Brick and make sure the circuit's modified LEGO electric wire is attached to output port A, pointing downward. Check the code for incorrect *printf* text statements. After correcting the errors, repeat this step to assure partial functionality of the two-switch hand controller.
8. Rotate the contact arm back to the center position. After 2 seconds, the seven-segment LED display should turn off. The mini-LCD screen should toggle *1 ON* and *2 OFF* messages every 2 seconds.
9. Remove the modified LEGO electric wire from output port A. Place it on output port B of the LEGO P-Brick. Make sure the connector is pointing downward for proper circuit operation.
10. Rotate the contact arm to the left touch sensor of the two-switch hand controller. The mini-LCD screen should toggle *2 ON* and *1 OFF* messages every 2 seconds. The character *A* on the seven-segment LED display should be present.
11. Rotate the contact arm back to the center position. After 2 seconds, the seven-segment LED display should turn off. The mini-LCD screen should toggle *1 OFF* and *2 OFF* messages every 2 seconds.
12. Edit the IC4 code using the following timing value for each if-else statement:

```
Sleep (5.0);
```

13. The toggle time between *ON* and *OFF* messages should be 5 seconds instead of 2.
14. Stop the program on the LEGO P-Brick.

The Smart Hand Controller Demonstrator Laboratory Project Summary The laboratory project of the smart hand controller demonstrated the ease in which the two-switch hand controller can be programmed to display text messages and control two electric loads using the LEGO P-Brick and IC4 computer programming language. The software code used to control the text messages on the mini-LCD screen of the LEGO P-Brick was managed using a *printf* statement. Output control of the high side seven-segment LED display driver circuit was operated by the *fd* function within the IC4 programming language. The *sleep* IC4 function controlled the time delay for toggling the two messages and deactivating an electrical load.

Additional Thoughts and Experimentation of Smart Controllers

I hope the information presented in this chapter demonstrates the hardware and software capabilities of a mechatronics-based HMI device. By using mechanics, electronics, and software, a flexible programmable HMI input device was built using the following components: a LEGO P-Brick, a few bricks, two LEGO electrics, an electronic circuit interface, and user-friendly computer programming. The last project can be modified for the robotics-control of mobile platforms and telerobots. The smart hand controller enables you to experiment with a low-cost HMI device for controlling simple or complex LEGO Mindstorms mechatronics devices and systems.

CHAPTER 7

Hybrid Robots

The LEGO Mindstorms *programmable bricks* (P-Bricks) can be used with other motorized construction kits, enabling mechatronics-based robots, smart devices, and machines to be created by the amateur roboticist. The prototyping techniques for hardware and software development of the LEGO Mindstorms robots discussed in the previous chapters of this book can be used to build hybrid machines. Hybrids consist of combining two different components into one operational unit. Although the components may look or function differently, they should have at least one common attribute.

This design requirement makes system integration development somewhat achievable in the performance and operation of the final product. For example, hybrid cars work using electricity and gasoline to power and provide propulsion for the vehicle. Electricity and gasoline are different energy elements, but they can be managed to work together for common goal of improving fuel economy and vehicle performance. Therefore, the selection of components must be made in a manner in which they can be integrated somewhat seamlessly into the hybrid robot.

In this chapter, several motorized construction sets, along with *engineering validation boards* (EVBs) and electronics prototyping kits, will be examined on how to create hardware and software controls for the electric drives and sensors to be used in hybrid robots. We will revisit the electronic circuits and software used in earlier chapters of this book to illustrate the engineering design practice of the reuse of ideas for expediting product development.

The following topics will be discussed in the chapter:

- Controlling Capsela Motors and Models with LEGO Mindstorms P-Bricks
- Reverse Battery Protection
- Hybrid Robots Concept: Interfacing Non-LEGO Electric Parts
- Measuring the Current of the Motor Capsule
- Why Take Data?
- Capsela Robot Project
- Other Motorized Construction Kits for Hybrid Robots.

Controlling Capsela Motors and Models with LEGO Mindstorms P-Bricks

Capsela takes principles of physics, mechanics, and engineering and packages them in easy-to-use capsules. The Capsela Construction System is built around a series of interlocking capsules, each having a distinct mechanical or electrical function. Each kit provides easy-to-follow assembly instructions and activities that introduce basic science and engineering concepts. Figure 7-1 shows some of the mechanical and electrical capsules that make up the Capsela Construction System.

Several construction set themes exist. Each focuses on a particular area of physics, mechanics, or engineering. For example, the MX Racer is a Capsela set for building a wild car, racing chariot, or real working fan. The MX Trooper set is used to construct a car, a revolving robot, or a speedboat that uses big yellow pontoons to float. Figure 7-2(a) and 7-2(b) show the respective models.

A hybrid robot can easily be created with either one of these Capsela models, by connecting their electric motors to a LEGO P-Brick and adding sensors to the input ports of the programmable brick for environmental sensing and object detection. The *robot command explorer* (RCX) code, Robolab, *Interactive C Version 4* (IC4), or *Not Quite C* (NQC) programming languages can provide the behavior and intelligence the hybrid robot needs to maneuver a course. The Capsela electric motors will be regulated and driven at a safe +5V source, using the driver circuits discussed previously in this book. The possibilities are only limited by your imagination.

The following laboratory project will demonstrate how to connect a LEGO P-Brick to a Capsela electric motor or light element and how to control it using IC4 programming language.

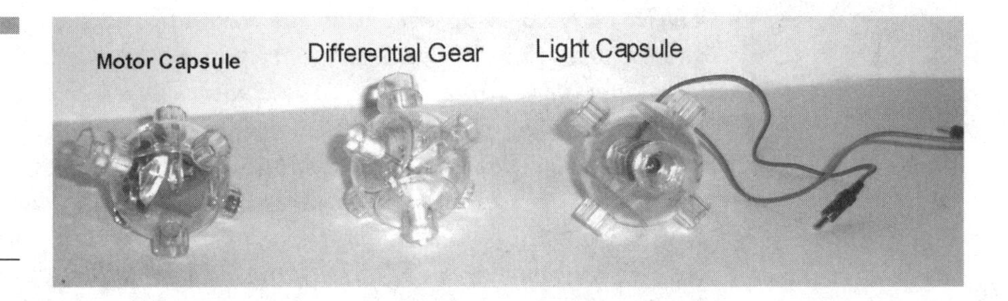

Figure 7-1 The mechanical and electrical capsules of the Capsela Construction System

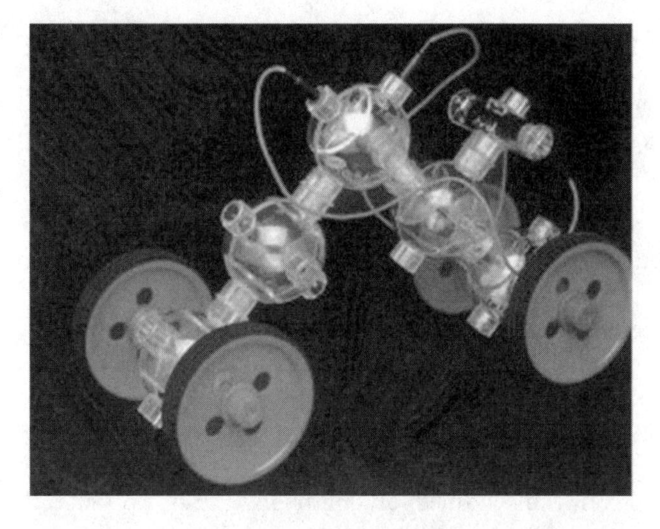

Figure 7-2(a) A Capsela MX Racer (courtesy of www.constructiontoys.com)

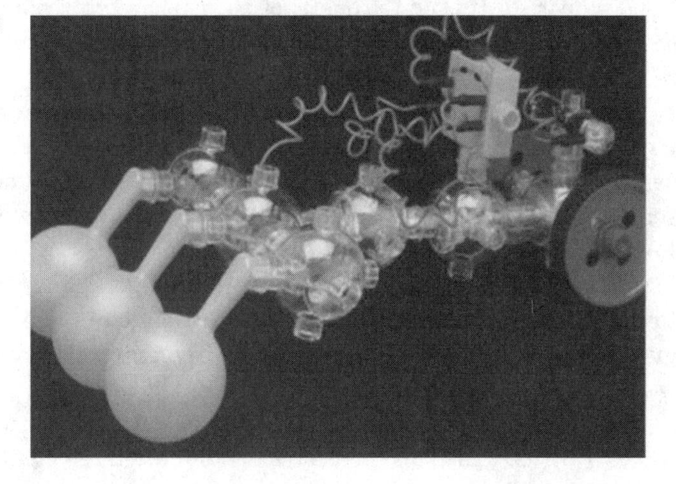

Figure 7-2(b) MX Troop (courtesy of www.constructiontoys.com)

TECH NOTE: *Capsela is a wonderful building set that also can be used for learning science. The motors and working gears are encapsulated, hence the name Capsela. You can work with crown gears, an electrical switch box, the transformation of energy, and other mechanical principles. The Capsela motorized building system provides the amateur roboticist with hands-on experiences. You can learn about energy and forces by actually handling batteries, creating motors, and seeing for yourself how energy changes, with the help of visible gears and wheels.*

Smaller kits explore energy changes, electric circuits, motion energy, speed, and torque.

Larger kits also provide introductory lessons on the following concepts: friction and traction, chain drive, inclined plane, crown gear, electrical switch box, buoyancy, electric circuitry, force, transformation of energy, vacuum, wheel-and-axle, propellers, and Newton's Third Law (to every action, or force, there is equal and opposite force).

Using fun and intriguing Capsela sets, you will be fascinated and quickly master valuable science principles.

LEGO P-Brick-Controlled Capsela Motor Capsule

An important element to controlling Capsela electric capsules using a LEGO P-Brick (RCX or Scout) is the output interfacing circuit. Extreme care must taken when driving external electromechanical loads because of the high inrush and steady state currents drawn by them. The solid-state driver that is used to drive LEGO electrics (motors, lights, and tone generators) has a maximum sourcing current of 500 mA (milliamperes). Therefore, any electrical device in addition to the LEGO electrics attached to the output ports of the P-Brick must not exceed this output current value. The circuit in Figure 7-3 is capable of meeting these electrical output requirements. Use it to manage the current level properly and to provide some protection to the internal solid-state driver of the P-Brick.

The DC-to-DC converter high side driver was discussed and experimented with in Chapter 6, "Smart Hand Controllers." The DC-to-DC conversion of the driver circuit consists of the +8V being stepped down to +5V, using the 7805 voltage regulator *integrated circuit* (IC). The +5V output provides a positive rail voltage for the 2N3904 NPN transistor to use as a

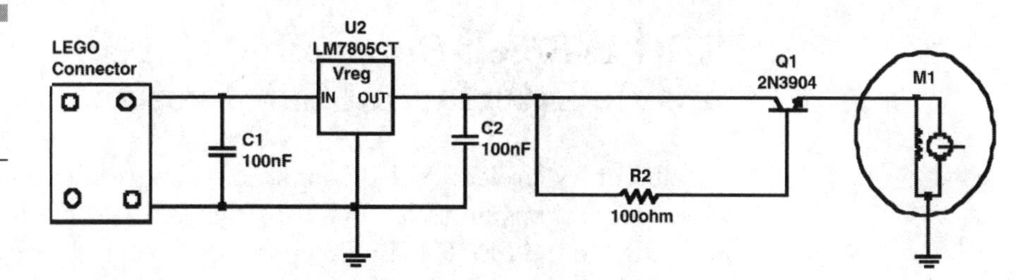

Figure 7-3 A DC-to-DC converter high side driver circuit

mechanism to source the current to the electrical or electromechanical load attached to the emitter lead of the solid-state component.

The 2N3904 NPN transistor is a general-purpose device used in small audio amplifiers and electronic switching circuits. This transistor is capable of sourcing a maximum of 200 mA with a maximum gain of 300. In order to assure sufficient sourcing current to the Capsela motor capsule for proper electromotive operation, the 470 Ω resistor was removed from the collector lead of the transistor.

TECH NOTE: *Reuse of software, mechanical subsystems, and electrical electronic circuits is an efficient way of expediting the mechatronics development process of robots and smart machines. This technique is illustrated in this project by making a minor change to high side driver circuit used in Chapter 6 laboratory projects. By building a bookshelf of proven hardware and software components, you eliminate a lot of debug time.*

The *bill of materials* (BOM) for this project is as follows:

One 7805 voltage regulator IC

Two 100 nF capacitors

One 100 Ω, 1/4-W resistor,

One 2N3904 NPN transistor

One Radio Shack Electronics Learning Lab or equivalent circuit prototyping breadboard station

Two alligator test leads

One Capsela motor capsule

One two-switch hand controller

One LEGO P-Brick

One *digital multi-meter* (DMM) with PCPC interface

One modified LEGO electric wire

LEGO P-Brick-Controlled Capsela Motor Capsule Interfacing Lab Project

This laboratory project will demonstrate how to electrically interface a Capsela motor capsule to a LEGO P-Brick. An IC4 program will then be downloaded to the LEGO P-Brick, enabling you to activate the motor using the two-switch hand controller built in Chapter 6.

Assembly Instructions and Test Procedure

Use the following steps to complete this lab project:

1. Build the DC-to-DC high side driver circuit, shown in Figure 7-3, on the Radio Shack Electronics Learning Lab kit or equivalent circuit prototyping breadboard station.
2. Attach the DC-to-DC high side driver circuit to the LEGO P-Brick, using the modified LEGO electric wire.
3. Attach the Capsela motor capsule to the DC-to-DC high side driver circuit using the two alligator test leads, as shown in Figures 7-4 and 7-5.
4. Open the IC4 software. Go to the program window editor and type the following code:

```
void main()
{          while(1){
        if (digital(1)==1){
            printf("1 ON");
            sleep(2.0);
            fd(1);

        }
        else {
            printf("1 OFF");
            sleep(2.0);
            brake(1);
        }
        if (digital(2)==1){
            printf("2 ON");
            sleep(2.0);
            fd(2);

        }
        else {
            printf("2 OFF");
            sleep(2.0);
            brake(2);

        }

    }

}
```

5. Place the infrared tower in front of the LEGO P-Brick.
6. Turn on the LEGO P-Brick and download the software into the P-Brick.

Figure 7-4 Attaching alligator test leads to the Capsela motor capsule

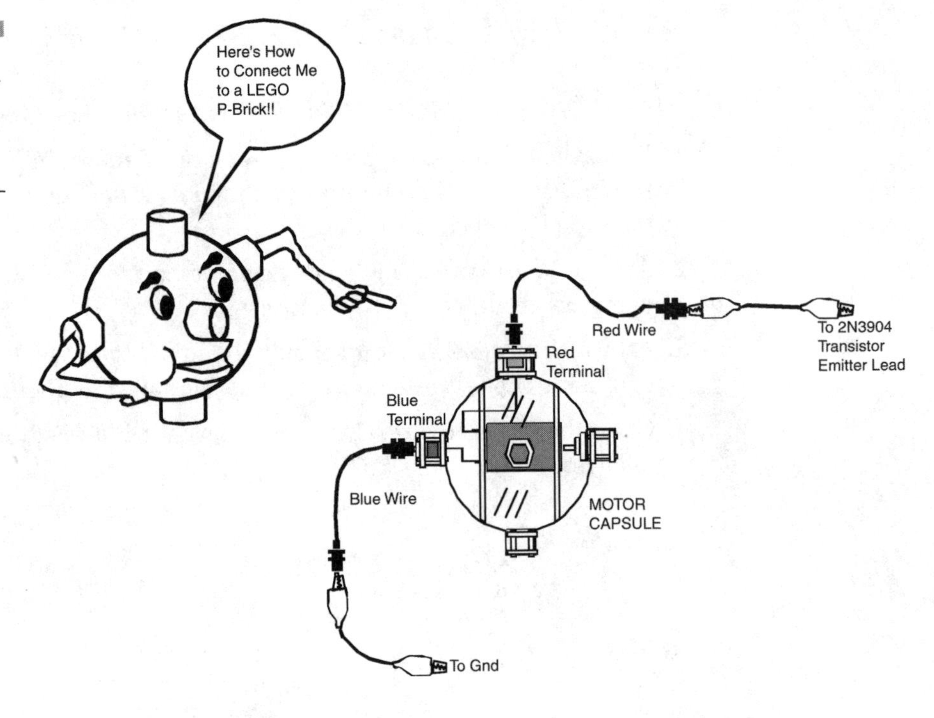

Figure 7-5 Alligator test lead attachment to the motor capsule

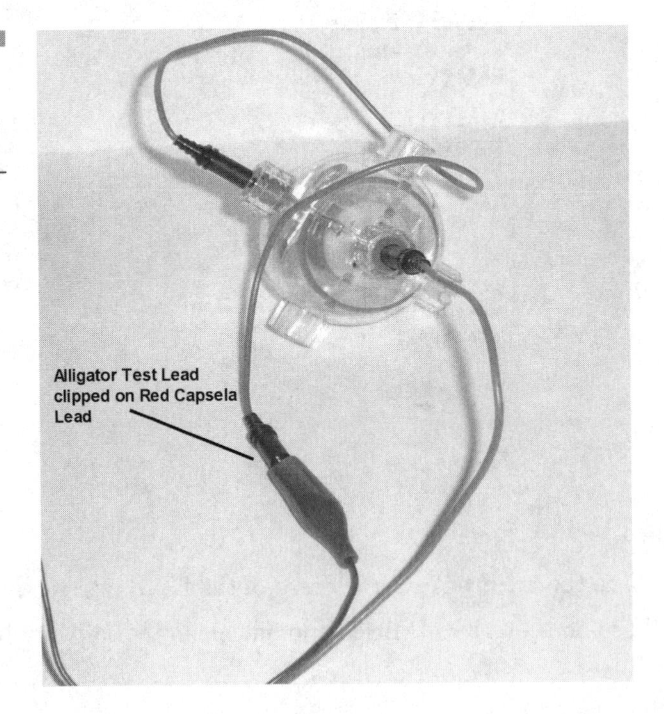

7. Place the contact arm of the two-switch hand controller in the center position of the LEGO electromechanical switch.
8. Press the *RUN* to execute the code. The mini-LCD screen should toggle *1 OFF* and *2 OFF* messages every 2 seconds.
9. Rotate the contact arm to the right touch sensor of the two-switch controller. The mini-LCD screen should toggle *1 ON* and *2 OFF* messages every 2 seconds. The motor capsule should be turned on at this time. If not, deactivate the LEGO P-Brick and make sure the circuit's modified LEGO electric wire is attached to output port A, pointing downward. Check the DC-to-DC converter circuit wiring on the breadboard. Check the code for incorrect *printf* text displays. After correcting the errors, repeat this step to assure partial functionality of the two-switch hand controller.
10. Rotate the contact arm back to the center position. After 2 seconds, the 7-Segment LED display should turn off. The mini-LCD screen should toggle *1 ON* and *2 OFF* messages every 2 seconds.
11. Repeat steps 9 and 10 several times to assure the driver circuit is working properly.
12. Stop the two-switch hand controller program running on the LEGO P-Brick by pressing *RUN*.
13. Modify the hand controller program within the IC4 *integrated development environment* (IDE) editor to demonstrate reverse battery protection of the 7805 voltage regulator IC, using the following code modification.

```
void main()
{           while(1){
        if (digital(1)==1){
            printf("1 ON");
            sleep(2.0);
            fd(1);


        }
        else {
            printf("1 OFF");
            sleep(2.0);
            brake(1);
        }
        if (digital(2)==1){
            printf("2 ON");
            sleep(2.0);
            bk(2);


        }
```

```
        else {
            printf("2 OFF");
            sleep(2.0);
            brake(2);

        }

    }

}
```

14. Download the modified software into the P-Brick.
15. Place the contact arm of the two-switch hand controller in the center position of the LEGO electromechanical switch.
16. Press *RUN* to execute the code. The mini-LCD screen should toggle *1 OFF* and *2 OFF* messages every 2 seconds.
17. Rotate the contact arm to the right touch sensor of the two-switch controller. The mini-LCD screen should toggle *1 ON* and *2 OFF* messages every 2 seconds. The motor capsule should be turned on at this time.
18. Rotate the contact arm to the left touch sensor of the two-switch hand controller. The mini-LCD screen should toggle *2 ON* and *1 OFF* messages every 2 seconds. The motor capsule should not turn on.
19. Temporarily stop the program by hitting *RUN* on the LEGO P-Brick.

LEGO P-Brick Controlled Capsela Motor Capsule Interfacing Lab Project Summary The objective of this laboratory project was to allow you the opportunity to interface a non-LEGO electric part to the LEGO P-Brick. The Capsela motor capsule was the chosen component because of ease of wiring and low current draw from the LEGO P-Brick's output port.

The first part of the laboratory project consisted of modifying the high side driver circuit used in one-switch hand controller project discussed in Chapter 6. The modification consisted of removing the 470 Ω resistor from the collector lead of the 2N3904 transistor and replacing it with a straight wire. The 7-Segment LED display was removed from the original circuit and replaced with the Capsela motor capsule, as shown in the circuit schematic of Figure 7-3. Figures 7-4 and 7-5 provided additional information on how to wire the motor capsule to the DC-to-DC converter high side driver circuit.

With the two-switch hand controller attached to input ports 1 and 2, the software from Chapter 6 used for mechanical switch detection of two touch sensors was downloaded into the P-Brick using the infrared tower. Rotating the contact arm of the two-switch hand controller to the right touch sensor validated the embedded software. The motor capsule wired to input port 1 via the high side driver circuit turned. When the contact arm of the hand controller moved away from the touch senor, the motor stopped.A modification in the code was made to the following lines of instruction:

```
if (digital(2)==1){
    printf("2 ON");
    sleep(2.0);
bk(2);
}
```

The *bk(2)* highlighted in boldface type enables the motor capsule not to reverse shaft rotation when the left touch sensor is on, captured by the *if(digital(2)==*1 function statement. The LEGO P-Brick now has the limited capability to disable control of the attached motor capsule with the aid of the two-switch hand controller and 7805 voltage regulator IC.

Reverse Battery Protection

Reverse battery protection is an important safety system that protects electronics from the potential scenario in which the power supply leads are inadvertently reversed then attached to the device under test. The laboratory project demonstrated, using the *bk(n)* IC4, provided reverse battery protection for the transistor high side driver circuit with the aid of the 7805 voltage regulator IC.

Traditional reverse battery circuits consist of a silicon diode placed in series between the positive battery node, with the anode attached and input power lead of the electronic circuit-cathode wired.

Under normal attachment conditions, current will flow through the anode to cathode pins of the diode to the input power leads of the electronic circuit under power. Therefore, a minimal resistance path is created for the electrons to flow freely from the battery, through the attached electronic circuit, to ground (forward-biasing of the silicon diode). Upon reversing the battery leads, a high resistance is created because the current is not able to flow freely (reverse-biasing of the silicon diode). Thus, an open circuit condition exists across the silicon diode, enabling all the battery supply voltage to be across the semiconductor device.

TECH NOTE: *If directional control is desired using a LEGO electric motor, the bk(n); function will reverse the shaft of the electromechanical component. By removing the DC-to-DC converter high side driver circuit and attaching the LEGO electric motor directly to the output port, the direction of the motor can be changed using the two-switch hand controller and* back *IC4 function.*

Hybrid Robots Concept: Interfacing Non-LEGO Electric Parts

Hybrid robots are basically kludged-up machines using parts from a variety of sources that collectively work together in a functional and maintainable system. The key to creating a good hybrid robot is to understand the requirements of the target machine. An approach to obtaining the requirements is to understand the core subcomponent or system responsible for the managing most of the *input/output* (I/O) processing of the target robot. In developing LEGO Mindstorms mechatronics-based robots, the RCX P-Brick is the core subcomponent for managing most of the I/O processing.

To begin interfacing non-LEGO electric parts, you must understand the internal circuits used in the RCX P-Brick. You must understand which circuits capture input data from passive and active sensors, and after processing the digital or analog signals received from the outside world, will drive the electrical load attached to its output port. For example, the 500-mA maximum current available to each of the output ports was obtained from discussion on the *LEGO Users Group NETwork* (LUGNET) on the Internet. With this critical piece of information, the appropriate small DC motor can be selected, along with the driver circuit. If the amateur roboticist has some circuit design knowledge, a custom driver circuit can be developed as well.

Measuring the Current of the Motor Capsule

To assure the current draw of the motor capsule is less than the 500-mA maximum output value of the LEGO P-Brick, a measurement can be easily made. A DC milliammeter can be inserted in series between the emitter lead of the 2N3904 transistor and the motor capsule. Figure 7-6 shows a partial wiring schematic of the DC-to-DC high side driver circuit and a DMM in series between the two specified parts.

After connecting the DC milliammeter in series between the 2N3904 transistor emitter lead and the motor capsule, the two-switch hand controller (timed delay on) IC4 program can be executed. One item to note about the flexibility of using a DMM over an analog unit is the reverse current value can be read and will be identified with a negative value. The analog milliammeter will deflect to the left, but peg to the left because of the reverse current flow and the mechanical stop that prevents the needle from spinning like a motor.

The following laboratory project will outline a test procedure for measuring current as well as displaying it on a desktop PCPC or notebook computer.

Measuring the Current of the Motor Capsule Lab Project

This laboratory project will allow you to characterize the Capsela motor capsule using a DMM with a serial PCPC interface. This measurement technique can also be used with the LEGO electric motor.

The BOM for this lab project is as follows:

One DMM with a serial PC interface (Radio Shack Cat. No. 22-168 or equivalent measurement meter)

One Desktop PC or notebook computer with serial port

One motor capsule

One LEGO P-Brick with DC-to-DC high side driver circuit

Motor Capsule Current Measurement Test Procedure Use the following steps to complete the test procedure:

1. Download the following IC4 code into the LEGO P-Brick:

```
    void main()
{           while(1){
        if (digital(1)==1){
            printf("1 ON");
            sleep(2.0);
            fd(1);
            sleep(10.0);
            brake(1);

        }
        else {
            printf("1 OFF");
            sleep(2.0);
```

```
            brake(1);
        }
        if (digital(2)==1){
            printf("2 ON");
            sleep(2.0);
            fd(2);

        }
        else {
            printf("2 OFF");
            sleep(2.0);
            brake(2);

        }


    }

}
```

2. Attach the DMM with serial PC interface to the DC-to-DC converter high side driver, using Figure 7-6. Note: To take current-measurements

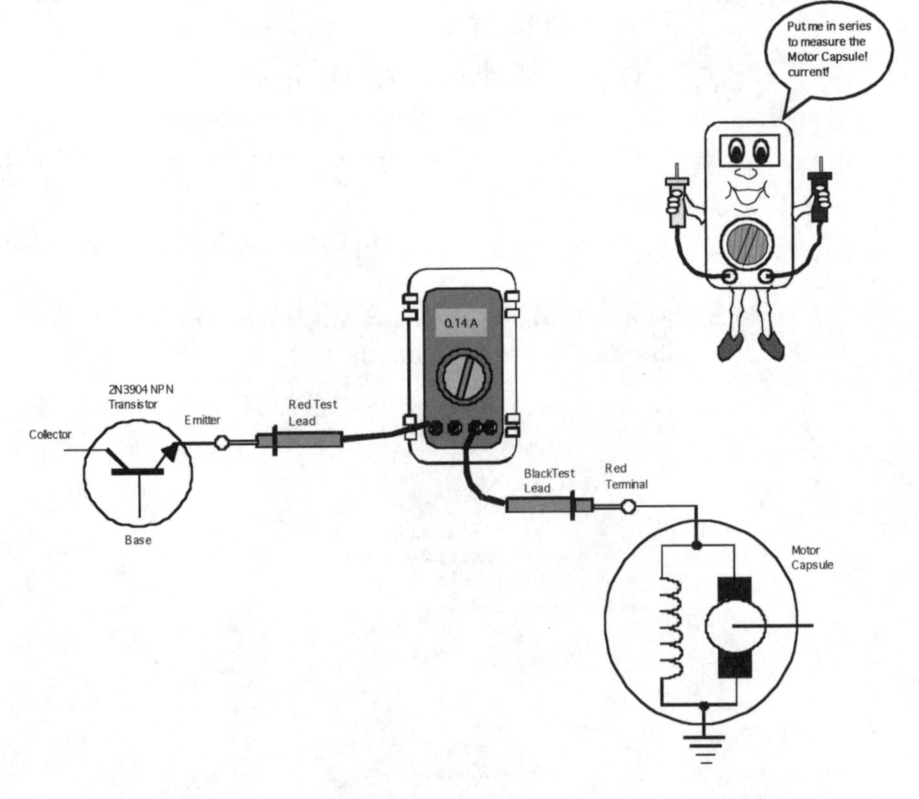

Figure 7-6 Taking a current-measurement from the DC-to-DC converter high side driver circuit

using a DC milliammeter, the measuring device must be inserted in series with the electrical load circuit that is driving it.

3. Attach the DMM with PC interface to the serial port of the desktop PC or notebook computer. Note: Prior to attaching the DMM to the serial port of the desktop PC or notebook computer, install the software that comes with the measuring device. The ScopeView® software that comes packaged with the Radio Shack DMM (Cat. No. 22-168) will be used as the measuring system example throughout the laboratory project.
4. Turn on DMM and run the ScopeView software.
5. Turn on the LEGO P-Brick and run the two-switch hand controller timed delay program.
6. Start a logging session within ScopeView, using Figure 7-7 as guide for setting up the software control panel.
7. Click *Scope* on the control panel.
8. Rotate the contact arm on the two-switch hand controller to the right touch sensor, latching it closed.
9. Click *RUN* on the ScopeView output window to start a data-viewing session. A series of pulses should be visible on the output window of ScopeView. The motor capsule will turn on for 10 seconds and off for 2 seconds. This pulsating signal will continue as long as the right touch sensor of the two-switch hand controller is latched closed.

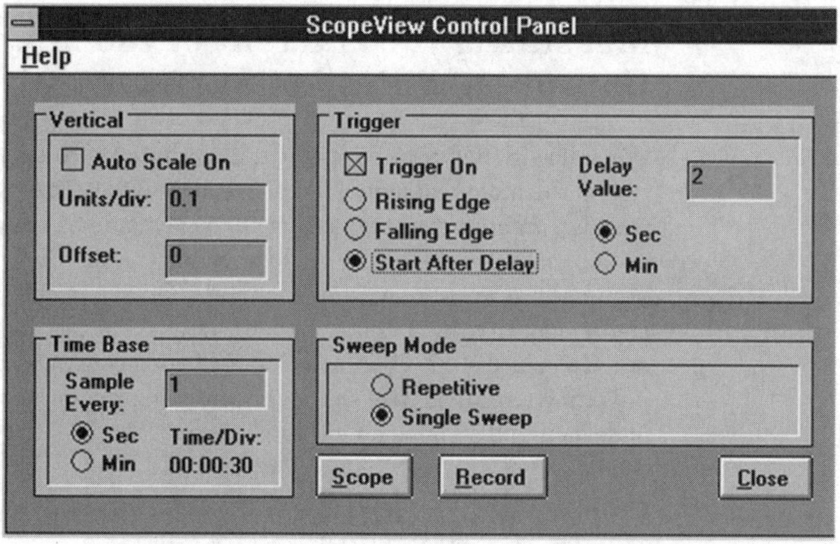

Figure 7-7 The control panel settings for the ScopeView data-logging session of the motor capsule current

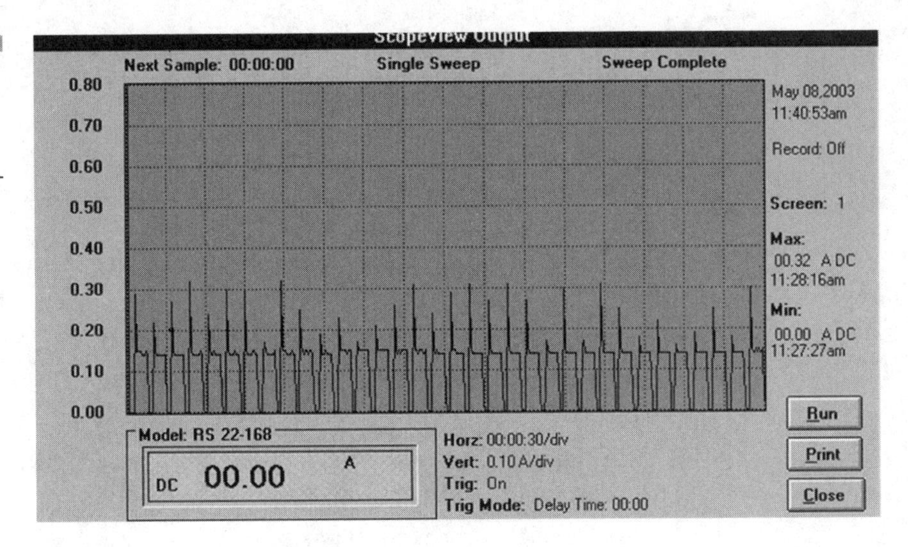

Figure 7-8 The motor current data captured by the author

10. After completing one sweep, click *START* to stop the data-logging session. Figure 7-8 shows the plotted data taken by the author. Record the maximum motor current (Imax) and the steady state value (I), using the plotted data shown on the output window. Imax_______ A Steady State(Isteady)________ A.
11. Stop the timed delay program in the LEGO P-Brick by pressing *RUN*.

Measuring the Current of the Motor Capsule Lab Project Summary

The following lab project outlined a test procedure for measuring the current of the motor capsule. The initial current of 0.32 A (320 mA) is based on the high amount of energy required to overcome the mechanical resistance of the shaft. Therefore, for a few seconds the 320 mA current is initiated with each activation cycle of the motor. On the graph, sharp peaks (transients) identify the 320-mA inrush current. The steady state current measured was at 0.14 A (140 mA). The following code modification created the pulsating output control signal for switching the motor capsule in a 10 seconds on and 2 seconds off cycle:

```
    {             while(1){
if (digital(1)==1){
    printf("1 ON");
```

```
                sleep(2.0);
                fd(1);
        sleep(10.0);
        brake(1);
        }
```

As seen in the code, the *sleep (10.0)* and *sleep (2.0)* are the IC4 functions responsible for the 10 seconds on and 2 seconds off cycles.

Recording Motor Capsule Current Data to a File

The following test procedure will outline how to record data onto a storage media device (3.5-inch floppy disk or hard drive).

1. Activate the DMM and run the ScopeView software.
2. Activate the LEGO P-Brick and run the two-switch hand controller timed delay program.
3. Start a logging session within ScopeView, using Figure 7-7 as a guide for setting up the software control panel.
4. Click *Record* on the control panel. The *Save File* dialog box should appear on the screen.
5. Select the appropriate drive for saving the data. The author used the A: drive to save data onto a 3.5-inch floppy.
6. Click the *Filename* box and type **mamp.txt**. Click OK.
7. Click *Scope*.
8. Rotate the contact arm on the two-switch hand controller to the right touch sensor, latching it closed.
9. Click *RUN* on the ScopeView output window to start a data viewing session. A series of pulses should be visible on the output window of ScopeView. The motor capsule will turn on for 10 seconds and off for 2 seconds. This pulsating signal will continue as long as the right touch sensor of the two-switch hand controller is latched closed.
10. Click *RUN* on the ScopeView output window to start a data viewing session. A series of pulses should be visible on the output window of ScopeView. The motor capsule will turn on for 10 seconds and off for 2 seconds. This pulsating signal will continue as long as the right touch sensor of the two-switch hand controller is latched closed.
11. After completing one sweep, click *CLOSE* to stop the data-logging session.

Reviewing and Playing Back the Logged Data

The logged data can be viewed or played back within the ScopeView as follows:

1. Click *CLOSE* on the control panel *graphical user interface* (GUI). The main menu should be displayed on the screen.
2. Click *Playback* on the main menu panel. The data file reviewer GUI should be displayed.
3. Click *Open File.* The *Read File* dialog box should be displayed.
4. Click the *mamp.txt* file. Click OK. The data file reviewer should start plotting the data on the screen. Figure 7-9 shows the results of the motor capsule current-measurement, played back within the ScopeView software.
5. Plot the ASCII data from the mamp.txt file in Excel or an equivalent spreadsheet application. See Figure 7-10. Compare the results with Figure 7-9.
6. Stop the timed delay program of the LEGO P-Brick.

Record and Playback Laboratory Project Summary

The objective of the Record and Playback lab project was to demonstrate how Capsela motor capsule data can be stored on storage media device like a 3.5-inch floppy disk or a hard-drive. The playback feature of the Scope-View software is convenient for you by enabling the ability to analyze the captured data later. Excel provides another analytical dimension to the

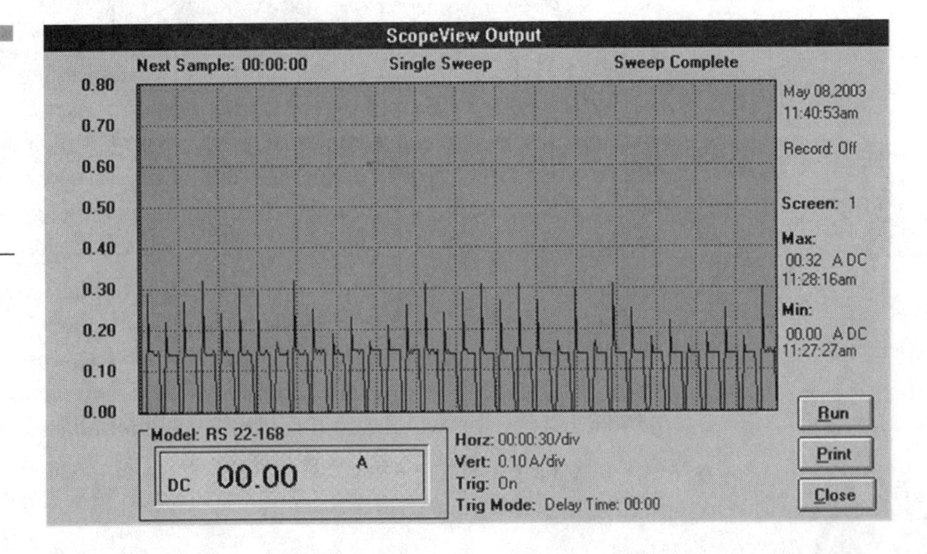

Figure 7-9 The measured data captured in Figure 7-8 playback within the ScopeView software

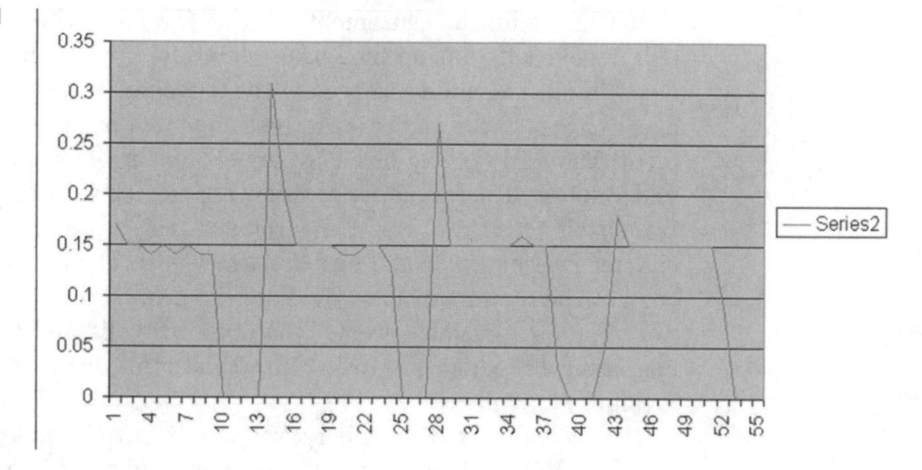

Figure 7-10 An Excel plot of the motor capsule data logged from the ScopeView software

measurement process because further analysis can be conducted on the data stored on the spreadsheet, using some of the embedded mathematical tools. The graph plotted from the captured data should be the same in terms of peaks (transients) and pulse as the data displayed on the ScopeView output window.

Why Take Data?

To understand the performance of a machine, you must take physical measurements of the mechanical and electrical systems. Test plans on both the mechanical and electrical systems are written and conducted by the designer to ensure the mechatronics-based machine meets all the functional requirements captured in the *systems design specification* (SDS). The two current-measurement projects outlined provided testing techniques that you can use to ensure the 200-mA requirement of the 2N3904 transistor is not exceeded.

The first measurements project outlined a test procedure for capturing one sweep of the current data from the LEGO P-Brick-driven Capsela motor capsule. The motor capsule was pulsed on and off in an attempt to capture transients associated with high inrush current. Therefore, the control panel of the software was set up to observe current transients on the output display window of the software. After adjusting the time/div and current/div scales with the appropriate values, the test measurement process could be executed by running the software capturing program as well as the IC4 code embedded within the LEGO P-Brick.

Before taking measurements, the designer of the device under test (DUT) generally has an idea or assumption of what the final outcome of test will be. The ability to predict the outcome of the test is based on the experience of the designer with the component or system under measurement. The author, who has 17 years of industry-based electronics circuit and systems design experience, knew the current limitation of the 2N3904 transistor. Therefore, the measurement test procedure outlined in this chapter was based on the author's knowledge of the appropriate level of current the motor capsule would sink to ground for proper circuit operation. The current-measurement lab results validated the proper circuit operation of the high side driver interface circuit to the motor capsule with a value of 140 mA.

The second part of the laboratory project introduced logging data using the ScopeView software onto storage media like a 3.5-inch floppy disk or hard-drive. The one complete sweep of a data-logging session was recorded and saved on a 3.5-inch floppy disk. By storing data on a storage media like a 3.5-inch floppy disk, the information can be reviewed at a later time. This method of collecting data is quite common in industry, and it allows the designer to analyze the captured information in the convenience of his or her office. The data can be viewed in Excel because of its ASCII (American Standard Code for Information Interchange) text-based format. You can conduct further analysis on the captured data from the motor capsule component using the embedded graphing tools within Excel.

The last laboratory project demonstrated how the data could be played back onto the output display window of the ScopeView software. By opening the data file stored on either a 3.5-inch floppy disk or hard-drive, you can review the data just as it was recorded during the measurement session. Using the ScopeView software enables you to quickly view and analyze the data captured from the motor capsule.The three laboratory projects on current-measurement techniques using the ScopeView software were presented to you as a robotics design tool for validating critical circuit or subsystem functions and features. The reason for taking data measurements is a matter of providing high confidence to you that your mechatronics-based machine will perform the required task with minimal chance of component or system failure occurring. The basic methods of collecting inrush or steady state current-measurements can be applied to collecting voltage as well as resistance from either analog or digital I/O devices wired to the LEGO P-Brick. Future chapters will explore some of these measurement techniques further.

Capsela Robot Project

The following assembly instructions are for building a Capsela robot. The completed robot will be wired to the LEGO P-Brick DC-to-DC converter high side driver controller. The robot will then move forward for 10 seconds, stop for 2 seconds, and will continue this motion until the contact arm on the two-switch hand controller is rotated away from the right touch sensor. After completing one cycle of motion, the robot will stop.

The BOM for this project is as follows:

One motor capsule (#1)
One worm gear capsule (#3)
One transmission capsule (#8)
One capsule (#9)
Eight octagonal connectors (#10)
Three couplers (#14)
Three coupler caps (#15)
Two leads (1 black,1 red) (#23)
Three large wheels (#25)
Two large tires (#28)

General Information About Capsules

The following information explains the electrical and mechanical function of the major *prime mover* capsule. A prime mover is the primary transmission actuator for providing motion in a mechanical system.

Motor capsule An external battery powers the built-in motor. The motor capsule is designed to run in both directions by reversing the voltage applied to the red and blue terminals.

Worm gear capsule The compound machine capsule combines the operation of speed reduction and crown wheel gears into a single unit. This capsule can reduce the shaft speed and change the direction of the rotating mechanical force through a right angle connection using an octagonal connector.

Transmission capsule This capsule can be engaged between any operating capsules to extend the gear action without changing the connecting rotating mechanical force.

Assembly Instructions

Use the following steps to complete this project:

1. Place the motor capsule on top of the transmission capsule using one octagonal connector. The red terminal should be pointing towards the back of the robot. See Figure 7-11.
2. Attach the motor and transmission capsule assembly on top of the worm gear capsule using an octagonal connector. Have the gray gear connecting the white gear of the transmission capsule. See Figure 7-12.
3. Attach a coupler at each side of the worm gear capsule using an octagonal connector. See Figure 7-13.
4. Attach two large tires around two large wheels. Place a tire and wheel assembly onto each coupler attached to each side of the worm gear capsule. See Figure 7-14.
5. Secure the tire and wheel assemblies in place, using two coupler caps. See Figure 7-15.

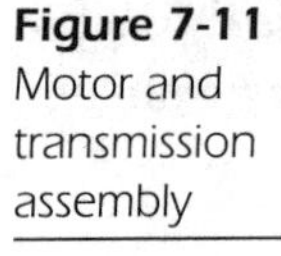

Figure 7-11 Motor and transmission assembly

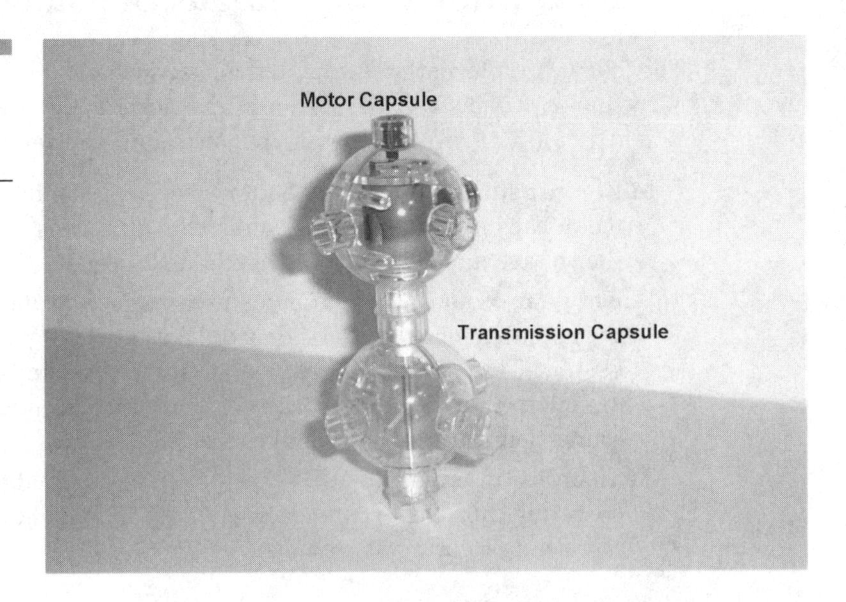

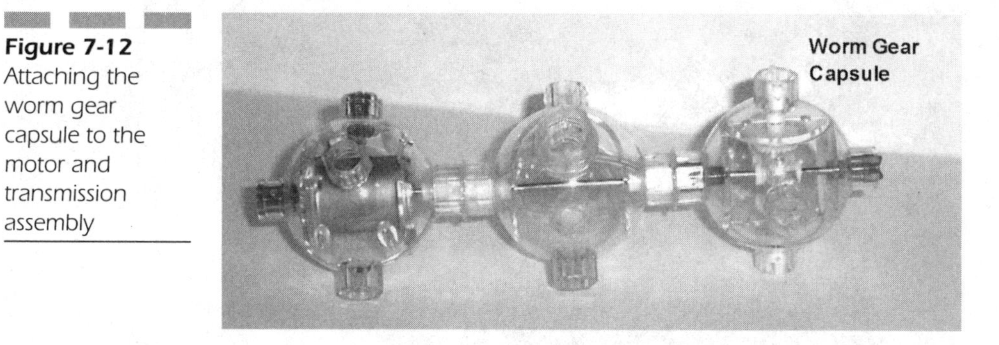

Figure 7-12 Attaching the worm gear capsule to the motor and transmission assembly

Figure 7-13 Couplers being applied to the worm gear capsule

Figure 7-14 Two large wheels being attached to the Capsela robot

Figure 7-15
Tire and wheel assemblies being secured with couplers

Figure 7-16
Attaching an octagonal connector to the back of the worm gear capsule

6. Place an octagonal connector onto the back of the worm gear capsule. See Figure 7-16.
7. Attach a coupler to the octagonal connector located on the back of the worm gear capsule. See Figure 7-17.
8. Place another large wheel to the attached coupler located on the back of the worm gear capsule. See Figure 7-18.
9. Secure the large wheel to the coupler using a coupler cap. See Figure 7-19.

Figure 7-17
Attaching a coupler to the octagonal connector

Figure 7-18
A third large wheel being mounted to the body of the Capsela robot

Figure 7-19
The large wheel being secured to the worm gear capsule using a coupler cap

10. Attach a half capsule to each side of the transmission capsule, using octagonal connectors. See Figure 7-20.

Figure 7-21 shows the completed Capsela robot.

Figure 7-20 Placing arms on the Capsela robot using half capsules and octagonal connectors

Figure 7-21 The completed Capsela robot

Measuring the Current of the Capsela /LEGO P-Brick Hybrid Robot Mechanical Full Load Condition

The following test procedure will outline measuring the full load condition of the Capsela/LEGO P-Brick hybrid robot using the measurement technique discussed on page 214 of this chapter. The objective of this measurement is to see the effect of having the motor capsule attached to a mechanical load and the amount of current required to drive it.

TECH NOTE: *The measurement lab procedure assumes you are downloading the IC4 code into the LEGO P-Brick for the first time. Therefore, the code is shown for this assumption.*

The BOM for this project is as follows:

One DMM with a serial PC interface (Radio Shack Cat. No. 22-168 or equivalent measurement meter)

One Desktop PC or notebook computer with serial port

One assembled Capsela robot

One LEGO P-Brick with DC-to-DC high side driver circuit

Mechanical Full Load Condition Current-Measurement Test Procedure Use the following steps to complete this test procedure:

Download the following IC4 code into the LEGO P-Brick.

```
   void main()
{          while(1){
        if (digital(1)==1){
            printf("1 ON");
            sleep(2.0);
            fd(1);
            sleep(10.0);
            brake(1);

        }
        else {
            printf("1 OFF");
            sleep(2.0);
            brake(1);
        }
        if (digital(2)==1){
```

```
                printf("2 ON");
                sleep(2.0);
                fd(2);

            }
            else {
                printf("2 OFF");
                sleep(2.0);
                brake(2);

            }


        }

    }
```

In order to take static measurements of the robot, place the Capsela machine on top of a half-capsule, using an octagonal connector.

1. Attach the DMM with serial PC interface to the DC-to-DC converter high side driver, using Figure 7-6. Note: To take current-measurements using a DC milliammeter, the measuring device must be inserted in series with the electrical load and the circuit that is driving it.
2. Attach the DMM with PC interface to the serial port of the desktop PC or notebook computer. Note: Prior to attaching the DMM to the serial port of the desktop PC or notebook computer, install the software that comes with the measuring device. The ScopeView software that comes packaged with the Radio Shack DMM (Cat. No. 22-168) will be used as the measuring system example throughout the laboratory project.
3. Turn on the DMM and run the ScopeView software.
4. Turn on the LEGO P-Brick and run the two-switch hand controller timed delay program.
5. Start a logging session within ScopeView, using Figure 7-7 as guide for setting up the software control panel.
6. Click *Scope* on the control panel.
7. Rotate the contact arm on the two-switch hand controller to the right touch sensor, latching it closed.
8. Click RUN on the ScopeView output window to start a data viewing session. A series of pulses should be visible on the output window of ScopeView. The motor capsule will turn on for 10 seconds and off for 2 seconds. This pulsating signal will continue as long as the right touch sensor of the two-switch hand controller is latched closed.
9. After completing one sweep, click *START* to stop the data-logging session. Figure 7-8 shows the plotted data taken by the author. Record the

maximum motor current (Imax) and the steady state value (I), using the plotted data shown on the output window. Imax_______ A, Steady State(Isteady)_________ A.

10. Stop the timed delay program in the LEGO P-Brick by pressing *RUN*.

Measuring the Current of the Capsela/LEGO P-Brick Hybrid Robot Mechanical Full Load Condition Summary The execution of the project was the same as the previous current-measurement exercises using one motor capsule. The setup of the ScopeView control panel, along with a one sweep data acquisition mode, were the requirements of measuring the inrush and steady state current values of the Capsela robot. The measured values the author captured during LEGO P-Brick IC4 software control of the robot were Imax at 330 milliamperes and Isteady of 290 mA. Therefore, under a mechanical full load condition, the steady state current value increased by 150 mA. This is typical of electric motors with a mechanical load attached to the shaft. Additional current is required from the battery to provide enough torque rotate the shaft with the attached mechanical load.

The 2N3904 transistor would not be able to handle this amount of current under a continuous mode of operation. Pulsing the mechanical load attached to the motor capsule's gear helps minimize the amount of time the 290-mA current is flowing through the 2N3904 collector to the emitter leads. To eliminate the possible failure mode of the transistor, this current-sourcing mode of operation must use another semiconductor switching device. The 2N4401 NPN transistor is a suitable replacement semiconductor device for sourcing 290 mA continuously. The transistor is able to handle a 600-mA maximum current sourcing continuously. Therefore, the 2N3904 transistor should be replaced with the 2N4401 device when used to drive motor capsules that are mechanically fully-loaded. Figure 7-22 shows the current-measurement plot of the Capsela/LEGO P-Brick hybrid robot.

TECH NOTE: *The Capsela 1000 Discovery set (Max Out) comes with 108 interlocking parts to construct over 100 land and water projects such as the following: a tug boat, water pump, crane, cable lift, generator, steam roller, tricycle, vacuum cleaner, and as many simple machines as your imagination can conceive! It includes a full-color Science Discovery Design Manual with easy-to-follow assembly instructions, as well as an illustrated basic Science Booklet to explore 18 physical science principles. This motorized construction set can be purchased from Brain Builders. Their web site is* `www.brain-builders.com/12001.html`. *See Figure 7-23.*

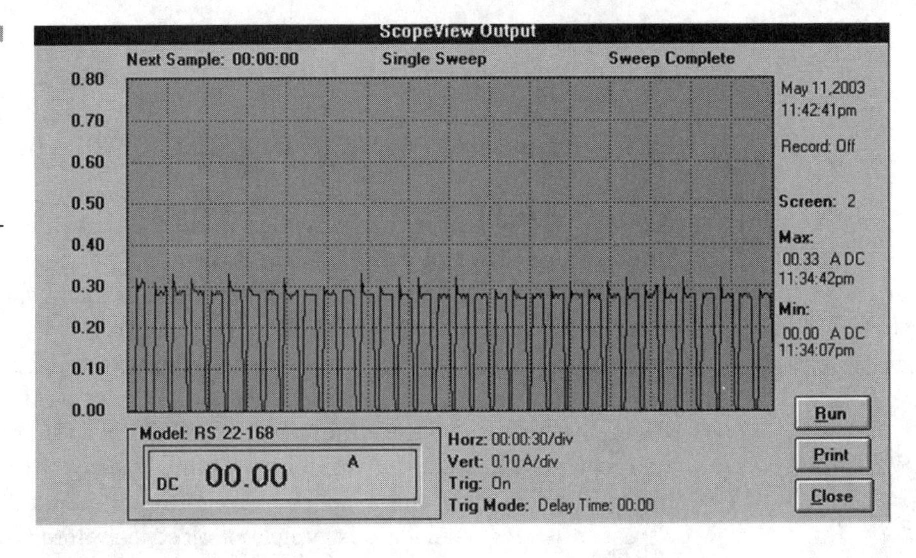

Figure 7-22 Current measurement plot of the Capsela /LEGO P-Brick hybrid robot

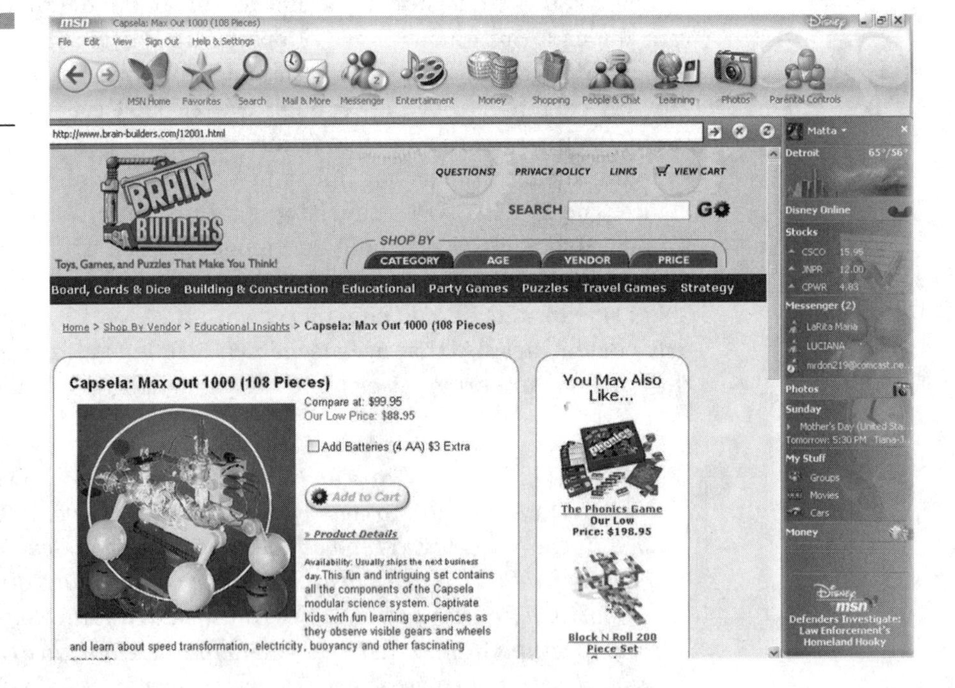

Figure 7-23 Brain Builder's partial web site

TECH NOTE: Simple Capsela Autonomous Robot *(SCAR) was an online robotics project documented by a robot hobbyist in January 2002. His journal details the why and how of SCAR development. This online journal captures some design concepts and lessons learned. The web site of S.C.A.R and its author is* `www.intp.us/richardbthursby/scar.html`.

Other Motorized Construction Kits for Hybrid Robots

Although the Capsela motorized construction set was discussed throughout this chapter, the same driver circuit techniques can be applied to other motorized construction kits. As demonstrated in the Measuring the Current of the Capsela/LEGO P-Brick Hybrid Robot Mechanical Full Load Laboratory Project, the current-sourcing requirement is the critical electrical parameter that must be taken in consideration when driving non-LEGO electric motors.

In the laboratory project, it was discovered the 2N3904 transistor was not suited to drive the motor capsule under mechanical full load. The current sourced to the transmission and worm gear capsules of the Capsela robot was 290 mA. This value is 90 mA above the maximum current level that the 2N3904 transistor can handle continuously. It was determined that the 2N4401 NPN transistor has an *Ice* (collector to emitter current) maximum of 600 mA. Therefore, the same high side driver topology (circuit geometry) can be used as a baseline output interface circuit, but with a different switching transistor. The bookshelf high side circuit that should be used for non-LEGO motors with a current requirement of 200 mA to 500 mA is a 2N4401 NPN transistor, instead of the 2N3904 semiconductor device.

With a baseline high side driver circuit design established, the LEGO P-Brick can now provide the programmable switching capable for controlling non-LEGO electric motors. The SAE A World In Motion (AWIM) Challenge 2 kit can be used as a mobile platform for building a hybrid robot. The AWIM Challenge 2 kit was designed by the SAE (Society of Automotive Engineers) to help children make the connection with math and science through a motorized construction kit. The kit has an assortment of wheels, gears, axles, a small electric motor, and plastic rectangle frame for attaching the components. The SAE AWIM Challenge 2 web site is www.sae.org/students/awim2.htm.

By adding a LEGO P-Brick along with an output electronic driver circuit for controlling the small electric motor, a hybrid mobile robot can be built. As with most laboratory projects discussed in this book, experimentation is the key to successful robotics architecture design. Attaching the small electric motor to the high side driver is simple. Attach the pair of alligator test leads used in the Capsela hybrid robot to the SAE AWIM, as shown in Figure 7-24.

The two-switch hand controller will then be able to turn on and off the small electric motor by rotating the contact arm to and from the embedded touch sensor. To measure the current of the electric motor, the measurement technique presented in this chapter can be used for characterizing the electromechanical load.

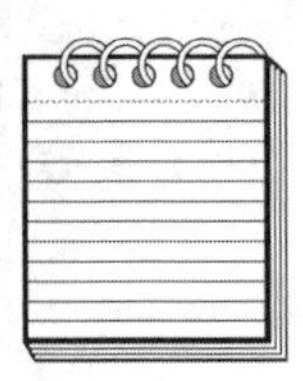

TECH NOTE: *The SAE AWIM Challenge 2 kit comes with a comprehensive lab manual. Although the material in the lab manual was written for a junior high school classroom, the experiments and projects will provide plenty of useful information needed to build a hybrid mobile platform robot.*

Figure 7-24
Attaching the SAE AWIM small electric motor to the LEGO P-Brick controlled DC-to-DC converter high side driver circuit.

The Erector Set and Hybrid Robots

The classic construction toy of the fifties provides an excellent mechanical framework for combining the LEGO P-Brick and the Erector set into hybrid robots. You can easily build an assortment of mechatronics-based robots and smart machines using the mechanical motorized construction set. The set contains steel girders, plates, brackets, gears, electric motors, screws, and nuts. The idea booklets packaged with the Erector Set can easily be transformed into mechatronics hybrid robots. Simply add a LEGO P-Brick, an electronic I/O interface circuit, and a behavioral-based software control interface for powering and moving the mechatronics-based machine.

Today's Erector Sets have come a long way since the days of A. C. Gilbert, the inventor of the mechanical construction kit introduced in 1914. Railroad workers assembling an electrical system out of girders and rivets inspired the Erector Set.

The original Erector Set enabled the hobbyist to build mechanical replicates of steam engines, ferris wheels, parachute rides, even a remote control robot. The metal pieces were a little heavy and the metal edged surfaces were not completely finished. This provided for a few scratches on the hand when using the set.

The new generation of Erector Sets has a combination of plastic as well as metal pieces. The metal edges have finished surfaces and the project books have an array of projects such as airplanes, robots, motorcycles, go carts, ferris wheels, windmills, and an airship. The electric motors are small and efficient in battery power consumption. These mechanized models can easily be converted in hybrid robots with the techniques discussed in this chapter.

Controlling an Erector Set Motor with a LEGO P-Brick Having an established control circuit expedites the building process of a hybrid robot project. The control method described in the SAE AWIM discussion is applicable to driving an Erector Set (ES) motor. To connect the ES motor to the high side driver circuit, simply take the alligator test leads from the breadboard controller and attach them to the wires of the electromechanical component. Figure 7-25 shows this electrical attachment method. The following minitest procedure will validate the control function of the two-switch hand controller to the ES motor.

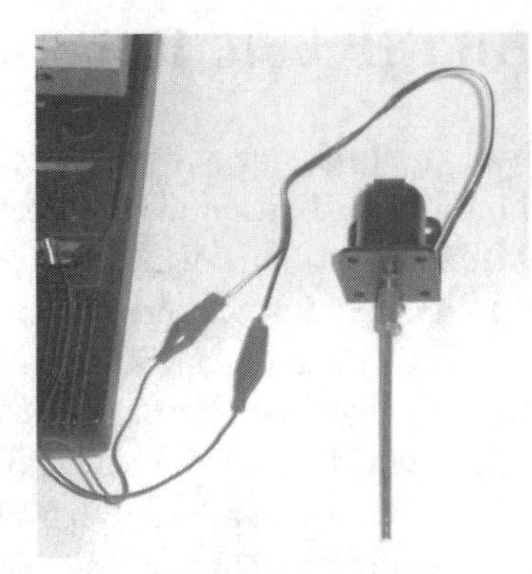

Figure 7-25 Electrical attachment of the Erector Set motor to the high side driver circuit erector

Set Motor Test Procedure Use the following steps to complete this test procedure:

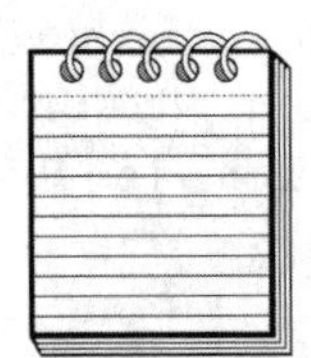

TECH NOTE: *The test procedure assumes the amateur roboticist is downloading the IC4 code into the LEGO P-Brick for the first time. Therefore, the code is shown for this assumption.*

1. Open the IC4 software. Go to the program window editor and type the following code:

```
void main()
{           while(1){
        if (digital(1)==1){
            printf("1 ON");
            sleep(2.0);
            fd(1);

        }
        else {
            printf("1 OFF");
            sleep(2.0);
            brake(1);
        }
        if (digital(2)==1){
            printf("2 ON");
            sleep(2.0);
            fd(2);

        }
```

```
            else {
                printf("2 OFF");
                sleep(2.0);
                brake(2);

            }

        }

    }
```

2. Place the infrared tower in front of the LEGO P-Brick.
3. Turn on the LEGO P-Brick and download the software into the P-Brick.
4. Place the contact arm of the two-switch hand controller in the center position of the LEGO electromechanical switch.
5. Press *RUN* to execute the code. The mini-LCD screen should toggle *1 OFF* and *2 OFF* messages every 2 seconds.
6. Rotate the contact arm to the right touch sensor of the two-switch controller. The mini-LCD screen should toggle *1 ON* and *2 OFF* messages every 2 seconds. The ES motor should be turned on. If not, turn off the LEGO P-Brick and make sure the circuit's modified LEGO electric wire is attached to output port A, pointing downward.
7. Rotate the contact arm back to the center position. After 2 seconds, the seven segment ES motor should turn off. The mini-LCD screen should toggle *1 ON* and *2 OFF* messages every 2 seconds.
8. Stop the IC4 program.

Erector Set Hybrid Robot, Final Thoughts The variety of projects that come with the Erector Set provides a wealth of hybrid robots for you to experiment with. The techniques presented in this chapter should provide enough information on software programming, building the experimental high side driver circuit, and testing the function of the ES motor controlled by the LEGO P-Brick. Try modifying the IC4 code listings to drive two non-electrical motors from the motorized construction kits discussed in this chapter.

The development of visual indicators for hybrid robot mechanical-movement status uses 7-Segment LED displays or regular LEDs. Set up the measurement control panel of the ScopeView software to trigger on various voltage or current signals produced by the electronics output driver circuit. Design, build, and test your LEGO P-Brick-based hybrid robots. Let your imagination be the motivational tool for discovery.

TECH NOTE: *On the Web, you can find an interesting article describing how the BRIO Corporation has brought back the Erector Set. The article provides a brief historical discussion on the Erector Set and the many corporations that helped evolve the 1950s construction kit for budding young engineers. To view the article, go to* `www.trains.com/Content/Dynamic/Articles/000/000/001/109 exais.asp`.

CHAPTER 8

Mechatronic Bots

Mechatronics robots enable the amateur roboticist to explore the mechanics, electrical/electronics, and software technologies within one convenient component or system. The electronics hardware along with embedded software helps improve the mechanical performance and function of the target component or system because of the capability to obtain data, process, and perform the design task effectively.

The efficiency of the mechatronics-based component and system is derived from the constant polling of the data from the external hardwired sensors attached to the internal microcontroller of the target smart device. The microcontroller processes the externally received data and, based on its embedded code, makes a logical decision to process the received data. This decision drives a hard or soft communications serial bus to perform the specific programming task. The embedded code inside of the microcontroller uses a conditional logic method for making output driver decisions on what external circuits should be switched on or off at the corresponding port of the electronic module (black box).

The LEGO Mindstorms *programmable bricks* (P-Bricks), both *robot command explorer* RCX and Scout, are essentially the black boxes for the tabletop mechatronics bot. The Hitachi H8 microcontroller, inside of the yellow and blue bricks, processes received digital and analog signals at the input ports of the P-Bricks. Conditionally, these input signals are processed and applied to the driver circuit which switches the output ports on and off.

The electromechanical loads attached to the programmable output ports activate and deactivate based upon the conditional processing of the embedded microcontroller. Therefore, the input processor, core processor, and output processor work in unison to ensure the data from the sensors and transducers drive the correct electrical actuators. With each component working together, the data flows from the sensor, to the microcontroller, to the output. The mechanical interface function is improved because of the internal self-adjusting circuits and software controls embedded inside of the P-Brick.

This chapter will explore the fundamentals of a mechatronic robot using the hybrid technique discussed in Chapter 7. The hands-on robot construction project will explain mechanical fabrication techniques. Java programming language will test the function of the robot for correct *input/output* (I/O) functions. Programming software techniques, circuit blocks and mechanical basics from the previous chapters will be revisited and analyzed for mechatronics bot connections and applications. Step-by-step instructions for building the robotic systems will be provided. Although the Erector motorized construction set is illustrated during the discussion on mechatronics robots, the same techniques can be applied be to other construction sets as well.

Robo Pump Proof of Concept (POC) Study

Robo Pump is a LEGO P-Brick mechatronics-based Erector model of an oil-drilling machine. To operate the Erector oil-drilling pump, attach the small DC motor to battery source with an electric switch wired to interrupt the voltage being applied across the electromechanical component.

The POC is to replace the external battery and switch with a LEGO P-Brick. The LEGO P-brick will give the function of the oil-drilling pump a customized control via software programmability. Robo Pump can be controlled using either the Scout or RCX P-Brick for operation of the attached small DC motor to the output port. The robotics-based oil drilling pump will be operated using a Java *graphical user interface* (GUI) capable of testing the I/O of the mechatronics-based system. A LEGO Cam will be mounted on the drilling pump to observe the reciprocating mechanical motion of the smart machine. The Robolab Vision Center software will be your viewing console for observing Robo Pump in operation.

Before proceeding with the discussion on building and testing Robo Pump, an explanation on how to attach LEGO bricks to Erector components will be presented.

The LEGO Brick Meets the Erector Set

For many years, LEGO bricks and Erector Sets have been in competition. The LEGO Company has developed many theme and skill builder sets to capture a wide range of construction kit hobbyists. BRIO-Erector has several construction kits as well, to meet the needs of their existing and potential metal plate-builder enthusiasts. For the amateur roboticist with a limited budget, building robots and smart machines using different construction mediums is the ultimate goal and a daily challenge. Therefore, this chapter will present a simple solution that should meet the challenge of building high quality bots at a low to medium cost investment.

The Erector Set has a variety of metal plates, girders, and brackets for building a host of mechanical machines. Using brackets to attach the two construction kit mediums is one technique that will enable the meshing of Erector pieces to Lego bricks. By drilling one small hole into the LEGO brick, an Erector bracket can be attached, enabling the two construction kit mediums to be connected together using small screws and nuts. Figure 8-1 shows the mechanical attachment technique.

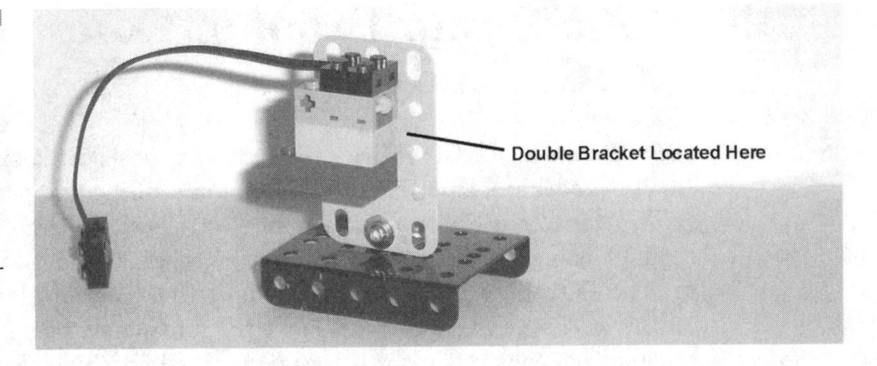

Figure 8-1 Mechanical attachment of LEGO Bricks to Erector plates using small drilled holes, brackets, screws, and nuts

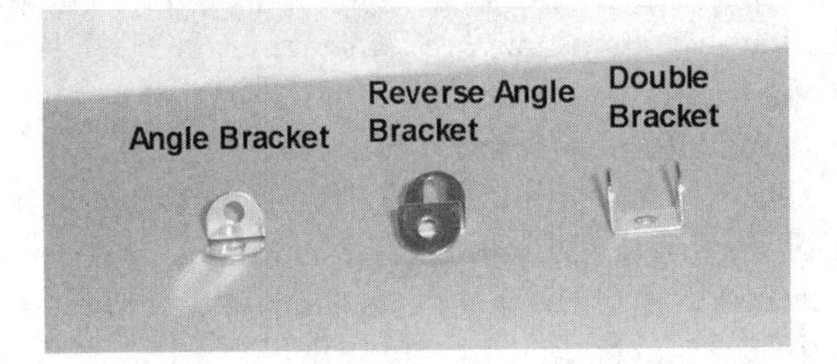

Figure 8-2 Various brackets used in Erector Sets for attaching parts

A LEGO brick will be core mechanical interface part for attaching a LEGO assembly or small electric to the Erector model. A 2x4 brick is a good choice for mechanical interfacing because it enables LEGO assemblies to be attached to the Erector model with good rigidity. Drill a small hole at an appropriate location on the brick to accommodate an Erector screw and nut. This hole will be used to attach the LEGO assembly to the Erector model in correct orientation. The appropriate Erector bracket should be selected based upon how the LEGO assembly will attach to the Erector model. Figure 8-2 shows several brackets used in attaching Erector parts together. Once the Erector attachment bracket is selected, the final assembly of the LEGO-Erector machine model can be built.

This technique enables the LEGO assembly to be mounted on an Erector mechatronics model in many positions and locations. The interfacing of LEGO bricks to Erector parts demonstrates the hybrid design and build concept discussed in Chapter 7, "Hybrid Robots." Using Erector parts as the framework of the robot provides a structurally-sound mounting base for adding LEGO bricks and assemblies to the Erector mechatronics model.

The LEGO core bricks (drilled blocks) enable the Scout and RCX P-Bricks and electrics to be easily mounted to an Erector metal-based platform. The Erector platform can be turned into a mobile device by attaching wheels and a small DC electric motor. Touch sensors, light sensors, LEGO Cams, and motors can be mounted to the Erector mobile platform using this mechanical attachment technique. To attach several LEGO bricks to the Erector model, several blocks need to be drilled for mechanically interfacing to the hybrid machine. A mechanical interfacing laboratory project will enable you to experiment with this hybrid component attachment technique.

TECH NOTE: *By drilling small holes into LEGO bricks, you can build up an ample supply of mechanical interfacing blocks. Drilling four small holes on each side of the brick enables you to mount additional blocks onto the core mechanical interface in any orientation.*

Hybrid Component Mechanical Interfacing Laboratory Project

The purpose of the Hybrid Component Mechanical Interfacing Laboratory Project is to enable you to attach a LEGO brick to an Erector plate using the attachment technique discussed earlier. This mini-laboratory will enable you to experiment with various mounting locations and orientations of the LEGO brick onto the Erector plate.

The *bill of materials* (BOM) for this project is as follows:

One touch sensor
One 2x3 brick
One 2x4 brick
One electric wire
Two double brackets
One flexible plate
One flanged plate
Four square nuts
Four screws
Four washers
One Erector Allen key
One Erector open-ended wrench
One pair of small needle-nose pliers

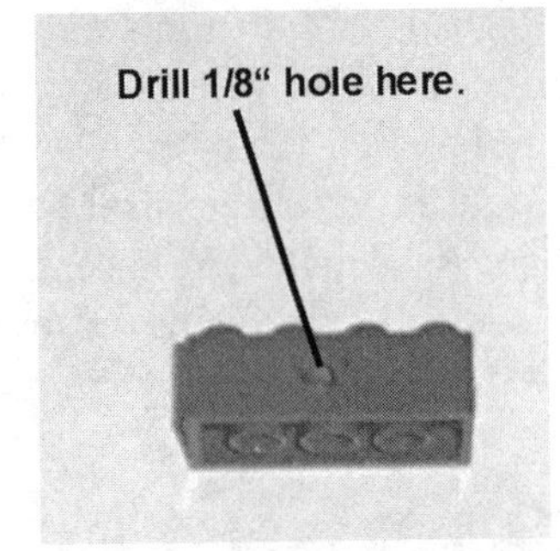

Figure 8-3
A 1/8-inch hole drilled into the center of a 2x4 LEGO Brick

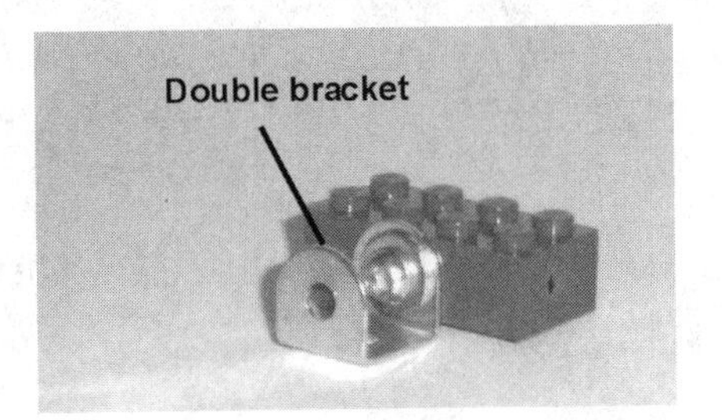

Figure 8-4
A double bracket attached to the 2x4 LEGO brick using a screw, washer, and nut

Assembly Procedure Use the following steps to complete this project:

1. Take the 2x4 LEGO brick and drill a 1/8" hole into the center using a small drill. See Figure 8-3.
2. Place a double bracket to the 2x4 LEGO brick.
3. Align the holes of both pieces and place a washer onto the double bracket.
4. Pass a screw through the three holes (washer, bracket, and LEGO brick).
5. Secure the double bracket to the 2x4 LEGO brick using a small square nut placed on the screw. See Figure 8-4.
6. Tighten using an Allen key and an open-ended wrench.
7. Place a 2x3 brick on top of the 2x4 plastic piece. Have the 2x3 brick off-set where two of the studs are showing. See Figure 8-5.
8. Place a touch sensor on top of the 2x3 LEGO brick.
9. Place an electric wire on top of the touch sensor.
10. Take the touch sensor assembly and attach it to the flexible plate, using a washer, screw, and nut. Tighten using a pair of small needle-nose pliers and an opened-end wrench. See Figure 8-6.

Figure 8-5
A 2x3 brick on top of the 2x4 part (notice the 1x2 stud offset)

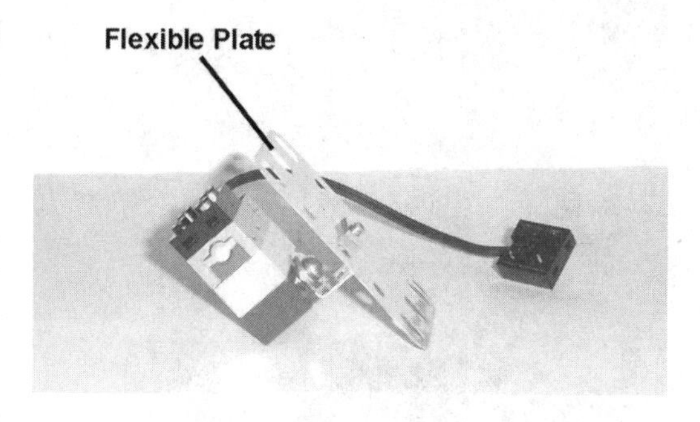

Figure 8-6
The touch sensor assembly attached to the flexible plate

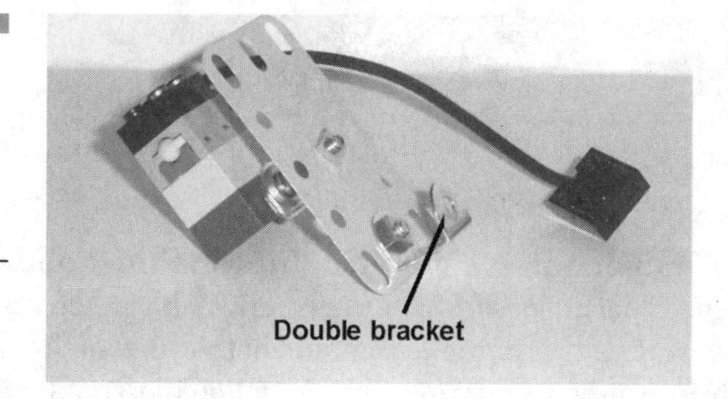

Figure 8-7
A double bracket secured to the touch sensor/flexible plate assembly

11. Mount the touch sensor/flexible plate assembly to a double bracket, using washer, screw, and nut. Tighten using an Allen key and an opened-end wrench. See Figure 8-7.
12. Take the touch sensor/flexible plate assembly and mount it on top of the flanged plate. Secure it with a washer, screw, and nut. Tighten using an Allen key and an opened-end wrench. See Figure 8-8.
13. The mechanical interface laboratory project is complete. Figure 8-9.

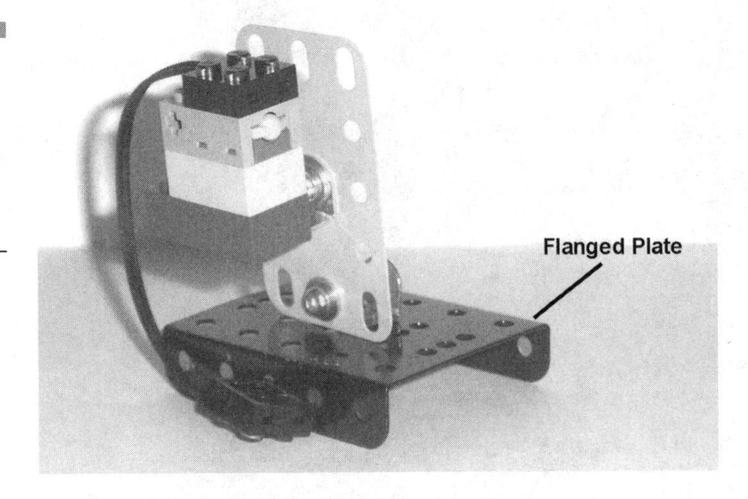

Figure 8-8 The touch sensor/flexible plate assembly mounted on top of the flanged plate

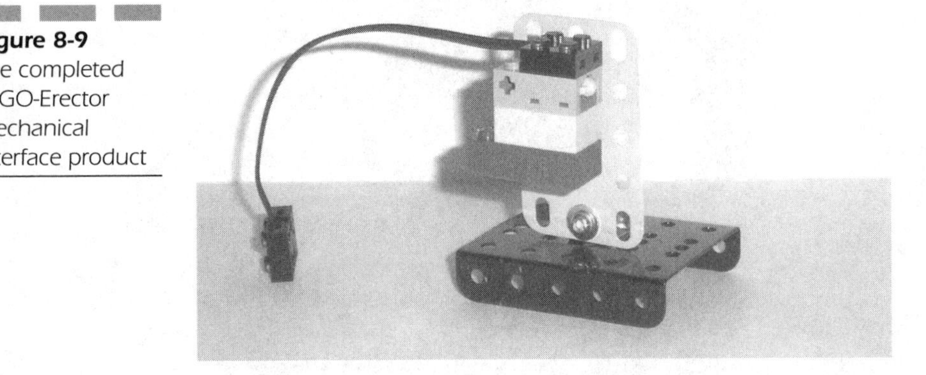

Figure 8-9 The completed LEGO-Erector mechanical interface product

Hybrid Component Mechanical Interfacing Laboratory Project Summary The purpose of the mechanical interface laboratory project was to illustrate a technique to combine both construction kit pieces to create a hybrid component or system. By following the assembly procedure outlined in the laboratory project, you should begin to see how other LEGO parts can be added to provide a structurally sound low prototype machine for robotics competitions in both performance and exhibition events.

Adding a Scout or LEGO P-Brick to an Erector-motorized model, along with small electrics-like sensors, a mini-camera, and motors, enhances the function of the mechanical-based machine tremendously. Therefore, any Erector model shown in the assembly manual or instruction sheet can now have a small black box capable of providing a programmable customization feature to the electrically motorized machine.

The following laboratory project will outline a test procedure for validating Robo Pump. Robo Pump is used as illustration, to demonstrate the process and technique of building a mechatronics-based hybrid robot. If you have an Erector Set or equivalent motorized construction set, the same build and test techniques can be applied.

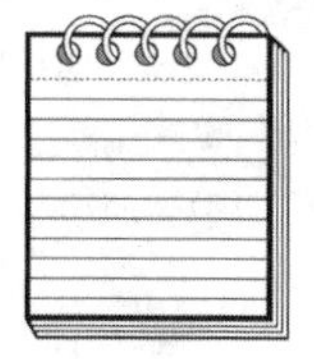

TECH NOTE: *An online web page called* ErectorWorld *has wealth of information on the Merkur Erector Set. It has information on the history of the metal construction kit, as well as several assembly instruction sheets that can be downloaded. The mechanical interface technique demonstrated in the previous laboratory project can be easily applied to these Erector models. You can find this interesting web site at* `www.erectorworld.com/efiles.html`.

Robo Pump's System Architecture

Robo Pump consists of mechanical and electrical interfaces for operation. Robo Pump is really a hybrid telerobot because of the mini-camera and virtual control panel console that assist you in operating the mechatronics-based bot. In order for the mechanical and electrical/electronics components of Robo Pump to work harmoniously, a system block diagram will need to be developed to assist in the assembly of the mechanics, electronics, and software components. Figure 8-10 shows the systems block diagram of Robo Pump.

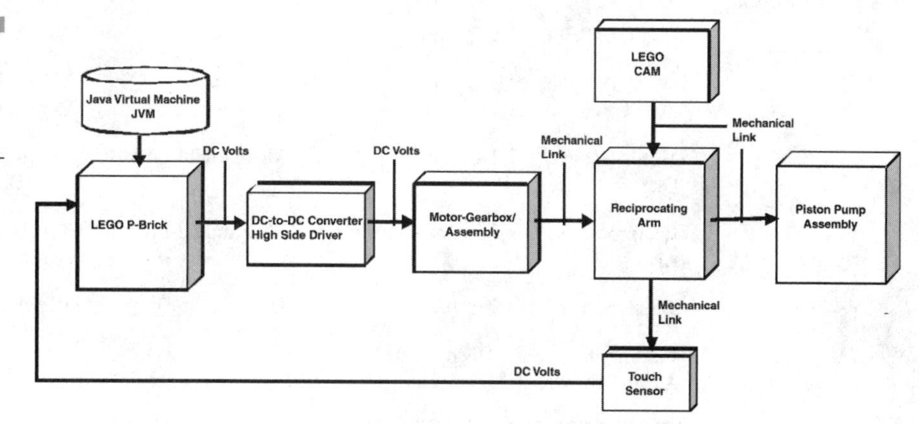

Figure 8-10 The system block diagram for Robo Pump

The arrows on the diagram show the data flow signals and the interaction between the mechanical and electrical/electronic components and assemblies. The mechanical connection technique demonstrated earlier will greatly aid you in building the hybrid bot. The system block diagram will also aid in the debugging process of Robo Pump's development by providing a synthesis (whole) view of how each subcomponent relates to each other by either DC volts or a mechanical link. If the component is electrical in design, it requires DC voltage for proper operation. If the component requires a push, pull or rotate task, a mechanical link will assist in producing the specified motion. Table 8-1 summarizes the interconnections between the subcomponents of Robo Pump.

The component interconnection table will help manage the mechanical and electrical/electronics data flow associated with the operation of Robo Pump. The table will also help in building the mechatronics bot by enabling each component to be built and validated before the next part is added. This approach minimizes the amount of debug time because each component will be validated prior to the overall robot being built. Trying to identify and repair all known design and function problems is a challenge you probably would rather live without.

Motor-Gear Drive Robo Pump's reciprocating arm is dependent on the motor-gear drive mechanism. The motor-gear drive is a critical electromechanical component of robots because it provides the locomotive motion required to perform some specific physical task. Therefore, the electro-

Table 8-1

Component interconnection table

Component	Input connect	Output connect	Input connect enabler
LEGO P-Brick	DC volts	DC volts	Touch Sensor/JVM
DC to DC—Hi Side Driver	DC volts	DC volts	LEGO P-Brick/DC to DC Converter High Side Dr.
Motor-Gear Drive	DC volts	Mechanical Link	DC to DC Converter High Side Dr.
Reciprocating Arm	Mechanical Link	Mechanical Link	Touch Sensor/Drill Bit Assembly
Touch Sensor	Mechanical Link	DC volts	Reciprocating Arm
Piston pump Assembly	Mechanical Link	NA	NA

mechanical functional integrity should be validated to ensure proper operation. The following laboratory project will investigate testing a LEGO P-Brick controlled Erector motor-gear drive. The testing methodologies presented in this chapter have been discussed previously in past chapters of this book. They have been provided again as a convenience to the reader.

Motor-Gear Box Test Procedure The following test procedure will outline the validation process for checking out the motor gearbox for Robo Pump. This same testing method can be applied to other mechatronics-based hybrid robots of your choice.

Use the following steps to prepare for the test procedure:

1. Go to Sun Microsystems web site (`http://java.sun.com/j2se`) to download and install J2SE version 1.3.1_01 software onto your notebook or PC hard-drive. Chapter 1, Figure 1-22, shows the home page for J2SE software.
2. Go to Jose Soloranzo's web site (`http://lejos.sourceforge.net/download.html`) to download and install leJOS version 2.0 software onto your notebook or PC hard-drive. Chapter 1, Figure 1-23, shows the download page for leJOS software.
3. Go to Tim Rinkens' web site (`http://rcxtools.sourceforge.net/`) to download and install the RCX Tools *integrated development environment* (IDE) software onto your notebook or PC hard-drive. Chapter 1, Figure 1-24, shows the download page for RCXTools software.

With the main software components installed on your notebook or PC hard-drive, the leJOS firmware is ready to be installed to the RCX P-Brick.

The leJOS Software Install Procedure Use the following steps to install the leJOS software:

1 Go to the *RCXTools_1_4* directory created in the step 3 download/install. Open the folder and double-click the *RCXDownload.bat* file. After a few seconds, the IDE tool should be on your notebook or PC video screen.
2. Click *Preferences* and check the pathway and COM port for your system. Make the appropriate changes and click *Accept*.
3. Turn on the LEGO P-Brick and have the infrared tower in front of it.
4. To send the leJOS firmware to the LEGO P-Brick, click *Download Firmware*. A dialog box will appear on the IDE tool, asking if you would

like to download the firmware. Click *Yes*. See Figure 1-25 in Chapter 1, "Software Tools for LEGO Mindstorms Mechatronics Embedded Systems Development," for details on the dialog box and the firmware download confirmation.

5. After downloading the firmware to the LEGO P-Brick, close the Download RCXTool.

Assembly Procedure The following steps are for the construction and control of the Erector motor gearbox, using the Direct Mode RCXTools IDE:

1. Build an Erector motor-gear box like the one shown in Figure 8-11, or spin your own design. Attach the motor-gear box to the output of DC-to-DC converter high side driver circuit. The input of the 7805 voltage regulator is attached to input port A of the LEGO P-Brick.
2. Turn on the LEGO P-Brick. Have the infrared tower in front of the P-Brick, for direct control of Robo Pump.
3. Download the *Direct RCX* Java program to the RCX P-Brick by clicking the button. The download time will take a few seconds to complete.
4. Click *Forward* on the IDE tool and the gears should turn.
5. Click *Stop* on the IDE tool and the motor-gears should stop.
6. Adjust the speed of the fan using the slide control A on the IDE tool. The fan blades should turn slow at 0 and fast at 7. As the slide control A is adjusted from power levels 0 through 7, the LCD screen of the LEGO P-Brick displays the same information as well.

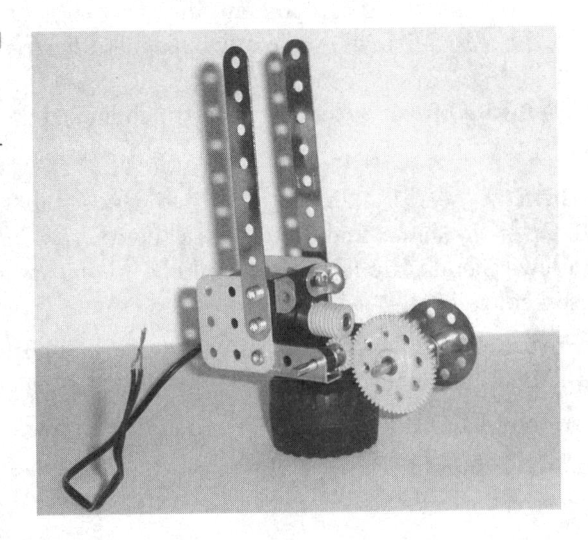

Figure 8-11 Robo Pump's motor-gearbox

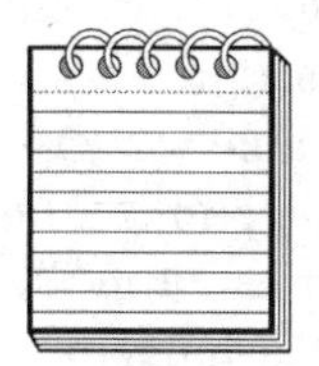

TECH NOTE: *To improve the rotational speed of the Erector motor-gear box, apply one drop of 4-in-One Household oil or equivalent small appliance lubricant to each gear. The oil helps reduce friction while increasing the speed of the electromechanical drive.*

The Author's Journal on Building Robo Pump

Robo Pump is based on extensive research reviewing an old Erector Set instruction booklet for various mechanical models that would assist in demonstrating the hybrid systems. The key requirement is that the chosen mechanical design must allow itself to be modified using standard LEGO Mindstorms components, attaching them to the Erector model using the hybrid component technique discussed earlier in this chapter. With this mechanical interface requirement in mind, each Erector model in the instruction booklet was carefully scrutinized for ease in attaching LEGO Mindstorms component to it. The instruction manual used in the hybrid robotics system development of Robo Pump was copyrighted 1964 by the A. C. Gilbert Co., the inventor of the Erector Set.

By the way, this instruction manual was used during my childhood days of tinkering with basic electric motors and Erector Set model building. Seventeen years later, I'm revisiting my childhood building days looking for a base Erector model to use in writing this chapter on mechatronic bots. Another reference book that helped in my choice of an Erector Set model is the 1993 *Meccano Erector Instruction Booklet #2*.

As I thumbed through the pages of the instruction booklet, I came across an interesting Erector model on page 4. The Meccano Erector Company decided to give the model on page 20 of the A. C. Gilbert Co. booklet a makeover. The model I'm referring to is a *walking-beam* engine. The design on page 20 of the original instruction booklet looked like something from the Smithsonian Museum. The design was quite big and looked cumbersome in terms of its mechanical up and down motion, which was based on a reciprocating arm.

The model on page 4 of the Meccano instruction booklet was smaller and had fewer mechanical construction parts than its distant cousin. This design would be a good demonstrator to include in the chapter on mechatronics and hybrid robotics system development. The POC of adding a LEGO P-Brick for operator control and intelligence along with a LEGO Camera and a touch sensor would enhance the electromechanical machine's function and appearance perfectly. Once the basic design of the walking-beam engine oil well drilling pump was realized, the software controls, electrical sensors, and vision components were selected for mechatronics bot conversion. The matrix shown in component interconnection table was created to help in design of the hybrid-based telerobotic oil well pump. The table allowed each connection of the components, either electrical or mechanical, to be captured within the design matrix for the target mechatronic robot.

After the completion of Table 8-1, I began construction of Robo Pump by building and testing the motor-gear box. With the electromechanical driver working properly, the rest of the Erector model can be built using the exploded view drawings shown on pages 4 through 6 of the Meccano instruction booklet.

Patience Is a Virtue when Building Erector Models It's been seventeen years since I built an Erector model, and for five hours I definitely validated the lapse in metal construction building. The exploded views for building the oil well drilling pump required several minutes of studying to identify which screw passes through hole number x on perforated strip a. The red arrows showing the process of part assembly and mechanical attachment using screws, spacers, washers, and nuts proved to be more challenging then I had expected. The small spaces intended for use with the open-ended wrench provided with the Erector construction kit made me realize that my fingers are not as small as they were when I was kid mechanically hacking on them seventeen years ago. Trying to keep the parts aligned while tightening the screws really was a test of patience for this amateur roboticist.

The approach I took in building the oil well drilling pump was to build the model in three sections: the motor-gear box, piston pump assembly, and the reciprocating arm. Figure 8-12 shows the complete hybrid Erector oil well drilling pump with each of the three assemblies highlighted.

The motor-gear box proved to be the easiest of all the three assemblies. The piston pump unit was quite challenging because of the small amount of space available, requiring me to use a surgical pliers to hold the nut while tightening the screw with an Allen wrench. With each turn of the screw, the flexible plates for building the sidewalls of the piston pump assembly would become misaligned. The trick to ensuring good alignment while tightening the screws is to turn the open-ended wrench in the opposite direction, forcing the plates to rotate in their secured mounted positions. This trick proved valuable during the remainder of the oil well drilling pump assembly.

After attaching the piston pump assembly to the motor-gearbox, the reciprocating arm was the last unit to build. Although the unit looks simple in

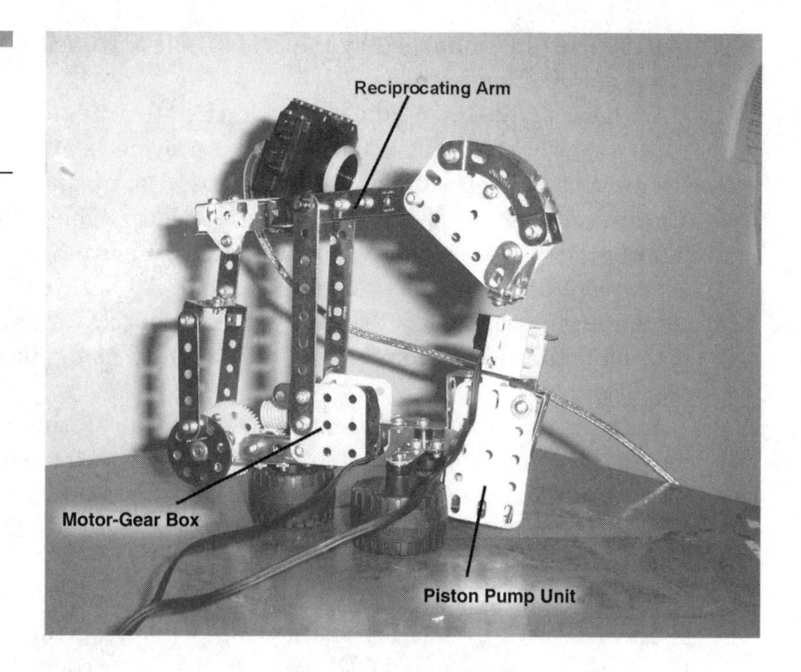

Figure 8-12 The three assemblies of Robo Pump

design, building it was an entirely different matter. Several areas in building the extended arm that required the use of the surgical pliers to assist in holding the nut while tightening the screw because of the limited amount of space. After 15 minutes of working in small spaces, the reciprocating arm was complete.

Final Assembly and Mechanical Adjustment The final step before testing Robo Pump was to attach the reciprocating arm to the motor gearbox. Although this sounds simple as a physical construction concept, the execution was more difficult. The reciprocating arm had to be placed on top of two 1x9 perforated strips attached to each side of the motor-gearbox. The center of the arm must be rotated on a 6 centimeter (2½) axle rod and secured with spring clips.

Properly adjusting the screws of the 1x3 and 1x5 perforated strips of the LEGO P-Brick-controlled pump was the ultimate test in mechanical construction patience. The screws were adjusted to ensure proper rotation to vertical up-down motion of the reciprocating arms. The Java-based RCX Direct Mode control panel GUI assisted in the adjustment of the screws by turning on output port A, which provided an 8V input voltage to the DC-to-DC converter high side driver. The driver circuit provided the appropriate sourcing current, managed by the 2N4401 NPN transistor to activate the motor of Robo Pump.

With each adjustment, the panel was used to activate the motor to see if the screws were too tight or too loose. This provided enough torque from the electric motor to move the perforated strips in an up and down motion. The arm's pivot consists of the 6 cm axle rod that enables the head of Robo Pump to move in a similar motion pattern. Figure 8-13 shows the two-adjustment area for proper arm movement. After several trial and error adjustment attempts on the screws, the reciprocating arm was able to respond to the various input speed values issued using the slide control for output port A.

With the mechanical functions of Robo Pump operating correctly, all that remains for final completion of Robo Pump is to attach the touch sensor and LEGO Cam to the hybrid mechatronic bot. Figures 8-14(a) shows the touch sensor attached to the piston pump unit. Figure 8-14(b) shows the LEGO Cam attached to the reciprocating arm of Robo Pump.

A final check for mechanical freedom of motion was carried out on the hybrid telerobot by sending speed commands using the Java-based RCX Direct Mode control panel GUI. Varying the slider control from 0 to 7 and noticing the movement (in relation to the speed of the motor-gear box) of the

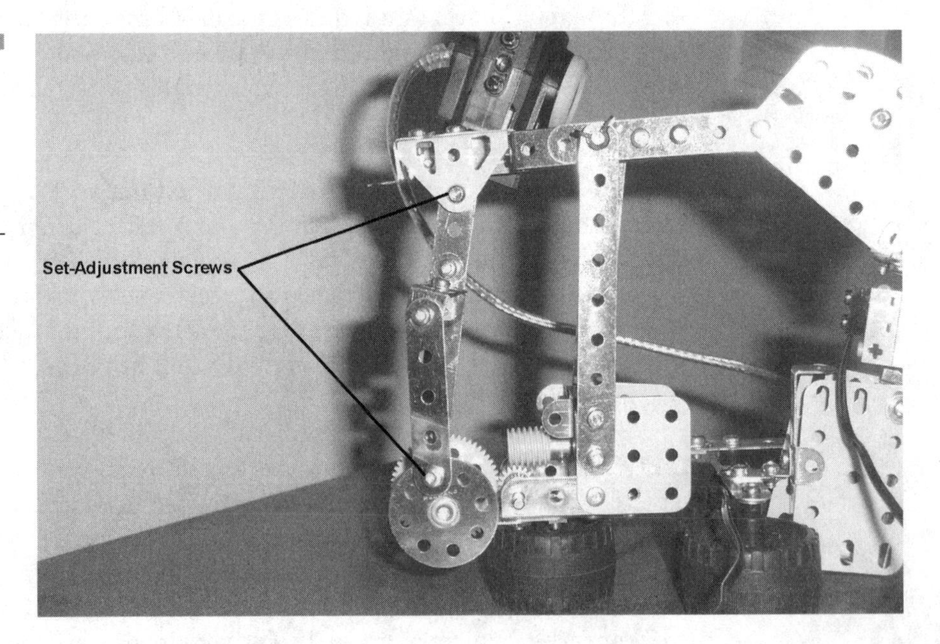

Figure 8-13 Adjustment of screws for proper reciprocating arm motion of Robo Pump

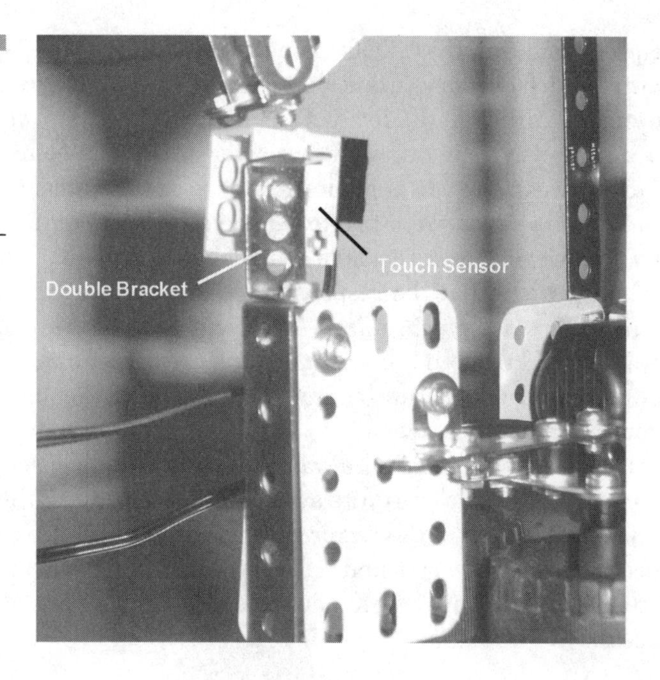

Figure 8-14(a) Touch sensor attached to the piston pump unit using an Erector double bracket

Figure 8-14(b) LEGO Cam attached to the reciprocating arm using an Erector double bracket

reciprocating arm, the LEGO Camera, and the touch sensor validated the input speed. The camera had no adverse effect on the reciprocating arms movement. Although the head of the reciprocating arm was contacting the touch sensor's yellow button, the response time for seeing the data values switch between 0 to 100 percent was not as consistent as I had hoped. Occasionally, the values would toggle between these two data values within the sensor value section of the Java control panel, but very sporadically.

The reason is that the response time of the software and the motion of the arm were not synchronized. The reciprocating arm moved faster than the code could read and process the touch sensor's data. To solve this problem would mean writing a new program with a quicker read-process response time.

Although the visual of data was not consistent, the mechanical actuation of the electric sensor was quite accurate. The location of the sensor, relative to the moving head of the reciprocating arm, was within the actuation distance required to depress and release the yellow button on the small LEGO electric. The overall design and mechanical test of Robo Pump was a success!

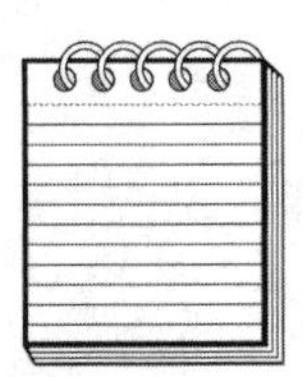

TECH NOTE: *The first walking-beam engine modeled using the Erector Set consisted of a wheel attached to a small electric motor. The reciprocating arms were attached to the wheel, providing direct transmission control of the rotating device. Go to* `www.erectorworld.com/efiles.html` *to see a PDF of this incredible machine.*

Measuring the Current Draw of Robo Pump

To ensure the DC-to-DC converter high side driver is sourcing the appropriate amount of current to turn to Robo Pump's motor-gear box, the following test procedure will measure this electrical parameter. Note: This test procedure was introduced in Chapter 7. In keeping with the reuse philosophy of this book, it has been modified for this laboratory project.

Test Procedure Use the following steps to complete this procedure:

1. Attach the *digital multimeter* (DMM) with serial PC interface to the DC-to-DC converter high side driver. Refer to Figure 7-6 as a reference circuit for attaching the DMM to the electronic driver circuit. Note: To take current-measurements using a DC milliammeter, the measuring

device must be inserted in series with the electrical load of the circuit that is driving it.

2. Attach the DMM with PC interface to the serial port of the desktop PC or notebook computer. Note: Prior to attaching the DMM to the serial port of the desktop PC or notebook computer, install the software that comes with the measuring device. The ScopeView® software that comes packaged with the Radio Shack DMM (Cat. No. 22-168) will be used as the measuring system example throughout the laboratory project.
3. Turn on the DMM and run the ScopeView software.
4. Turn on the LEGO P-Brick and run Direct Mode Java program.
5. Have the RCXTools Direct Mode GUI running.
6. Start a logging session within ScopeView, using Figure 7-7 as guide for setting up the software control panel.
7. Click *Scope* the control panel.
8. Click the slider control for output port A to level 7. Robo Pump' s motor-gear box should be running at the highest speed.
9. Click *RUN* on the ScopeView output window to start a data viewing session. A continuous current signal should be visible on the output window of ScopeView. The motor-gear box should continue to run for one complete sweep.
10. After completing one sweep, click *START* to stop the data-logging session. Figure 8-15 shows the plotted data taken by the author. Record the maximum motor current (Imax) and the steady state value (I) using the plotted data shown on the output window. Imax_______ A, Steady State (Isteady)_________ A.
11. Repeat test procedure steps 1 through 12 for each output level of the LEGO P-Brick. Compare the plots, noting any differences in current levels.
12. Stop the Direct Mode Java program in the LEGO P-Brick by pressing *RUN.*

Robo Pump Current-Measurement Summary This measurement laboratory project enabled you to characterize Robo Pump's performance by displaying its current draw while in operation. The test setup enabled one sweep of continuous measurement while the mechatronics pump was on at power level 7. The Direct Mode Java program enabled the output power level to be adjusted, using the slide control displayed on the virtual control

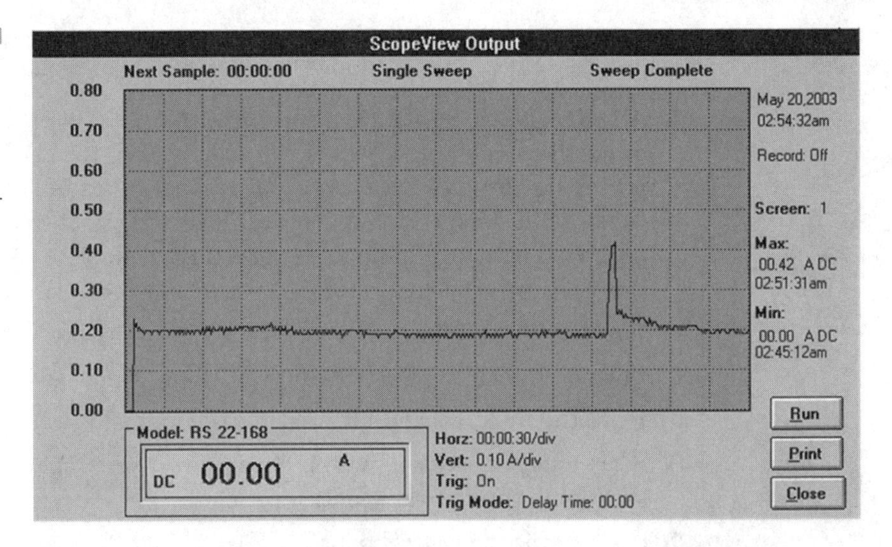

Figure 8-15 Robo Pump's current signal level, taken by the author

panel. This tool is very handy for providing a simple *human machine interface* (HMI) for testing the mechanics of a hybrid robot under development.

The plot of Figure 8-15 displays a large spike. The author created the transient by momentarily holding the gear of Robo Pump's motor-gear assembly. The current flowing through the collector-emitter lead of the 2N4401 transistor at the motor-stalled condition was 420 mA. Once the gear was release, the current level dropped to 190 mA. 190 mA is the steady-state current for properly drivingRobo Pump's reciprocating arm. As compared to the Capsela motor-capsule drive, Robo Pump's motor-gear box does not put much electrical current strain on the NPN transistor of the DC-to-DC converter high side driver.

Activating the LEGO Cam During Functional Testing of Robo Pump

The LEGO Cam can provide a visual perspective of Robo Pump while in reciprocating motion. The position of the small digital camera will provide a unique view when the arm and head assembly is in a constant up and down travel motion. Chapter 5 of this book discussed the test procedure for initiating the camera using Robolab Vision Center software. As added convenience to the reader, this section presents the test procedure for reference.

Opening the Robolab Vision Center The Vision Center is an advanced tool within the Robolab programming environment. Therefore, an Inventor 4 programming session must be opened to gain access to the vision software. The following steps will guide you to obtaining the vision center software. Prior to running the following vision center procedure, connect the LEGO Cam to the *universal serial bus* (USB) port of the desktop PC or notebook computer. Place the power stand telerobot in a good viewing location and mechanical movement area.

Vision Center Procedure Use the following steps to complete this procedure:

1. Run the Robolab software by double-clicking its icon on the Windows desktop.
2. Click *Programmer*.
3. Double-click *Inventor 4* under *Inventor*.
4. Within the Robolab *Untitled 1.vi diagram* main tool bar, click *Tools*. A drop-down menu box will be displayed.
5. Scroll down to *Select Camera* and click it. A *Select Camera* box will appear.
6. Click the scroll down arrow once. Microsoft WDM Image Capture (Win 32) will appear within the *Select Camera* box.
7. Click the check mark . The *Select Camera* box will disappear from the screen.
8. Click *Tools*.
9. Scroll down to select *Vision Center* and click it. The *Vision Center* window will appear on the screen. Note the image within the *Vision Center* window.

With the camera activated, Robo Pump can be operated using the Direct Mode Java-based control panel. The motion of Robo Pump's reciprocating arm can be seen on the vision center window of Robolab. The response time of the camera to the bot's reciprocating arm is somewhat slow. The visual is pretty much blurred at level 7, but as the speed is adjusted using the slide control, the images seem to improve. An interesting item to note is the angle of the camera relative to the left side of the head of the reciprocating arm. A small portion of it is visible within the vision center window. Of course, by changing the mounting orientation of the LEGO Cam, you could see more or less of the metal assembly.

TECH NOTE: *On the Web, there is a site that has 233 posted messages regarding the LEGO Vision Command software and other image processing applications. Subjects range from seeing in the dark using infrared diodes, to a robot that sorts parts, to LEGO-supports XP. The wonderful resource web site is the LEGO Users Group Network* (LUGNET), located at `www.lugnet.com/`.

TECH NOTE: *With each infrared control request made using the RCX Tools Direct Mode GUI, information relating to packets of data being received and sent by the LEGO P-Brick is available. The information is located at the bottom of the virtual control panel. This active toolbar can also be used as a real-time diagnostics tool when debugging circuit block add-ons to the P-Brick.*

TECH NOTE: *The Build-It-Yourself web site* (`www.built-it-yourself.com/`) *has a good examples of developing hybrid mechatronics-based robots using purchased or ordinary products found around the home. Creator John Galinato has developed an online site where kids in third through twelfth grade can learn about math, science, and art using the LEGO Mindstorms Robotic Invention System* (RIS) *and common household products. You can find a wealth of hybrid-based ideas on this site free of charge.*

Robot Construction: A Multimedia Approach to Robotics Development

The technique of building mechatronics-based robots using various motorized construction kits has a multimedia technology influence. Taking two distinct motorized constructions kits and merging them to produce a unique robot has a multimedia appeal in mechatronics. Multimedia is the combination of various art forms using text, graphics, and sound to produce a unique sensory and interactive product. Robotics development has taken this multimedia approach, based on the various control methods using virtual control panels, electromechanical switch boxes, voice and sound, and light detection.

You can combine these sensory control technologies to uniquely manipulate the smart machine to perform the given task. The motorized construction kits discussed should help you to develop unique subparts and systems for LEGO Mindstorms mechatronics products. The additional information on hybrid robot construction using the Erector Set as the core construction media should illustrate the importance of diversifying mechatronics-based robots and machines in design and development projects.

The hybrid component mechanical interfacing laboratory project, explored earlier in this chapter, was developed to enable you to experience the use of two different motorized construction kit media for the creation of unique experimentally developed smart machines. The off-the-shelf industrial and consumer-based robots produced by companies like Asea-Brown Boveri (ABB) and iRobot use various mechanical, software, and electronics technology media to build their mechatronics-based machines.

Mechatronics in itself is a multidisciplinary field combining mechanics, electronics, and software to solve society problems using smart components and machines technologies. Through the remainder of this book, the author will continue to explore how you can push the envelope of mechatronics-based development devices and smart machines using hybrid construction and multimedia techniques for LEGO Mindstorms robotics development.

Human Machine Interface (HMI) and a Smart Switch Design

In Chapter 6, "Smart Hand Controllers," the concept of HMI and LEGO Mindstorms was explained. The HMI concept is the capability to manipulate a machine using an input control device. The electrical switch is a simple HMI component for operating a LEGO Mindstorms mechatronics bot. Primarily, the electrical switch has two states: on and off. Therefore, any electromechanical device wired to the electrical switch will function in an on and off manner.

The method of human interaction with the electrical switch is dependent on how the internal contacts are mechanically manipulated by a *rock*, *push*, *slide*, or *momentary-push* actuation. The switching of states is somewhat immediate after the human initiated event has occurred. The electrical switch, when wired to a P-Brick, enables you to adjust the response of the input request to the internal contact switching state.

The electrical switch internal contact state can be programmed for either *timed delay off* or *timed delay on* events. The electromechanical load

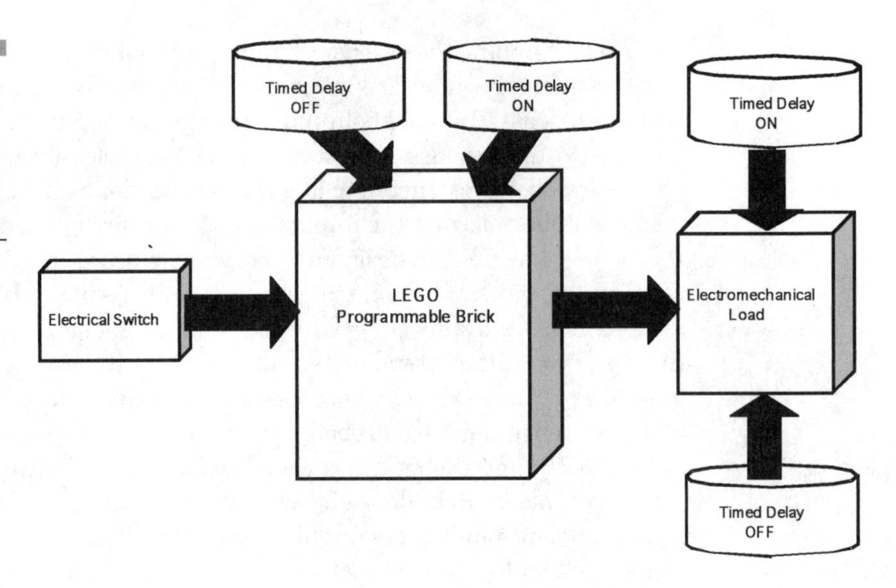

Figure 8-16 A programmable HMI input device (electrical switch) for controlling a mechatronics bot

attached to target output port of the P-Brick will respond according to the programmed input request. Now, the basic HMI input device is capable of having some programmed intelligence, based on it being hardwired to the P-Brick.

With a mechatronics bot seamlessly integrated into the HMI input device and the P-Brick, the input event-triggering component will initiate the bot's behavior. Figure 8-16 shows a system block diagram of the electrical switch HMI interface to a P-Brick.

In essence, a smart HMI input device has been created for the purpose of automatically turning an electromechanical load on or off. The capability to initiate, as well as complete, a task to be done at a specified time is an important requirement for a robot or smart machine. The smart switch described is capable of initiating such timed requests for Robo Pump. The following discussion will explore the smart switch POC and the switched control function to Robo Pump. The design of the smart switch will be described, along with the software requirements for controlling Robo Pump.

Smart Switch Design: Hardware and Software Requirements

Robo Pump can be controlled to either operate in a *time delay on* or a predetermined *time delay off* when an input event has occurred. The software can either keep Robo Pump operating for a few seconds when the switch is pressed and released, or the machine will turn on later, after the switch has been triggered. The software should enable a text message to be displayed on the LCD screen of the P-Brick.

The key components, used by the author with Robo Pump, capable of meeting these hardware and software requirements are the following:

- ***Interactive C Version 4* (IC4) programming language** For time delayed on and off control of Robo Pump's motor-gear assembly. An alphanumeric generator is provided with the software language as well.
- **DC-to-DC converter high side driver circuit** An electronics circuit capable of giving the necessary current.
- **Simple touch sensor** For input event triggering to the P-Brick.
- **LEGO P-Brick** The P-Brick can be programmed to display a short message on its LCD screen.

With the components identified, the next step is to build a simple prototype for testing the HMI controls to operate Robo Pump. The software will enable Robo Pump to continue to operate briefly when the touch sensor is pressed and released. The following laboratory procedure will test the smart switch design for creating a *timed delay on* feature for the LEGO P-Brick.

HMI Smart Switch for Controlling Robo Pump Laboratory Project

The objective of the laboratory project is to validate the HMI smart switch *timed delay on* control feature. Robo Pump will continue to operate for an additional 10 seconds after the touch sensor has been pressed and released.

The BOM for this lab project is as follows:

One touch sensor

One LEGO P-Brick

One standard LEGO electric wire

Robo Pump

DC-to-DC converter high side driver circuit (wired)

IC4 programming language

Assembly Instructions and Test Procedure Use the following steps to complete this procedure:

1. Attach the electrical wires of Robo Pump's motor-gear assembly to the DC-to-DC converter high side driver circuit. One wire of the motor-gear assembly should go the emitter lead of the 2N4401 NPN transistor. The remaining lead will go to ground.
2. Attach a touch sensor to input port 1 of the LEGO P-Brick, using a standard electric wire.
3. Place the infrared tower in front of the LEGO P-Brick.
4. Turn on the LEGO P-Brick.
5. Open the IC4 software programming language. Type the following program into the text editor window:

```
void main()
{           while(1){
        if (digital(1)==1){
            printf("P ON");
            sleep(10.0);
            fd(1);

        }
        else {
            printf("P OFF");
            sleep(2.0);
            brake(1);
        }
    }
}
```

6. Download the code into the LEGO P-Brick using the infrared tower.
7. Press *RUN* on the LEGO P-Brick.
8. Press and release the touch sensor. Robo Pump should be active for 10 seconds and then stop. On the LCD screen, *P ON* should be displayed when the touch sensor is pressed, and *P OFF* when the touch sensor is not pressed. If there is no mechanical movement from the hybrid robot, stop the program and recheck the code. After correcting any software errors, repeat steps 6 through 8. If Robo Pump is working correctly, repeat the procedure several times to ensure proper operation of the mechatronics-based machine.
9. Stop the program by pressing *RUN* on the LEGO P-Brick.

10. Turn off the LEGO P-Brick
11. Place the HMI smart switch project aside.

HMI Smart Switch for Controlling Robo Pump Laboratory Project Summary This laboratory project demonstrated how to create a smart switch by coding a *time delay on* feature of 10 seconds for Robo Pump. Once the touch sensor was pressed and released, the hybrid mechatronics bot will run for 10 seconds. After running for 10 seconds, Robo Pump's mechanical movement would stop. On the LCD screen, *P ON* and *P OFF* would display while the robot was in motion for 10 seconds, or with no movement, respectively.

TECH NOTE: *The LEGO Cam can be moved to help reduce the load of Robo Pump's reciprocating arm. By placing the digital camera to the center of the reciprocating arm, as shown in Figure 8-17, the load can be evenly distributed through the mechatronics bot mechanical arm.*

Figure 8-17 Redistributing the weight by moving the LEGO Cam closer to the center of Robo Pump's reciprocating arm

Programming the One-Switch Hand Controller for an HMI Smart Switching Function

Reusing components for building mechatronics bots helps you to quickly prototype a robotic design with the high confidence that the major features and functions should work with minimal problems. To illustrate the reuse design practice, the *timed delay off* software program for Robo Pump will be initiated using the one-switch hand controller discussed in Chapter 6 of this book. The laboratory procedure will be reintroduced, with some minor modifications to the testing steps required for integrating Robo Pump to the basic switch hand controller.

Test Procedure Use the following steps to complete this procedure:

1. Build the one-switch hand controller, using Figures 6-4 through 6-15. The following steps will test the electrical switching function of the hand controller.
2. Attach the hand controller to input port 1 of the LEGO P-Brick.
3. Turn on the LEGO P-Brick.
4. Push *RUN* on the LEGO P-Brick.
5. Turn the steering wheel clockwise on the one-switch hand controller. The input port 1, as well as Robo Pump, will be on.
6. Turn the steering wheel on the one-switch hand controller counterclockwise. The input port 1 LED will be off. Robo Pump will be in operation for 10 seconds. After the 10 seconds has elapsed, Robo Pump will stop running.
7. Turn the steering wheel on the one-switch hand controller clockwise and counterclockwise. The input port 1 will toggle on and off, with Robo Pump following the input event request of the HMI smart switch.

Using the Scout P-Brick to Control Robo Pump

The Scout P-Brick can easily control Robo Pump by using one of the SAC programs. The input and output ports of the P-Brick can be used to provide an input trigger and motor-gear drive control, respectively. The following laboratory project will illustrate the conversion method of replacing a LEGO P-Brick with a Scout programmable device.

Test Procedure Use the following steps to complete this procedure:

1. Attach the driver circuit to output port A of the Scout P-Brick, using the modified LEGO electric wire. The wire attached to connector should be pointing downward for correct polarity interface to the driver circuit.
2. Program the Scout P-Brick to operate as an inverting logic gate, using the following *Stand Alone Code* (SAC) program.

```
Motion          Touch
Forward

                Brake
```

3. Run the code in SAC mode
4. Rotate the contact arm of the one-switch hand controller clockwise, using the steering wheel, locking it into place. The Robo Pump should be off.
5. Rotate the contact arm of the one-switch hand controller counter-clockwise from the locking position, using the steering wheel. Robo Pump should be in full reciprocating motion. If not, turn the Scout P-Brick off and check for wiring errors, wrong resistor values, misplaced transistor and integrated circuit leads to bread holes. Once the errors are found and corrected, repeat steps 5 and 6 to obtain the working results.

Summary of the One-Switch Hand Controller and the Scout P-Brick Laboratory Projects This laboratory project reuses the one-switch hand controller introduced in Chapter 6, to illustrate the ease with which an existing component can be used in a different robotics application. The basic switching function of the one-switch hand controller was modified using software to provide a *timed delay off* feature for operating Robo Pump. After attaching the hand controller to input port 1 of the LEGO P-Brick and running the IC4 code, the *timed delay off* feature was initiated. When the switch was rotated from right to left, Robo Pump would remain running until the 10 second time had elapsed. The operation was repeatable upon each clockwise/counterclockwise rotation of the hand controller's contact arm.

The replacement of the LEGO P-Brick with a Scout P-Brick was the second part of the HMI smart switch design evaluation. The Scout P-Brick was wired to the one-switch hand controller at input port 1. Output port A was wired to the DC-to-DC converter high side driver circuit. An inverter logic function program was built using the SAC embedded within the Scout P-Brick. Once the code was executed, Robo Pump would respond to inverted actions of the one-switch hand controller. For example, with the no switch on, Robo Pump would be active. If the switch was activated via the contact arm rotating clockwise using the steering wheel, Robo Pump would be inactive. The inverting operation using the Scout P-Brick's SAC was operable with each clockwise/counterclockwise rotation of the contact arm.

Additional Thoughts on Mechatronics Bots

In this chapter, additional information on hybrid robots was illustrated using a Meccano Erector Set, LEGO Mindstorms electric elements, and the P-Brick. The hybrid component mechanical interface laboratory was provided so you could practice the technique of combining LEGO bricks to Erector structures. The technique of hybrid mechanical construction was illustrated by the Robo Pump project. The various control switching projects for operating Robo Pump were designed to illustrate how the touch sensor or an electric switch on/off function could be enhanced using software.

I hope that this hybrid robotics project will provide useful information to the amateur roboticist interested in creating mechatronics-based robots.

CHAPTER 9

Walking Robots

The concept of walking robots have been around for over a 200 years. Since 1893, walking robots have been a fascination with the amateur roboticist. The Steam Man, a biped built by George Moore, was powered by a 0.5 hp gas-fired boiler and reached a walking speed of 9 mph (14kph). The Steam Man's capability to walk was helped by a swinging arm that guided the robot circles. The biped's traction was aided using heel spurs created by the smoke flowing from its head and nose. The Steam Man's was made from a mechanical gauge used to monitored pressure and mounted on the neck of robot. The concept of walking machines can easily be explored using basic mechanical blocks that will be explored in this chapter.

The LEGO Mindstorms *Robotic Invention System* (RIS) can enable you to explore this interesting area of robotics using mechatronics techniques. The approach that will be used in the exploration of walking robots is to develop a modular design in which the walking robot's mechanical drive system can be used with either a Scout or *Robot Command Explorer* (RCX) *programmable brick* (P-Brick).

The electronic method of control for the walking robot will be developed with a heuristic approach. Heuristic development is the act of experimentally building a system or component for product creation. During the heuristics development phase, various remote-switching applications will be investigated with the experimental walking robot. The modular walking mechanism dynamics will be measured using a graphics calculator as a portable data-acquisition device.

The Concept of Walking Robots

The locomotion of walking robots can be classified into three categories: slide, lever, and wheel-or-track. The slide and lever mechanisms provide a reciprocating motion; whereby the robot will be propelled by a wobble-thrust movement. The wheel-or-track is the most common locomotion technique used in mobile-based robotics.

The requirements of for a robot to successful negotiate soft uneven terrain are better mobility, stability of the walking platform, energy efficient, and smaller impact on the ground. The leg design used with walking robots is usually capable of accomplishing these requirements because of the light structural materials used in order for the robot to transverse its terrain, along with a microprocessor-based controller for motion and stability controls. Sensors and electromechanical actuators assist the walking robot by monitoring and regulating power to drive the legs with high efficient DC-to-DC converters, and electric motors, and smart power *metal-semiconductor field effect transistor* (MOSFET) drivers. The software embedded inside of the microprocessor is responsible for the management of stabilizing the legs over different terrain, using sensors for object and force detection.

Choosing a locomotion system for a walking robot is based on the terrain it will traverse. Operational flexibility, power and energy efficiency, payload requirements, stability, and impact on the environment are other factors effecting the selection of a locomotion system. The robot must be able to move on the target terrain while maintaining proper management of the electronic systems payload it carries, using the mechanically driven legs that propel it.

The motion controls of a walking robot should operate the body mechanics so that the leg movements automatically produce the desired mechanical movement of the mechatronics-based machine. The control system needs to control the gait of the robot. The gait is a sequence motion of the legs that can be defined as the time and location of the placing and lifting of each foot. Gaits can be divided into two classes: periodic and nonperiodic. Periodic gates repeat the same sequence of leg-foot placement. Periodic gates programmed into the microprocessor use a phase variable for indicating the stage of leg placement. The walking robot can also be controlled using nonperiodic gates that have no sequence in their leg-foot placement. The number of gaits is based on the total legs used with the mechatronics-based machine. For example, four legs will have four gaits or leg-foot placement sequences. Figure 9-1 shows a gait diagram for a four-legged walking robot.

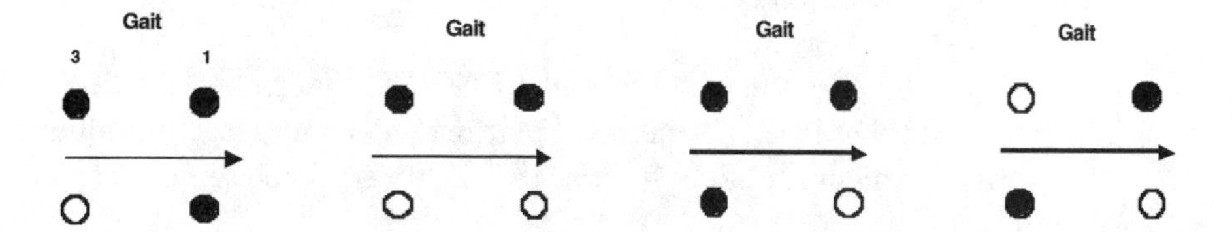

Figure 9-1 Gait (foothold) diagram of a four-legged walking robot

The black dots represent a foot touching the ground during the machine's movement through the terrain. Free gaits are controlled using logic-based rules that determine when the legs are put into a phase variable. The logic-based rules are usually established on certain criteria that depend on the design requirement of the robot's walking motion. One of the design criteria for logic-based rules is maximizing the foot-leg stability of the walking machine. The other free gait criterion is transferring the legs in motion that are at the end of the walking-movement stroke.

The control system will manage the body movement of the walking robot using sensors and a dedicated microprocessor for leg-motion stabilization. In essence, the control system will manipulate foot placement of the robot through the motion coordination of the nonsupporting legs with the operation of the moving legs in order to find the next foothold (gaits).

The stability of walking robot is the capability to maintain the mechanical body at the standing position of its legs. There are two forms of stability for a walking robot: statically stable and dynamically stable. A statically stable system means that the legs are stationary and the motion of the robot can be stopped at any time without the robot losing its standing position. Statically stable machines can be controlled using kinematic models. A dynamically stable system of a walking robot can be maintained only in active motion, when the legs are moving. Dynamically stable systems are managed using dynamical models embedded within the controller's microprocessor memory registers.

The concepts presented are based on mathematical models programmed into the controller of the walking machine. A *hardware-in-the-loop* (HIL) method of an operating the walking robot uses the math model running on microprocessor-based embedded controller. The data read from the machine's sensors are processed by the microprocessor. The received data are adjusted to minimize the error signal within the controller's microprocessor. Minimizing this signal enables stabilization of the leg of the robot.

Now that the technical concepts of walking machines have been discussed, applying these design guidelines will be explored using the LEGO Mindstorms RIS.

Building a Modular Walking Platform Mechanism

The key to a successful LEGO Mindstorms mechatronics-based walking robot is the mechanical platform mechanism. The walking platform mechanism should be built so that it can easily be removed from the P-Brick with minimal effort. The reason for modularity in building walking robots is it makes it simple to transfer the target robot's walking platform mechanism between the Scout and the RCX P-Bricks. Modularity also enables the walking mechanism to be tested prior to installing it on the target P-Brick. Therefore, the walking platform mechanism will be an example of *recycling of mechanical assemblies* (ROMA).

The ROMA concept reduces the amount of development time required in building Mindstorms mechatronics-based robots. ROMA enables mechanical building blocks to be used in new designs by taking proven existing systems, tweaking the modular unit for the application, and using it in the target robotics subsystem. Figure 9-2 shows the ROMA concept diagram for the mechatronics-based walking platform mechanism.

To create a ROMA diagram, begin by using the locomotion components of the walking platform as the inputs to the target machine. In the walking platform mechanism, the electromechanical parts consist of the gears (driven/driver) and the electric motors. The mechanical body of the walking platform will be the convergent point where the input parts are mounted. The end result of joining these components together is the completed walking platform. The ROMA diagram should help you build subcomponents that will eventually be attached to each mechanical block, completing the walking robot's platform mechanism.

A walking platform can easily be built using the *Constructopedia* instructions for building a Pathfinder 1 robot and a walking mechanism. Following the assembly instructions in Figures 9-3 through 9-5 enables you

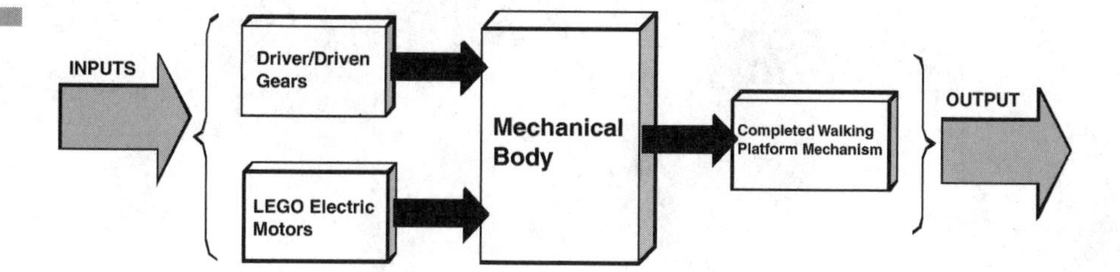

Figure 9-2 ROMA concept diagram for mechatronics-based walking platform mechanism

Figure 9-3(a) Building base of walking platform

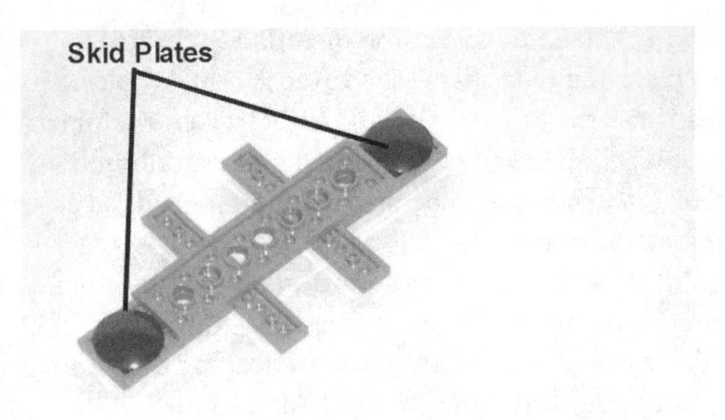

Figure 9-3(b) Adding skid plates to bottom of base

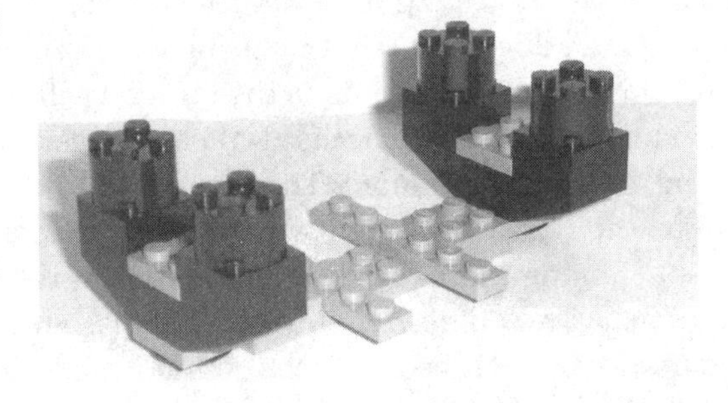

Figure 9-4 Adding LEGO P-Brick mounting supports to walking platform base

Figure 9-5 Adding two LEGO motors with electric wires to the walking platform base

to build the walking platform body. This modular mechanical structure provides the foundation for the addition of the walking mechanism. Figures 9-6 through 9-7 show the assembly of the walking mechanism. The walking mechanism consists of the driver and driven gears, the legs for the gait-based robot, and supporting bricks and plates for attaching the mechanical subunits to the body of the smart machine. Figure 9-8 shows the completed walking platform.

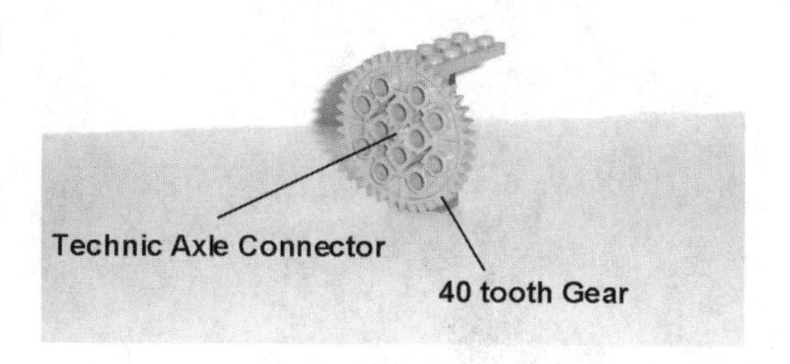

Figure 9-6(a)
A 40-tooth gear with axle-connector support

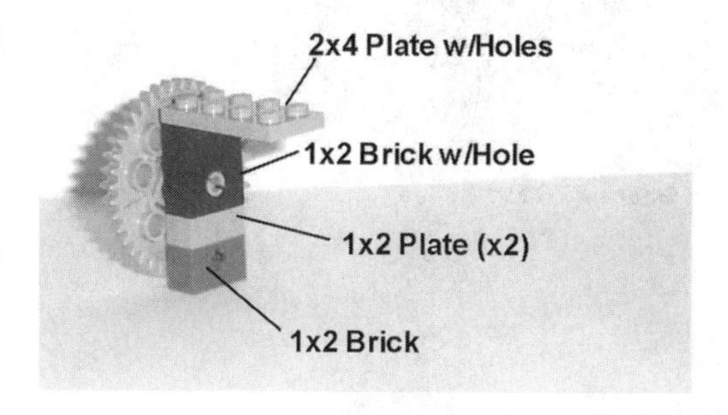

Figure 9-6(b)
A 40-tooth gear with axle-connector support (back view)

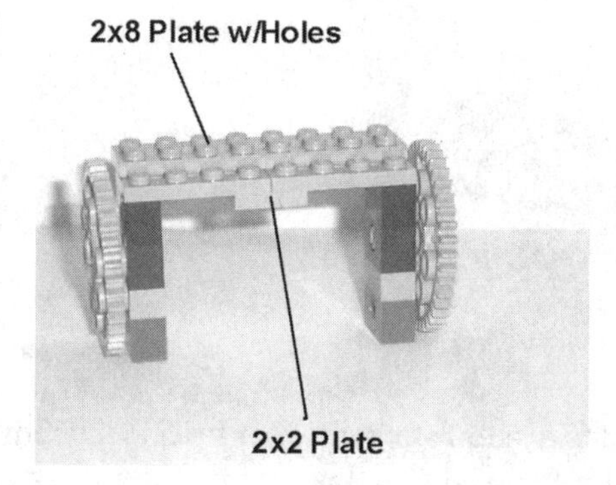

Figure 9-7
Building the front and rear driven gear assembly set #1

Assembly Instructions for the Modular Walking Platform Mechanism The following assembly instruction steps explain how to build the modular walking platform:

1. Build the base of the walking platform, using Figure 9-3(a). Use the following LEGO Pieces: two 2x4 plates with holes, one 2x8 plate with holes, two 1x6 plates, and two skid plates.
2. Assemble the P-Brick mounting supports to the walking platform base, as shown in Figure 9-4. The LEGO pieces required to build these four

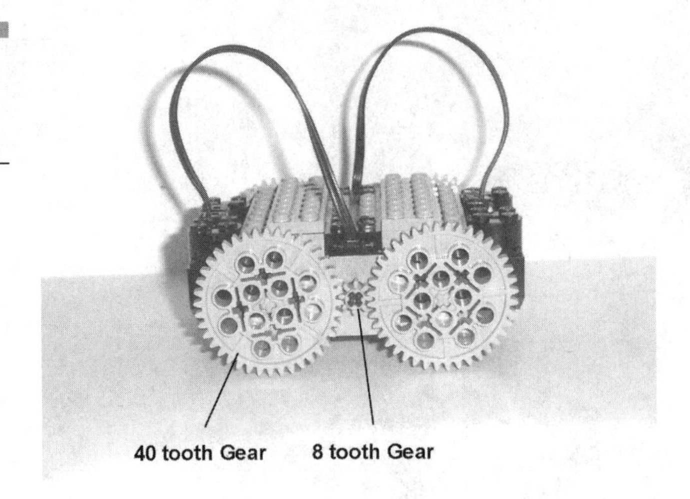

Figure 9-8(a) The completed walking platform mechanism

Figure 9-8(b) Plane view of walking platform mechanism

mounting supports are four 2x2 circular bricks, four inverted roof bricks, four 2x2 bricks, and two 2x2 plates.

3. Add two LEGO motors with electric wires to the walking platform base, as shown in Figure 9-5.

This completes the walking platform body assembly instructions. Next, the walking mechanism building instructions will be discussed.

Adding the Walking Mechanism to the Modular Platform Body
Use the following steps to complete this procedure:

1. Build the 40-tooth gear with axle-connector support, using Figures 9-6(a) and 9-6(b). Four supports are needed.

2. Take two of the four gear and axle-connector supports and attached them together using a 2x8 plate with holes, as shown in Figure 9-7. Build another set of this assembly, thereby having two complete sets.
3. Attach the two support assemblies built in step 2 to the walking platform body. The final unit is shown in Figure 9-8(a). Figure 9-8(b) shows the plane view of the modular walking platform mechanism.

To provide additional mounting support to the P-Brick of the walking platform mechanism, add 2x2 plates to each corner of the electromechanical unit, as shown in Figure 9-9.

The LEGO P-Brick can now be attached to the modular electromechanical device. Carefully place the P-Brick on top by sliding the embedded controller slowly on top of the modular walking platform until the P-Brick simply stops. Press the P-Brick onto the modular unit, as shown in Figure 9-10. The smart electromechanical device is now ready for a quick check of its gearing mechanism.

Testing the Walking Platform Mechanism The final step before adding the legs to the walking platform is to test the mechanisms gear

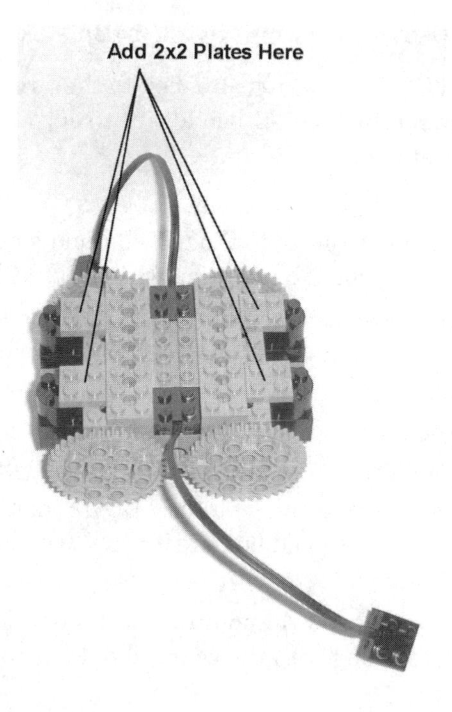

Figure 9-9 2x2 plates added to walking platform mechanism for P-Brick mounting strength support

Figure 9-10 The LEGO P-Brick mounted on top of the walking platform mechanism

drive. This simple test is to assure the gears are not binding while in rotation. It is important that this testing step is carried out before adding the walking robot's feet. Troubleshooting and mechanical adjustment is somewhat challenging when the robot is completely assembled. To test the robot's gear drive, simply to use the LEGO remote control and test each of the output ports individually and together, observing the rotation of the mechanical drive components.

Testing Procedure Use the following steps to complete this procedure:

1. Connect the LEGO electric wires attached to the two motors of the walking platform mechanism. Attach a wire to output port A and the other one to output port B.
2. Turn on the LEGO P-Brick.
3. Press and hold the up arrow *A* on the LEGO remote control. The motor, along with the gears, should be rotating without binding.
4. Press and hold the down arrow *A* on the LEGO remote control. The motor, along with gears, should be rotating in the opposite direction of step 3, without binding.
5. Press and hold the up arrow *C* on the LEGO remote control. The motor, along with the gears, should be rotating without binding.
6. Press and hold the down arrow *C* on the LEGO remote control. The motor, along with gears, should be rotating in the opposite direction of step 5 without binding.
7. Press and hold the up arrow *A* and up arrow *C* on the LEGO remote control. The motors, along with the gears, should be rotating without binding.

8. Press and hold the down arrow *A* and down arrow *C* on the LEGO remote control. The motors, along with the gears, should be rotating in the opposite direction of step 7 without binding.

Now that the walking platform mechanism is working correctly, we can now add the feet to complete the robot.

Completion of the Walking Robot The foot-leg assembly is the last component to add to the walking robot. The assembly consists of using four foot-leg units attached to two 1x8 bricks with holes. The four subunits are connected to the mechanism of the walking platform using four connector pegs. The connector pegs are attached to each 40-tooth gear and the ends of the 1x8 brick with holes. Figure 9-11 shows the leg-foot assembly, and its attachment to the walking platform mechanism.

After experimenting with the modular design of the robot, it was discovered that the 1x4 bricks located at bottom of each gear-axle connector were structurally inadequate. The bottom supports mounted to the P-Brick were collapsing. To provide additional reinforcement to the robots gear-axle connector assembly, the four 1x2 bricks should be replaced with two 1x8 plates. Figure 9-12 shows the replaced parts in their respective locations on the bottom of the robot.

Alignment of the leg units to the gears is critical because movement can become bound. Carefully adjust each 40-tooth gear so that each connector pin is parallel to the next one. The end holes of the 1x8 brick can then be placed on top of the connector pin of each gear. Slowly rotate one of the gears by hand and note the movement of the leg-foot assembly. There should no binding with the gears or no locking action of the leg-foot assembly. Complete this assembly process to the other leg-foot component on the

Figure 9-11 Attaching the leg-foot assembly to the walking platform mechanism

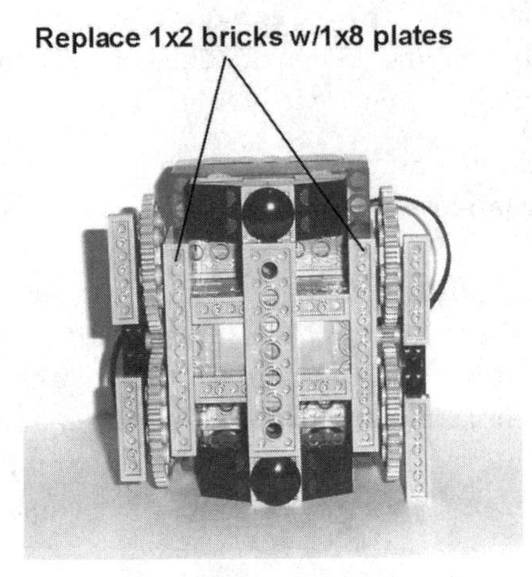

Figure 9-12 The reinforcement on the bottom of the walking robot

other side of the walking platform mechanism. With both leg-foot units attached, the walking robot can be tested.

Orientation of the LEGO Electric Wires Before testing the walking robot using the LEGO remote control, the electric wires should be orientated in such a way that the correct movement response is possible. The placement and orientation of the electric wires dictate the direction in which the walking robot will move. This orientation of the electric wires enabled the author's robot to move in the forward direction. Figure 9-13 shows the orientation to ensure correct movement while testing the robot.

Testing the Walking Robot The following test procedure, modified from the previous plan, will enable you to validate the correct movement of the bot, based on various motor drive requests. The LEGO remote control will be used to validate the movement and direction of the electromechanical platform of the walking robot.

1. Connect the LEGO electric wires attached to the two motors of the walking platform mechanism. Attach one wire to output port A and the other one to output port B. See Figure 9-13 for orientation of the electrical connectors.
2. Turn on the LEGO P-Brick.
3. Press and hold the up arrow *A* on the LEGO remote control. The motor, along with the gears, should be rotating without binding. The robot should be turning left.

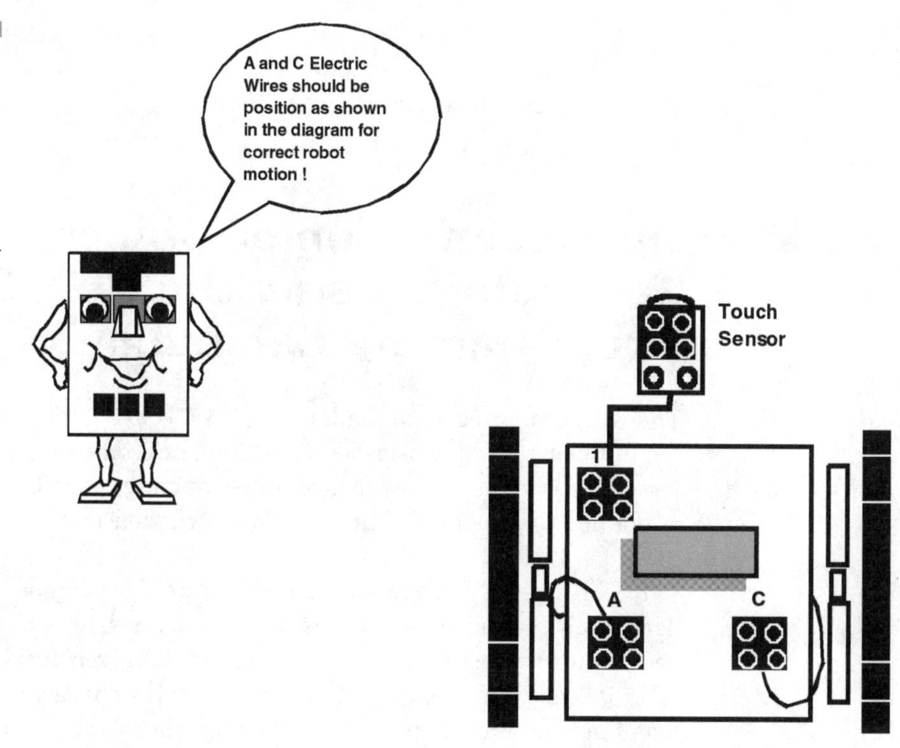

Figure 9-13 To ensure correct movement while testing, the connectors should be oriented as shown.

4. Press and hold the down arrow *A* on the LEGO remote control. The motor, along with gears, should be rotating in the opposite direction of step 3, without binding. The robot should be returning to its original position.
5. Press and hold the up arrow *C* on the LEGO remote control. The motor, along with the gears, should be rotating without binding. The robot should be turning right.
6. Press and hold the down arrow *C* on the LEGO remote control. The motor, along with gears, should be rotating in the opposite direction of step 5, without binding.
7. Press and hold the up arrow *A* and up arrow *C* on the LEGO remote control. The motors, along with the gears, should be rotating without binding. The robot should be moving forward.
8. Press and hold the down arrow *A* and down arrow *C* on the LEGO remote control. The motors, along with the gears, should be rotating in the opposite direction of step 7 without binding. The robot should be moving backward.

If the movements of the walking robot are incorrect, stop the bot and make the appropriate mechanical or electrical corrections. Repeat the test procedure again to assure a correctly working robot.

Implementation of Object Technology Using the RCX Code Programming Language

The RCX code programming language is a convenient tool for implementing automated and sensor-based robots. The icon-based programming follows the structure and procedure-based languages like C and C++ because of the *no jump to* subroutine or *goto* instruction methods for logic control flow.

RCX language has a top-down logic control flow for operating robots. The data received from outside events through a touch or light sensor is embedded within the behavior programmed for the given task. The task of the robot is really a defined *object* that describes the characteristics of the intelligent machine. Therefore, RCX code language is a simple object technology tool for quickly and conveniently programming a robot or intelligent machine. Although the complexity of the program might be large based on the robot's behavioral requirements and in the management of the code, the capability to combine data and function into one programming structure makes it a good object technology tool for the rapid software development of LEGO Mindstorms robots.

RCX Programming and State Diagrams

State diagrams are a convenient way of capturing the behavioral requirements of a LEGO Mindstorms robot. State diagrams provide a graphical representation of how the robot will behave based on a triggered event. Object technology is an important element to developing robotics software because the mechatronics-based machine behavior is embedded within the object-oriented structure. Data associated with robot's behavior is also part of the structure of the object.

The RCX programming language can be considered an application of object technology because the embedded data along with the behavior is part of the task-oriented structure. For example, the touch sensor watcher is an object, because of the data received by a triggered event and the output behavior that results from this external action. Combining other programming blocks increases the capability of the RCX code-based object to exhibit the target behavior the designer is developing.

The following lab project will explore the use of RCX code programming language for implementing a quick automated test program for validating the forward-reverse walk modes of the robot.

What Is a Smart Robot?

The following experimental project discussion will demonstrate the concept of a robot being able to stop activity after a predetermined amount of time has expired. The word smart, in this context of a robotics system, means to be able to turn on or off after an input event has been triggered. Therefore, the machine has the intelligence, based on its programming, to be able to start or stop a specific task or behavior upon receiving an input command from an external source. The robot is thereby reactive in trait because of the input request producing a corresponding output response.

A Smart Walking Robot

Traditionally, robots are smart machines capable of performing tasks based on the embedded code residing inside of its microcontroller. Autonomous robots use sensors to receive environmental data to transverse terrain without electrical tethers to confine it. To illustrate how RCX code is a great low-fidelity prototype programming language, this project will demonstrate a robot's capability to stop its movement after the touch sensor has been pressed and released.

Assembly and Software Test Procedure The walking robot will be able to move forward with an input event of a press and release request from a touch sensor. This walking motion will continue for 10 seconds, after which the robot will stop. To resume motion, the touch sensor will be pressed and released again, repeating the forward movement behavior as before.

The BOM for this procedure is as follows:

One touch sensor

One LEGO electric wire

LEGO Mindstorms RIS RCX code version 1.0, 1.5, or 2.0 software

One *experimental walking robot* (EW robot)

Use the following steps to complete this procedure:

1. Attach a touch sensor to input port 1, using a LEGO electric wire to the EW robot. See Figure 9-14.
2. Using LEGO Mindstorms RIS RCX code version 1.0, 1.5 or 2.0 software, program the EW robot for a *10 seconds on* forward function. See Figure 9-15.
3. Download the code into program slot 1.
4. Push *RUN* on the LEGO P-Brick.
5. Press and release the touch sensor. The robot should begin walking forward for 10 seconds, then stop. If not, stop the program and check the code for correct I/O ports and connector orientation. After correcting the problem or problems, repeat step 5.
6. Press and release the touch sensor. The EW robot should begin walking forward for 10 seconds, then stop. Each press and release action to the touch sensor will enable the robot to walk forward.

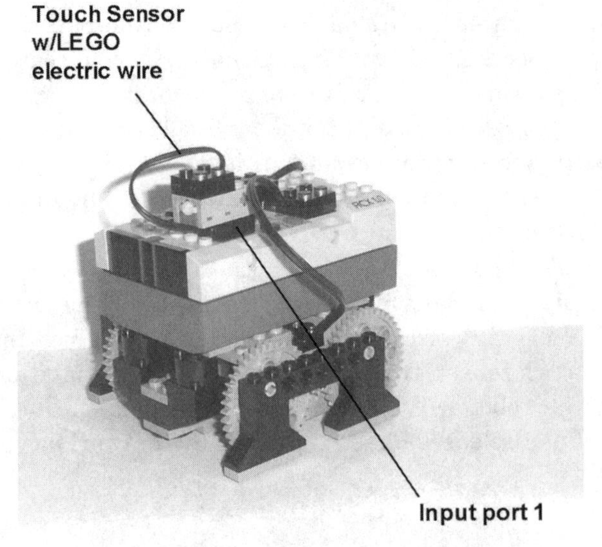

Figure 9-14 Adding a touch sensor to input port 1 using a LEGO electric wire

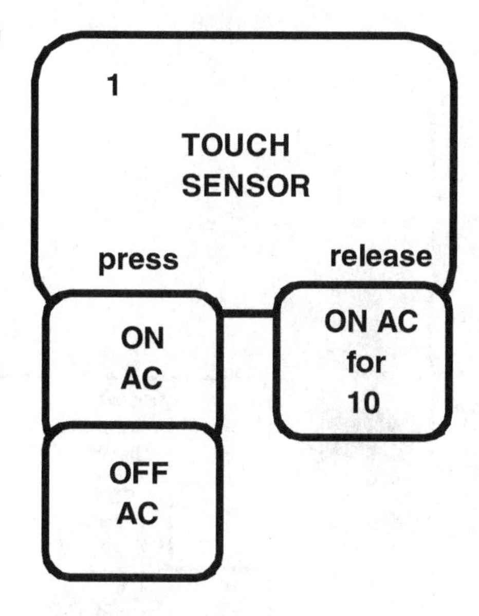

Figure 9-15
RCX code for the smart EW robot function

A Smart Walking Robot Project Summary The experimental project demonstrated the smart concept as discussed earlier by enabling you to write code for a *10 second on* function. The first press and release action to the touch sensor enabled the robot to be calibrated by walking forward for 10 seconds then stopping. Subsequent input requests of the robot to walk forward were commanded by the additional press and release of the touch sensor.

RCX Code Musing

The following project outlines a RCX code muse whereby the EW robot will advance forward for another 10 seconds after a 5-second halt. The function will be executed after a touch sensor press and release event.

Test Procedure Use the following steps to complete this procedure:

1. Attach a touch sensor to input port 1, using a LEGO electric wire to the EW robot. See Figure 9-14.
2. Using LEGO Mindstorms RIS RCX code version 1.0, 1.5 or 2.0 software, program the EW robot for an additional *10 seconds on* forward after *5 seconds halt* function. See Figure 9-16.
3. Download the code into program slot 2.

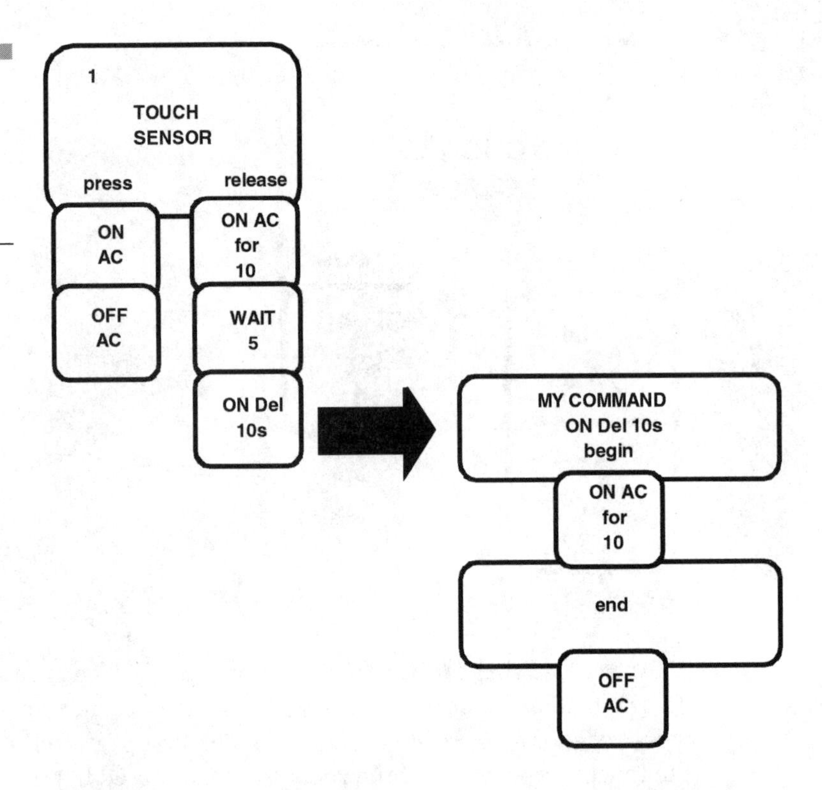

Figure 9-16 RCX code for additional 10-second forward motion after 5-second halt

4. Push *RUN* on the LEGO P-Brick.
5. Press and release the touch sensor. The robot should begin walking forward for 10 seconds, then stop. A 5-second halt period will be initiated followed by a 10-second walk forward function. If not, stop the program and check the code for correct I/O ports, timing values, and connector orientation. After correcting the problem or problems, repeat step 5.
6. Press and release the touch sensor. The EW robot should begin walking forward for 10 seconds, then stop. It will rest for 5 seconds, then resume walking forward for an additional 10 seconds before coming to a final stop. Each press and release action to the touch sensor will enable the robot to walk-halt-walk as specified in the embedded code.

RCX Code Musing Summary Writing code for what-if scenarios can be an enjoyable way to explore robotics. The test procedure outlined is an example of such technical musing. By adding a *wait 5* instruction along with a command block of a *10 seconds on* forward walking command, you had the opportunity to explore enhancing the RCX code through experimental play. Writing software should not have to be a brain-drain activity. The RCX code programming language enables you to have fun in creative invention and exploration through your musings.

TECH NOTE: *The modular design of the walking robot's platform mechanism is capable of being mounted to the LEGO RCX. All the gears and leg-foot assemblies and supports enable the P-Brick to mount on top of the mechanical package that will thrust it into forward or reverse modes of direction. The Scout P-Brick can also be used with this design as well. To mount it on top of the platform mechanism, move the brick slowly until it stops. Lightly press the Scout onto the mechanism, to ensure a secure fit to the electromechanical package.*

TECH NOTE: *The RCX code is a great low-fidelity prototype programming language for the quick development of robotics concepts. Initially, writing the robotics application in RCX code enables the mechatronics concept to be validated in a convenient and quick manner. After the concept has be validated, the RCX code can be translated into a robust programming language like Not Quite C* (NQC), *Java, or Robolab software for further enhancements and function expandability.*

RCX Code and the Palm Handheld Computer

The RCX code embedded in slots 1 and 2 of the LEGO P-Brick can be wirelessly awakened using the Palm handheld computer and the infrared tower. In Chapter 3, "Palm Diagnostics and Robots," an in-depth discussion on how to use the Palm handheld computer with an infrared tower was explained. The same project activities can be used to experiment with the EW robot-embedded RCX code functions. Therefore, a sample project using the test procedures outlined in Chapter 3 will be presented. Note: Information on the serial cable connections and software installation should be reviewed in Chapter 3, prior to performing the lab test procedure.

RCX Code Exploratorium Using the Palm Handheld Computer
Use the following steps to complete this procedure:

1. Attach the Palm handheld computer, using the cradle, to the LEGO Mindstorms infrared tower, as shown in Figure 3-13.
2. Place the infrared tower in front of the EW robot. Turn on the LEGO P-Brick.
3. In the center of the *BrickRemote* screen, touch your stylus to the left arrow. The left motor should be activated, turning the EW robot slightly to the right.
4. Touch the right arrow on the application screen with the stylus. The left motor should be activated, but turning in the reverse direction. The EW robot should turn slightly to the left to its original position.
5. Touch the up arrow on the application screen with the stylus. The right motor should be activated, turning the EW robot slightly to the left.
6. Touch the down arrow on the application screen with the stylus. The right motor should be activated, but turning in the reverse direction. The EW robot should turn slightly to the right to its original position.
7. At the top left-hand corner of the application screen, touch the stylus to the *Task On* down arrow. A drop-down box will appear. Touch *Task Off* then back to *Task On*.
8. Touch the *Pgm* arrow and a drop-down box will appear. With the stylus, select *Pgm 1*.
9. With the stylus, touch the number *0*, underneath *Task On*. Program 1 should be activated on the LEGO P-Brick, as indicated by the little man running on the LCD screen.
10. Touch with the stylus to change from *Task On* to *Task Off*.
11. With the stylus, touch the number *0*, underneath *Task Off*. Program 1 should have stopped running, as indicated by the little man standing on the LCD screen.
12. Repeat steps 7 through 11 to activate and experiment with program 2, using the Palm handheld computer.

TECH NOTE: *If you're looking for some cool and sophisticated walking robots, check out Colins' LEGO Robot page. The web page has three prototype walking robots built from LEGO Technic and Mindstorms RIS kits. Each prototype robot has clear colored pictures of these fascinating walking bots, and video clips show the robots movements. Colins' Web page address is* `http://filebox.vt.edu/users/cocampb2/LEGO/`.

LEGO Mindstorms Meets the 486 Notebook Computer

Old technology can sometimes find a new twist with the aid of the LEGO Mindstorms P-Brick. As discussed in Chapters 7 and 8 on the subject of hybrid robots, old computers can find new applications by using a bridge to enable the old technology to work with the new. For example, the 486 notebook computer that I had since the earlier '90s has unique software for turning a graphics calculator into a portable data acquisition system. The *Texas Instruments* (TI) Graph Link software enables any TI-based graphics calculator to be transformed into a scientific instrument for collecting and storing data. The collected data can be retrieved using the TI Graphics Link software to extract and display the data on the screen or to store the data onto the hard-drive or floppy disk for future analysis.

The method of using the old notebook with the sophisticated LEGO P-Brick kept running around in my head until a flash of inspiration was bestowed upon me. The TI *Calculator Laboratory* (CBL) could be used with the motion detector to record the EW robot's distance-versus-time profile. The walking robot could be wirelessly put into motion using the either Palm handheld computer or the infrared remote control. The motion detector would record the robot's movement, recording distance per time and sending it to the graphics calculator via the CBL. After the data had been stored inside the calculator, its numerical contents could be retrieved from the calculation tool using the 486 notebook computer. Using an interface cable, the calculator's data would be sent to the old notebook computer's serial port. The TI software would then be able to display the contents directly onto the screen.

The 486 notebook computer has been resurrected from the dead. Figure 9-17 shows a typical receive session of data from the TI-82 graphics calculator to the 486 notebook computer.

The CBL, when used as a signal conditioner, is capable of reading digital and analog signals. It has a sonic input for the motion detector as well. The graphics calculator serves as an HMI controller enabling you the ability to specify the appropriate physics data on the robot.

In this example, the TI-82 graphics was setup to detect and record the movements of the EW robot using the motion feature on the programmable unit. With the data displayed and stored in the calculator, it can be downloaded into the 486 notebook. With a little ingenuity and the appropriate software, the 486 notebook computer was resurrected by a little yellow LEGO brick.

Figure 9-17 Data being received from a TI-82 graphics calculator to the author's 486 notebook computer

Vernier Software and Technology and the EW Robot

Vernier Software & Technology is an educational company involved in developing software products for the Mac OS, Windows, and XP computer platforms. The company was started in 1981, in the garage of Dave Vernier, a high school physics teacher. In addition to developing software products for science education, the company also designs and manufactures sensors, digital control units, and laboratory manuals for science education.

The CBL is a legacy product, superceded by the sleeker CBL 2 module made by Texas Instruments. The CBL is a portable data acquisition unit for collecting math and science data. The motion detector is a small portable ultrasound unit capable of collecting data on a variety of movement oriented events like the following:

- Students walking toward and away from the sensor
- Carts rolling on a table or track
- Objects in simple harmonic motion, such as a weight hanging on a spring
- Pendulum motions
- Air track gliders
- Objects dropped or tossed upward

- A bouncing object
- LEGO Mindstorms EW robots

The motion detector is connected to the CBL using the sonic input port on the side of the unit. The physics software that comes on the 3.5-inch floppy has the CBL.82 (Calculator-based Laboratory TI-82 calculator version) program for executing the motion detection program. The motion detection software, in conjunction with the Palm handheld computer or infrared remote control, can be used to record the EW robot's distance-versus-time profile. This profile can be used as a design and development tool to improve the robot's walking platform mechanism by analyzing the peaks of the profile and determining which part of the mechanism needs improvement.

The following lab project will provide a test procedure for collecting such data from the EW robot.

Loading the CBL.82 Software onto a TI Graphics Calculator

Before proceeding with the lab project, the CBL.82 software needs to be installed on a TI graphics calculator. The following procedure provides the installation steps for the CBL.82 software onto the graphics calculator:

1. Open the TI Graph Link software by double-clicking the icon on the desktop of the notebook or PC computer. The TI-82 keypad with the *Untitled1* window will appear on the screen. See Figure 9-18.
2. Turn on the TI-82 calculator and press the *2nd-then LINK* key on the calculator.
3. With the right arrow key, move the cursor on the calculator to *RECEIVE*.
4. Push the enter key. The calculator should be displaying *Waiting . . .* on the screen.
5. On the notebook select *LINK* from the main menu. From the pull-down screen, select *Send*.
6. Select the *CBL.82P* from the *Exploring Physics and Math with CBL System* floppy. Note: CBL.82P program can be found on the CD-ROM that accompanies this book.
7. The CBL program will be displayed on the TI graphics calculator.

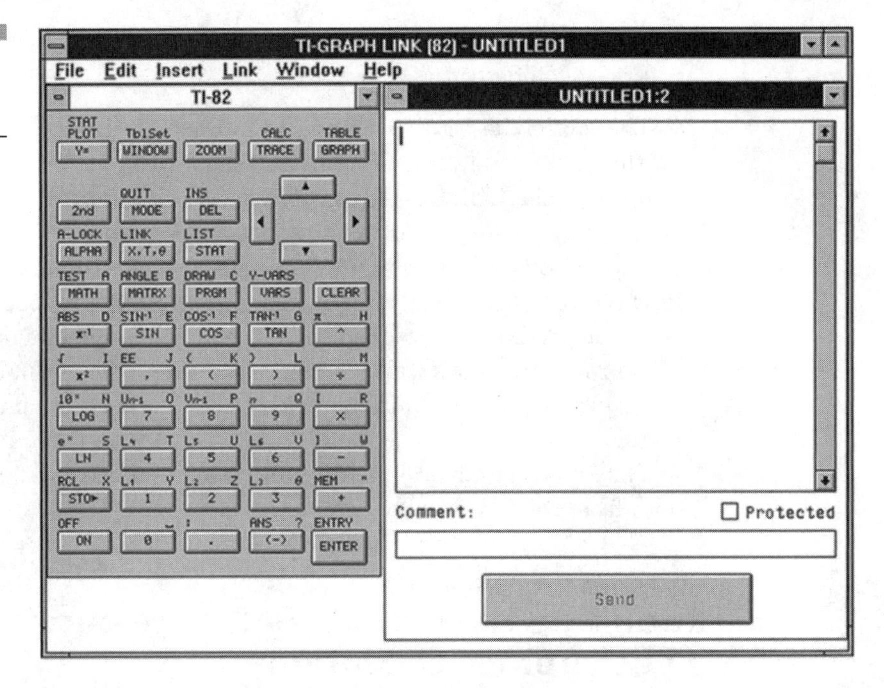

Figure 9-18 The TI Graph Link software IDE

Once the software is installed on the TI-82 graphics calculator, the EW robot's distance-versus-time profile can be obtained.

The TI Graph Link IDE Window Distance-Versus-Time Profile Laboratory Project

The distance-versus-time profile of the EW robot can now be obtained using the portable data acquisition equipment of the CBL, the motion detector, a Palm handheld computer or infrared remote control. The profile of the robot walking away from the motion detector will be the opposite of the ultrasonic characteristics of the robot walking towards it.

The BOM for this project is as follows:

One CBL or CBL 2

One Vernier motion detector

One Palm handheld computer

One Palm cradle

One LEGO Mindstorms serial cable

One null modem adapter

One gender changer (male-to-male)

One LEGO Mindstorms infrared tower

One LEGO Mindstorms infrared remote control

Assembly and Test Procedure Use the following steps to accomplish this procedure:

1. Attach the Vernier Motion Detector to the CBL or CBL 2 using the sonic input port. See Figure 9-19.
2. Attach the TI graphics calculator to the CBL using the small connecting cable. See Figure 9-20.
3. Place the EW robot in front of the motion detector, pointing away from the motion detector. See Figure 9-21.
4. Assemble the null modem adapter, gender changer, and Palm cradle serial cable, according to the instruction in Chapter 3, pages 100–102 .
5. Place the Palm handheld computer into the cradle. Turn on the unit and run the *LRemote* application.
6. Turn on the EW robot.
7. Turn on the CBL and the TI graphics calculator.

Figure 9-19
Attaching the motion detector to the CBL using the sonic input port

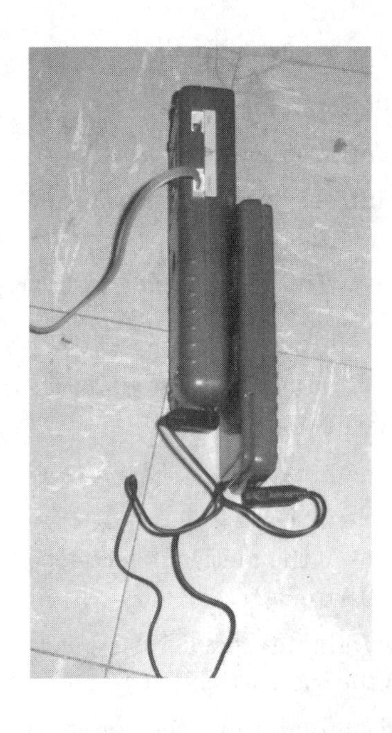

Figure 9-20 Attaching the TI graphics calculator to the CBL using a connecting cable

Figure 9-21 The EW robot in front of the motion detector

8. Press the *prgm* key on the calculator.
9. Press the enter key, after the *prgrm CBL* execute message.
10. Select *motion* from the main menu. Press enter. The motion program will be displayed on the calculator. Follow the prompts per the instructions on the calculator display.

11. The *Enter Collection Time in Seconds ?* prompt will appear on the calculator screen. Enter *10* using the calculator keys followed by the enter key.
12. Press the enter key again. The green LED on the motion detector will illuminate and the measuring device will emit a series of beeps.
13. Run the P1 program by tapping it using the stylus on the Palm handheld computer. The EW robot should walk away from the motion detector. It will continue to walk forward for 10 seconds, then stop.
14. After the unit has collected the data, the plot program will appear on the screen. Press the enter key.
15. Select the *Distance - Time* from the plot options. The calculator will then begin to plot the data.
16. Remove the calculator from the CBL by disconnecting the connecting cable.
17. Connect the TI graphics calculator to the notebook (486 for the author) or desktop PC.
18. Open the TI Graph Link software.
19. From the main menu, select *LINK* and then *Get LCD from TI-82*, followed by *Clipboard*. A *Receive* window will appear on the screen.
20. Click *Receive*. The *LCD SAVE* window will appear on the screen with a calculator plot. Save the image to clipboard. Using a paint program or Excel, the image can be pasted onto the application window. Figure 9-22 shows the data the author collected for the robot walking away from the motion detector.

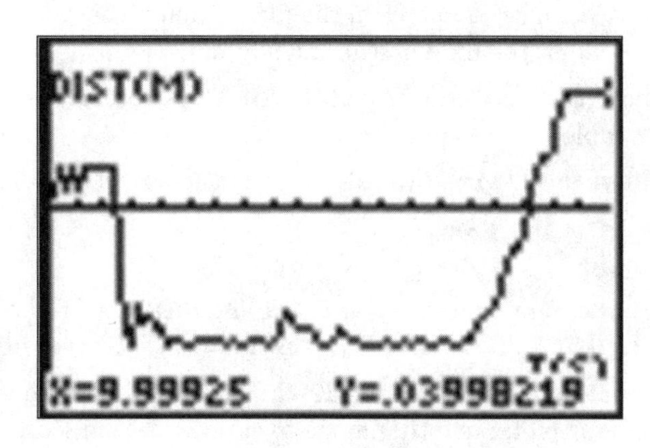

Figure 9-22 The author's data, collected and plotted showing the distance-versus-time for the robot walking away from the motion detector

The Distance-Versus-Time Profile: Robot Walking Toward the Motion Detector Here is the laboratory data collected for the robot walking towards the motion detector. Use the following steps to collect forward motion data of the robot using the following procedure:

1. Place the EW robot in front of the motion detector, pointing toward the motion detector.
2. Place the Palm handheld computer into the cradle. Turn on the unit and run the *LRemote* application.
3. Turn on the EW robot.
4. Turn on the CBL and the TI graphics calculator.
5. Press the *prgm* key on the calculator.
6. Press the enter key after the *prgrm CBL* execute message.
7. Select motion from the main menu. Press enter. The motion program will be displayed on the calculator. Follow the prompts per the instructions on the calculator display.
8. The *Enter Collection Time in Seconds ?* prompt will appear on the calculator screen. Enter *10* using the calculator keys followed by the enter key.
9. Press the enter key again. The green LED on the motion detector will illuminate and the measuring device will emit a series of beeps.
10. Run the P1 program by tapping it using the stylus on the Palm handheld computer. The EW robot should walk toward the motion detector. It will continue to walk forward for 10 seconds, then stop.
11. After the unit has collected the data, the plot program will appear on the screen. Press the enter key.
12. Select *Distance - Time* from the plot options. The calculator will then begin to plot the data on the calculator.
13. Remove the calculator from the CBL by disconnecting the TI Graph Link cable.
14. Connect the TI graphics calculator to the notebook (486 for the author) or desktop PC.
15. Open the TI Graph Link software.
16. From the main menu, select *LINK* and *then Get LCD from TI- 82*, followed by *Clipboard*. A *Receive* window will appear on the screen.
17. Click *Receive*. The *LCD SAVE* window will appear on the screen, with calculator plot. Save the image to clipboard. Using a paint program or

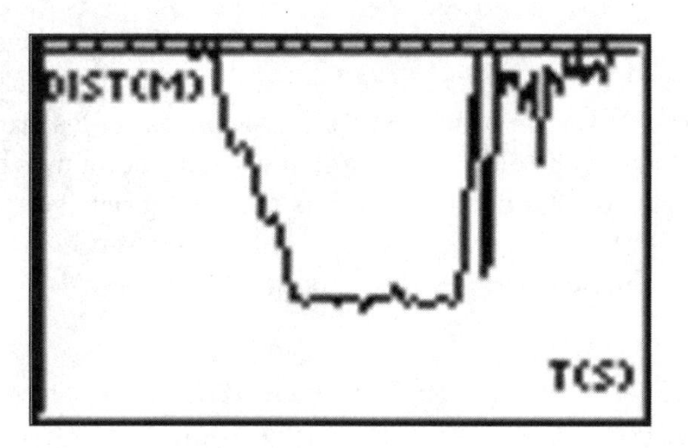

Figure 9-23 The author's data, collected and plotted showing the distance-versus-time for the robot walking toward the motion detector

Excel, the image can be pasted onto the application window for viewing. Figure 9-23 shows the data the author collected for the robot walking toward motion detector.

Distance-Versus-Time Profile Laboratory Project Summary

This laboratory project provided a unique way of measuring the dynamics of the EW robot, by capturing its distance-versus-time profile. The laboratory investigation looked at two motion profiles of the robot: walking away from the motion detector, and walking towards the motion detector.

The first laboratory project enabled the amateur roboticist to capture data on the walking robot moving away from the motion detector. The EW robot was placed in front of the motion detector and remotely activated using a Palm handheld computer. The L-Remote software installed on the handheld computer activated the smart walking robot program stored in P1 slot of the LEGO P-Brick. The robot advanced away from the motion detector, which was set up to record 10 seconds worth of movement data. The code embedded in the LEGO P-Brick enabled the robot to walk forward for 10 seconds, then stop. This beginning motion function is the calibration phase for initializing the touch sensor with a 10-second shutoff feature after releasing the small electric switch. After the TI graphics calculator recorded the robot's movement, the information was downloaded to a notebook computer (486 machine) for viewing of the profile.

The second part of the laboratory project consisted of repeating the motion measurement procedure but for recording the robot walking toward the motion detector for a 10-second window. The data was collected on the graphics calculator and the profile viewed on the notebook computer. Looking at the graphs in Figures 9-22 and 9-23, the profiles are typical of objects moving away and toward the motion sensor. Therefore, this measurement technique could be used as an experimental security system, where the movement of the robot can be recorded and viewed in order to analyze the bot's surveillance distance per a specified time.

TECH NOTE: *Vernier Software & Technology is a company driven by science education and computer-aided laboratory experimental methods. They have an extensive online resource of curriculum and laboratory experiments in the scientific disciplines of physics, physical science, chemistry, biology, and water quality. The physics measurement experiment outlined in the laboratory project can be implemented using the legacy CBL or CBL 2. Check out their web site at* `www.vernier.com/` *for additional sensors and robotics project ideas.*

CBL-DCU for EW Robot Control

On page 149 of *LEGO Mindstorms Interfacing*, several hands-on laboratory projects discuss the CBL-DCU for robot diagnostics. The *Vernier Digital Control Unit* (DCU) is a small box with a short cable that plugs into a digital connector on the CBL. The DCU can electronically switch six electrical/electromechanical loads independently of each other. The output current from each electronic driver is 600 mA and is managed using a ULN2003A transistor array integrated circuit. The DCU communicates with the CBL or CBL 2 through the short cable that plugs into the digital-out connector specified earlier. An external power supply or batteries are required to activate the DCU.

Inside the DCU are visible LEDs that indicate the status of the six output lines. A socket is provided for connecting electronic devices that you have built. A cable with bare wires on one end is provided for your first projects. The DCU provides an alternative control method for operating the EW robot using a graphics calculator as the HMI device. Figure 9-24 shows a picture of the DCU.

The following project will enable you to experiment with the DCU in a robotics control application.

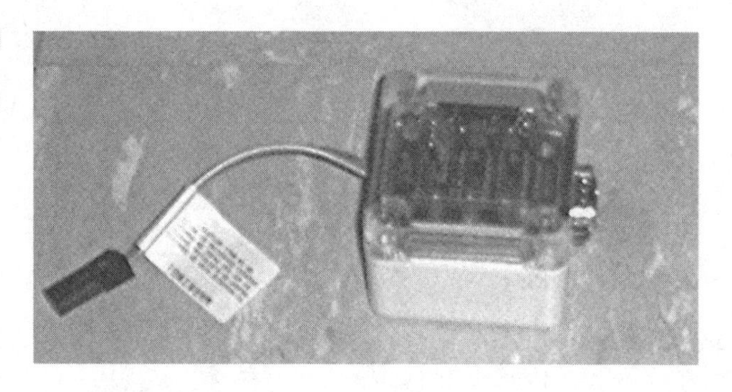

Figure 9-24
The DCU is a small electronic switching interface box with six output drivers for electronic control of electrical or electromechanical loads.

DCU Basics

By using the calculator as the main processor that manages the I/O for the target system, the DCU enables the computational device to have a real world interface for controlling external devices. The DCU is a six-output channel switcher capable of driving electrical, electronic, and electromechanical loads with a sourcing current of 600 mA. The output voltage is +6.08V DC, which makes it a sufficient electromotive force driver capable of switching externally-connected devices on and off.

From an electrical standpoint of view, to wire an external circuit or electrical load to the solid-state interface box is quite easy. The key is to identify the output circuits that are mapped with the corresponding LED inside of the plastic box. The electrical interface that enables the box to interface with the real world is a nine-pin D-shell connector. The output voltage signals are tied to each of the designated pins. The pins are also associated with one of the six output channels as well. Therefore, the following pins and their respective output channels represented by an LED are as follows:

- Pin 1= LED 1
- Pin 2= LED 2
- Pin 3 = LED 3
- Pin 4 = LED 4
- Pin 5 = LED 5
- Pin 6 = LED 6
- Pin 7 = Ground
- Pin 8 = Ground
- Pin 9 = +6.08V DC

Table 9-1

Pins, color-coded wire, LEDs, and output voltage for the multiconductor cable

Pin number	Color-coded wire	LED	Output voltage
1	Black	1	+6.08V DCDC
2	Brown	1	+6.08V DCDC
3	White	3	+6.08V DCDC
4	Blue	4	+6.08VD CDC
5	Orange	5	+6.08V DCDC
6	Red	6	+6.08VDC DC
7	Yellow	NA	0 VDC DC
8	Gray	NA	0 VDC DC
9	Green	+6.08VDC DC	+6.08V DCDC

Vernier Software & Technology has made it convenient for the user of the DCU to have better access to this signal by providing a multiconductor cable which has each wire color-coded. Table 9-1 shows the pins, color-coded wire, and the LED outputs for the multiconductor cable.

Pins 7 and 8 can be wired together, since they both provide common grounds for the output driver circuits of the DCU. Pin 9 provides a constant +6.08V DC. This constant output signal makes a great reference voltage for externally-wired electronic circuits.

The following lab project will demonstrate how to use the DCU and its control software for managing the six output voltage channels.

DCU Laboratory Project

This laboratory project will enable you to use the six output channels and validate the Table 1. This project will enable you to become familiar with the control software as well.

The BOM for the project is as follows:

One CBL or CBL 2

One Radio Shack Electronics Learning Lab or equivalent circuit prototyping station

One TI graphics calculator (Models -73, -82, -83, -83+, -86, -89+, -92, and -92+)

One DCU (Vernier Cat. No. DCU-CBL)

120V AC-to-6V DC output CBL or CBL 2 power supply (Vernier Software Cat. No. IPS)

Assembly and Test Procedure The following procedure outlines the installation of the control software required to operate the DCU with a TI graphics calculator:

1. Open the TI Graph Link software by double-clicking the icon on the desktop of the notebook or PC computer. The TI-82 keypad with the *Untitled1* window will appear on the screen. See Figure 9-18.
2. Turn on the TI-82 calculator and press the *2nd-then LINK* key on the calculator.
3. Move cursor to *RECEIVE* on the calculator with the right arrow key.
4. Push the enter key. The calculator should be displaying *Waiting . . .* on the screen.
5. On the notebook, select *LINK* from the main menu. From the pull down screen, select *Send*.
6. Select *DCUINIT* from the accompanying CD-ROM. The CBL program will be displayed on the TI Graphics Calculator.
7. Repeat the steps, installing the *DCUTOGGL* program onto the graphics calculator.
8. Attach the DCU into the Digiout port of the CBL or CBL 2.
9. Attach the multiconductor cable to the nine-pin D-shell connector of the DCU.
10. Attach the CBL or CBL 2 to the TI graphics calculator, using the connecting cable.
11. Plug in the AC adapter into the DCU
12. Turn on the calculator and the CBL. The green LED should illuminate.
13. Take the bare conductors of the multicable and push them through the holes of the Radio Shack Electronics Learning Lab breadboard or equivalent circuit prototyping. Use the convention where pin 1's color is placed first on the breadboard. Use Table 9-1 to layout the wiring of the multiconductor cable to the circuit breadboard.
14. Run the *DCUINIT* software first, using the *prgm* key on the calculator.
15. Run the *DCTOGGL* software. Text will appear explaining how to use the software for electronic control switching applications.

16. Take a DMM and set it up to read DC volts. Place the black test lead of the DMM to the yellow wire. The red test lead should go to the black wire.
17. Press the number *1* on the calculator keypad. The LED 1 on DCU should be on.
18. Press the number *1* on the calculator keypad again. The LED 1 on the DCU should be off.
19. Repeat steps 17 through 18 for each multiconductor cable until Table 1 has been validated.

This exercise will help the amateur roboticist to become familiar with the DCU and understand the output voltages produced by the DCU's transistor array drivers inside of smart interface box.

A DCU-Controlled Walking Robot

The diagnostics project discussed on page 150 of *LEGO Mindstorms Interfacing* will be presented, but modified to illustrate the remote control interfacing technique of operating a walking robot with the DCU and a graphics calculator.

The BOM for this procedure is as follows:

One 6 to 9V DC electromechanical relay
One 1N4001 silicon diode or equivalent
One 1 KΩ 1/4-W resistor
One CBL or CBL 2
One DCU (Vernier Cat. No. DCU-CBL)
One TI graphics calculator (Models -73, -82, -83, -83+, -86, -89+, -92, or -92+)
One Radio Shack Electronics Learning Lab breadboard or equivalent circuit prototyping station
One modified LEGO electric wire
120V AC-to-6V DC output CBL or CBL 2 power supply (Vernier Software Cat. No. IPS)

Assembly and Test Procedure Use the following steps to complete this procedure:

1. Wire the DCU output electromechanical switch on a Radio Shack Electronics Learning Lab breadboard or equivalent circuit prototyping station using Figure 9-25.
2. Attach the modified LEGO electric wire to input port 1 of the EW robot.

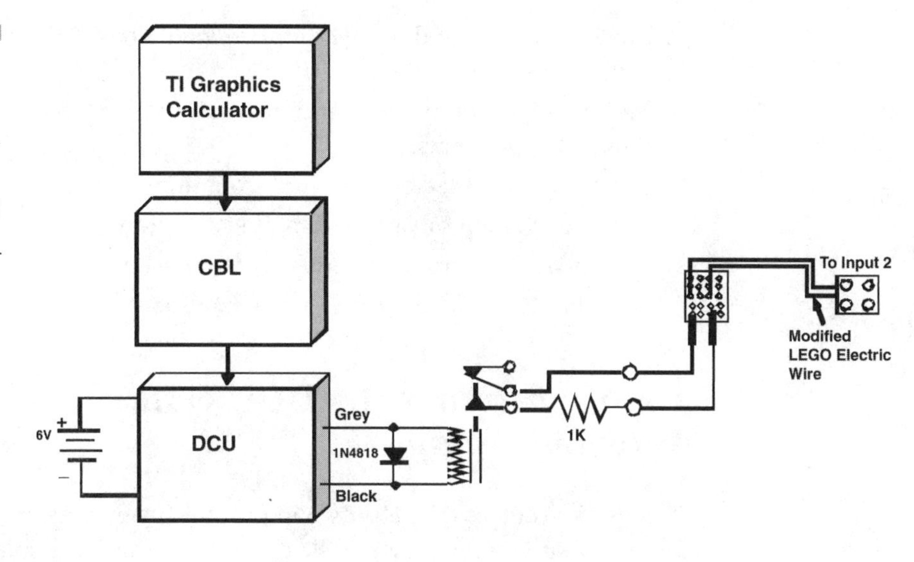

Figure 9-25 Functional block diagram and circuit interface for the DCU-controlled walking robot

3. Install the *DCUINIT* and *DCUTOGGL* programs into the TI graphics calculator using the following procedure:
   a. Open the TI Graph Link software by double-clicking the icon on the desktop of the notebook or PC computer. The TI-82 keypad with the *Untitled1* window will appear on the screen. See Figure 9-18.
   b. Turn on the TI-82 calculator and press the *2nd*-then *LINK* key on the calculator.
   c. Move cursor on the calculator to *RECEIVE*, using the right arrow key.
   d. Push the enter key. The calculator should be displaying *Waiting . . .* on the screen.
   e. On the notebook select *LINK* from the main menu. From the pull-down screen, select *Send*.
   f. Select the *DCUINIT* from the accompanying CD-ROM.
   g. The CBL program will be displayed on the TI graphics calculator. Repeat the steps, installing the *DCUTOGGL* program onto the graphics calculator.
4. Run both the *DCUINIT* and *DCUTOGGL* programs in the TI graphics calculator.
5. Run the walking robot program in slot location 1.
6. Press *1* on the calculator. The status should be *ON*. The DCU LED should be on. The robot should move forward for 10 seconds and then stop.

7. Press *1* on the calculator. The status should be *OFF*. The EW robot should be stopped as well.
8. Repeat steps 6 and 7 several times, noting the response of the robot to the input actuations.
9. Switch the modified electric wire from input port 1 to input port 2.
10. Run the walking robot program in slot location 1.
11. Repeat steps 6 through 8 for the new embedded code.
12. Stop the RCX code program on the EW robot.

Characterizing the DCU Control Interface Signal

This measurement lab activity will graphically display the 0.2V and 4.99V signals associated with the DCU control interface switching the close and open signals respectively. Connect the DMM with PC interface to the serial port of the desktop or notebook computer.

Test Procedure Use the following steps to complete this procedure:

1. Turn on the DMM and run the ScopeView software.
2. Attach the test leads from the DMM to the modified electric wires attached to the input port 1 of the EW robot.
3. Run the P1 program embedded inside the LEGO P-Brick.
4. Start a logging session within the ScopeView control panel window. Change the trigger value to *4.98* and the units/div to *1*.
5. Click *Record* on the software control panel. The *Save File* dialog box should appear on the screen.
6. Select the appropriate drive for saving the data. The author used the A: drive to save data onto a 3.5-inch floppy.
7. Click the *Filename:* box and type *ew_robot.txt*. Click OK.
8. Click *Scope*.
9. Run both the *DCUINIT* and *DCUTOGGL* programs in the TI graphics calculator.
10. Run the walking robot program in slot location 1.
11. Press *1* on the calculator. The status should be *ON*. The DCU LED should be on. The robot should move forward for 10 seconds and then stop.
12. Click *Run* on the ScopeView output window to start a data-logging session.

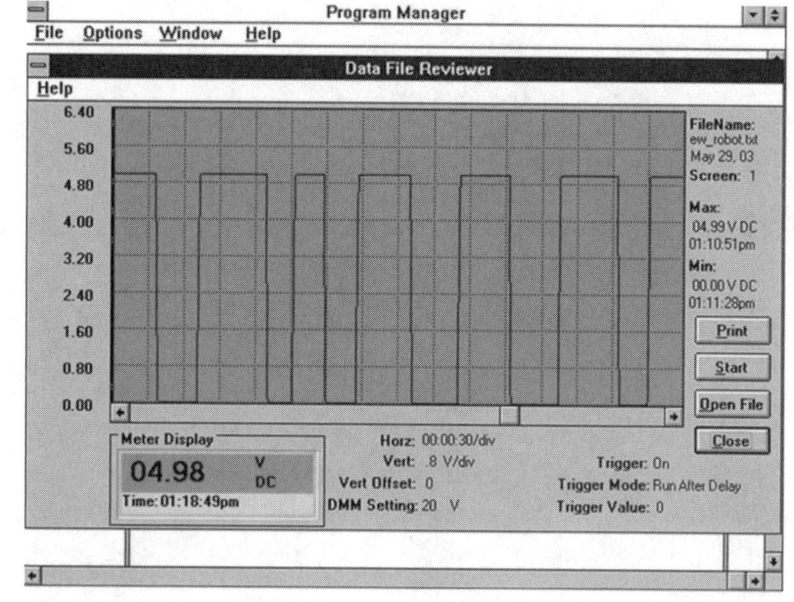

Figure 9-26 Output plot of the DCU-controlled interface signal collected with the ScopeView software

13. The software should start plotting a trace on the screen. Let the signal run for about 20 seconds.
14. Press *1* on the calculator. The status should be *OFF*. The EW robot should be stopped.
15. Repeat this for on/off DCU switching with the graphics calculator four times.
16. Click *Close* on the control panel to stop the data-logging session.

Playing Back the Logged Data Use the following steps to play back the logged data:

1. Click *Close* on the control panel GUI. The main menu should be displayed on the screen.
2. Click *Playback* on the main menu panel. The data file reviewer GUI should be displayed.
3. Click *Open File*. The *Read File* dialog box should be displayed.
4. Click the *ew_robot.txt* file. Click *OK*. The data file reviewer should start plotting the data on the screen. Figure 9-26 shows the results of the DCU-controlled interface signal data logged within the ScopeView software.
5. Plot the ASCII data from the *ew_robot.txt* file in Excel or an equivalent spreadsheet application.

CHAPTER 10

Audio Bot Concepts

The LEGO *programmable brick* (P-Brick) and the Scout P-Brick are capable of generating sounds and tones using software and a piezo-electric element. The software is used to create various sounds and tones using programming synthesis techniques. An embedded hardware circuit is used for amplification and driving the periodic signal to the piezo-electric buzzer.

This method of creating sound and tones internally and externally is an interesting topic in mechatronics-based robots. The audible signals provide information to the amateur roboticist about the output status of a mechanical function of a robot or the input event regarding sensory data it is receiving from the environment. Sound and tone generation can also be used in robot self-diagnostics or as an intruder-alert signal for an autonomous robotic security system.

This chapter will explore concept applications and techniques in creating sound using different tones. We will experiment with the techniques for creating sound and tones using internal and external methods. We will discuss circuits used for the creation of tones, and how the LEGO P-Brick can generate tones using hardwired circuit interfaces. Embedded software controls using programming techniques along *with graphical user interface* (GUI)-based tools will show you how to find a specific tone through experimentation.

Audio Bots: Creating a Sound Synthesis Laboratory

The final chapter in this book will provide concepts for you to explore through *Proof of Concept* (POC) studies. The objective behind audio bots is to illustrate concepts in creating sound synthesis machines using the techniques discussed in previous chapters of the book. The audio bot is primarily an experimental concept into developing sound and tone-generating robots that provide audible feedback based on the sensor data it receives from the environment.

Creating sound and tones using either a distributive bus or locally based audio generating robotic devices will be discussed in this chapter. By combining sounds and tone patterns, the amateur roboticist is able to develop LEGO Mindstorms robots that communicate using programmed audible indicators.

The material in this chapter should provide food for thought on how to create an autonomous sound synthesis laboratory using the mechatronic-based tools created with the LEGO Mindstorms *Robotic Invention System* (RIS) to provide audible feedback from the environment.

What Does Sound Have to Do with Robots?

Professor Fred G. Martin defined a robotic system as follows:

> Robotic systems are systems that interact with people, each other, and the world around them, using sensors, actuators, communications, and control programs. [1]

Expanding on his definition, robotic systems can interact with people, each other, and the world around them using sound as an audible form of communication. Tones with varying frequencies can be used as an audible warning or information system. For example, autonomous robots that roam in different environments can be used to alert security personnel of an intruder, using loud and distinctive sounds. Digital data can be interpreted using two unique tones, one for a logic *1* or logic *0*.

The LEGO P-Brick can produce sound or tone internally or externally. Internal sound or tone generation consists of using digital registers mapped to a sound or tone lookup table embedded within the microcontroller. The table has specific frequencies and duration for the various sounds and tones you can program the mechatronics bot to use within the specified task or behavior. Pulse width modulation techniques will provide the fundamental frequency, along with a duty cycle for enabling sound or tone generation to occur.

External sound is accomplished using external circuits or a tone generator block. By activating an output port, the hardwired sound/tone generator can be switched on or off by a specific input event. Figures 10-1 and 10-2 show the concepts of generating a sound/tone using either hardwired or software techniques.

In the hardwired approach to generating sounds and tones shown in Figure 10-1, the LEGO P-Brick is the front-end component. It drives a voltage regulator circuit that steps down the +7 to -8V DC output to a +5V signal level. The Basic Stamp module can use signal level for enabling the frequency generation process to occur. The frequency signal is then coupled to an 8 Ω speaker using hardwired electrical wires. The output port of the LEGO P-Brick is activated by the reception of the appropriate input signal.

---

[1]Fred G. Martin's Robotics 1 Class web site is www.cs.uml.edu/~fredm/.

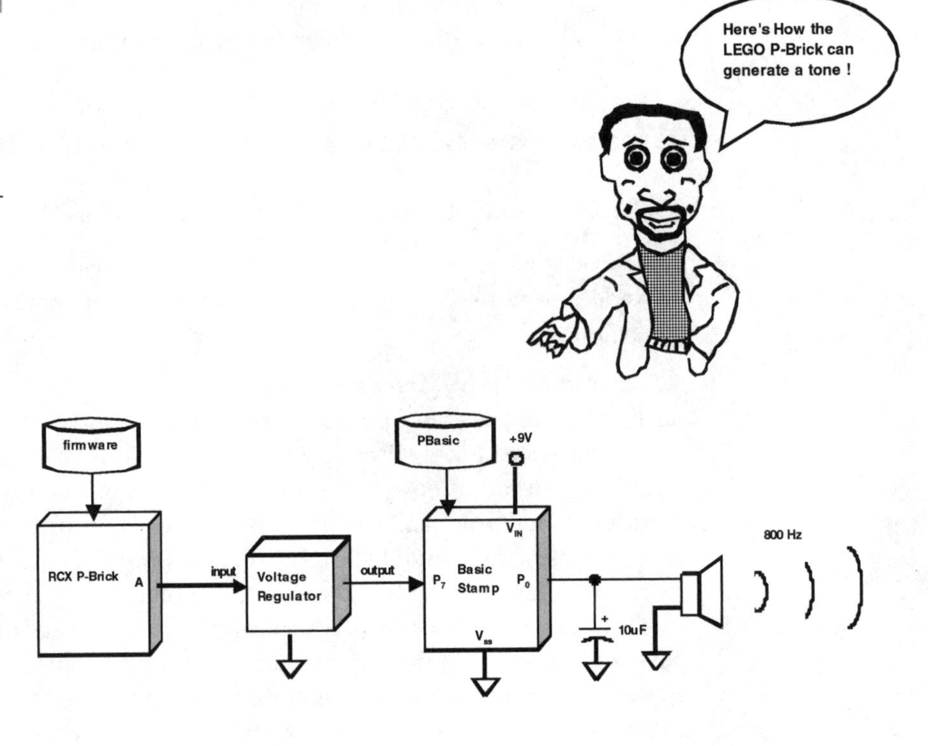

Figure 10-1 Hardwired approach to generating a tone or sound with a LEGO P-Brick

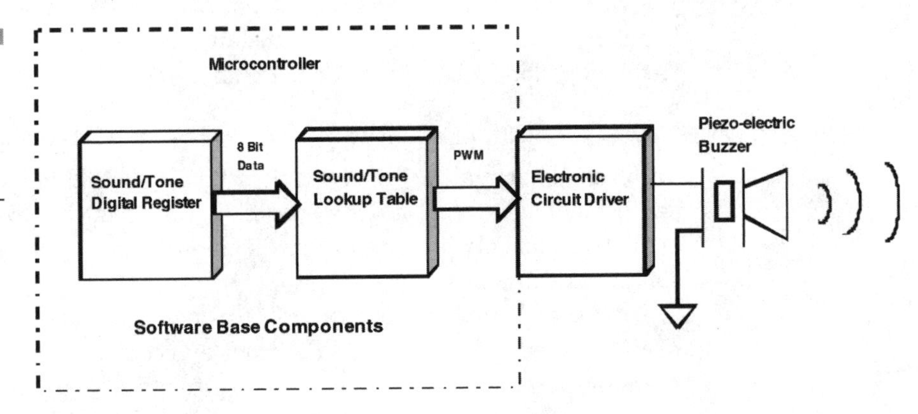

Figure 10-2 Software approach to generating a tone or sound

The data that the LEGO P-Brick can accept is digital (Boolean value), analog (percentage), or analog (raw). The type of data representation used is dependent on the robotics application you have developed.

The software approach shown in Figure 10-2 is based on using basic commands or instructions that the LEGO P-Brick *robot command explorer*

(RCX) code programming language has in its library. The commands or instructions are in binary data format. The microcontroller uses these commands for processing and producing the equivalent frequency that drives an electronics driver circuit. The electronic driver circuit, in turn, is hardwired to an internal piezo-electric buzzer that generates the audio sound or tone.

The two sound/tone generation concepts are effective in producing the audible signals for the particular robotics application you have developed. If you want a distributive robotics sound/tone control, then the Figure 10-1 concept would be the approach to use for the application.

Figure 10-2's audible concept enables a compact approach to sound/tone generation because the internal RCX code commands and instruction are used for generating LEGO Mindstorms-based audio applications, instead of additional circuits and hardware modules. The tones and beep command blocks follow the system block diagram shown in Figure 10-2.

The following discussions will explore the two fundamental approaches to sound-tone generation using software and hardwired electronic circuits. Some of the following topics will enable you to explore the development of the audio-based POC robots using some of the tools explained previously in this book.

Creativity is the main tool for exploration. Develop, experiment, and enjoy the possible applications and ideas for sound/tone generating robots.

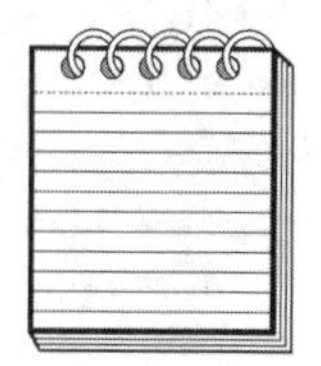

TECH NOTE: *Fred G. Martin teaches a robotics course at the University of Massachusetts-Lowell campus, where graduate students learn about design and construction of robotic systems. Course notes on various lab experiments convey his design philosophy on robotics. To find more info on Fred and his class, go to* `www.cs.uml.edu/,fredm/`.

The LEGO Beep-and-Tone Command Blocks: A Simple Software Approach to Creating Sound

A quick approach to creating sound using software is to program the internal piezo-electric buzzer of the LEGO P-Brick using the RCX code beep-and-tone command bricks. The beep command enables you to add one of six different sounds to your mechatronics-based robot. The tone command block tells the LEGO P-Brick to play a note of specific pitch, for a selected amount of time. You can choose the pitch of the note, based as a frequency in hertz (Hz). For example, to play middle F you would use 699 as the

Table 10-1

Tone command block matrix

Musical note	Very low	Low	Middle	High	Very high
C	131	262	523	1047	2093
C#	139	277	554	1109	2217
D	147	294	587	1175	2349
D#	156	311	622	1245	2489
E	165	330	660	1320	2640
F	175	350	699	1398	2797
F#	185	370	741	1482	2963
G	196	392	785	1570	3140
G#	208	416	832	1663	3326
A	220	440	880	1760	3520
A#	233	466	932	1865	3729
B	247	494	988	1976	3951

frequency. Table 10-1 shows the tone command block matrix for all musical notes and their corresponding frequencies.

TECH NOTE: *The BricxCC is an Integrated Development Environment* (IDE) *for the* Not Quite C *(NQC) programming language. Besides enabling you to write, edit, and compile NQC code, BricxCC also has a piano keyboard that can play and record tones and sounds. The sound synthesize tool is interactive and enables all the musical notes to play on the LEGO P-Brick. The piano keyboard can be found under the Tools drop-down menu of the BricxCC GUI.*

The LEGO Tone Generator Brick: A Simple Hardware Approach to Creating Sound

Attaching a LEGO tone generator brick to the output port of the P-Brick is a quick approach to creating sound using hardware. Figure 10-3 shows the small yet versatile tone generator block.

The tone generator brick is an electronic module capable of producing two tones, depending on the position of the rotating top. In the LEGO Basics kit (Cat. No. 735), the small 2x2 round brick provided audible tones

Figure 10-3 LEGO tone generator brick

for the various LEGO models shown in ideas booklet. The block was capable of producing two tones: a police siren and a fast-ringing chime.

The LEGO P-Brick can easily operate the tone generator by placing the audible block on top of the three output ports. Writing code to turn on the tone generator can be accomplished using programming languages like RCX code, NQC, Robolab or *Interactive C Version 4* (IC4). For example, the RCX code used in the EW robot Figures 9-15 and 9-16 software programming code can easily be used to drive the tone generator.

The Transistor Relay Driver Circuit: The Sound/tone Generator Applications Enabler

The transistor relay driver circuit shown in Figure 10-4 has been discussed in previous chapters of this book.

The transistor relay driver is a convenient way for the LEGO P-Brick to control external circuits and devices beside the basic electrics of small incandescent light bulbs and motors. By using any of the three outputs ports and turning them on or off depending on the input event, the same *input/output* (I/O) processing concept can control external sound/tone generator circuits, using the transistor relay driver. On page 144 of *LEGO Mindstorms Interfacing*, Figure 4-25 shows the system block diagram for the *Conveyor Robot* (C-Bot)-controlled audio tone generator. Figure 10-5 presents the system block diagram here, for your convenience.

The system block diagram has been modified to show the audio switching interface circuit components. The voltage regulator, as discussed previously in "What Does Sound Have to Do with Robots?" section of this chapter, steps down the output voltage of the LEGO P-Brick to +5V DC. This 5V signal

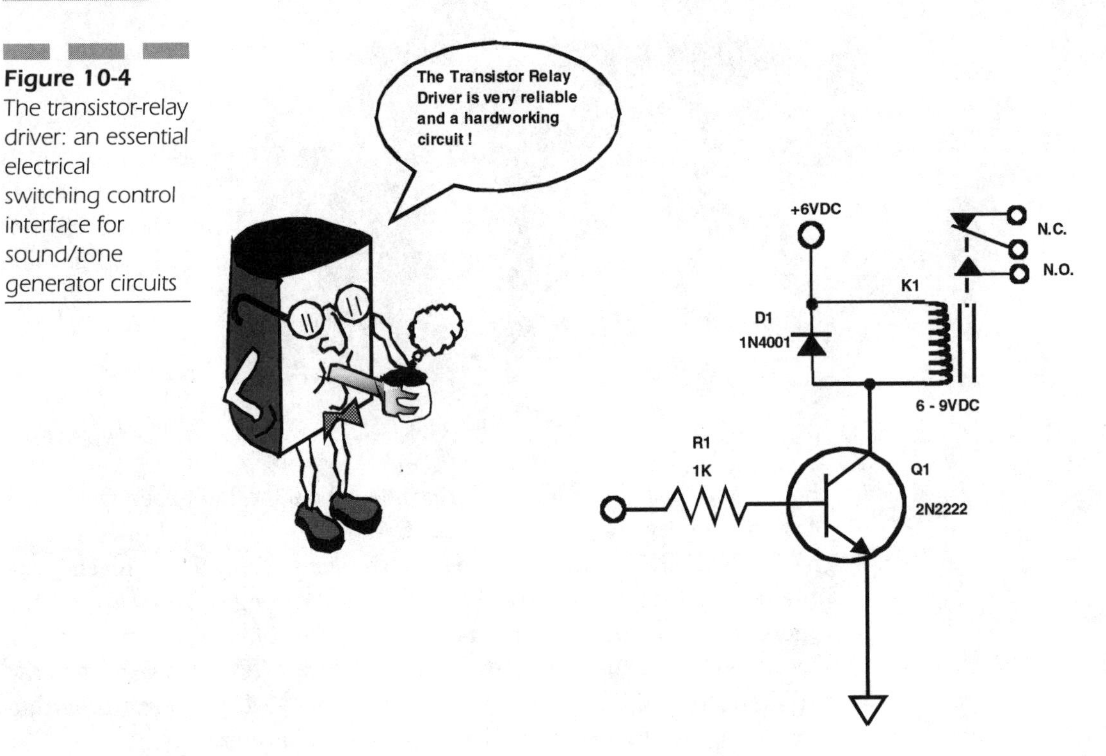

Figure 10-4 The transistor-relay driver: an essential electrical switching control interface for sound/tone generator circuits

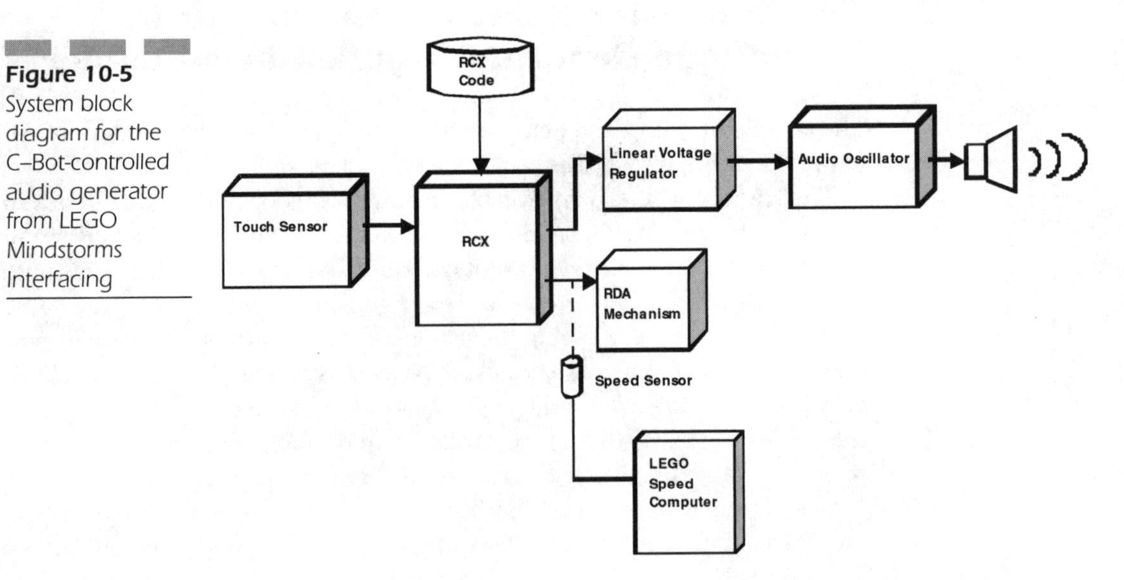

Figure 10-5 System block diagram for the C–Bot-controlled audio generator from LEGO Mindstorms Interfacing

then provides a voltage rail or bus for powering the audio oscillator circuit. The output port is activated by the appropriate input event required by the LEGO P-Brick.

The tone generator circuit used in this application enables you to change the oscillator's output frequency by selecting a different R (resistor) or C (capacitor) value. A 1 MΩ potentiometer provides frequency selection control by rotating the variable resistor clockwise or counterclockwise. The choice of tone is dependent on the robotics application you specify. Figure 10-6 shows the circuit schematic used in the C-Bot application.

In order to use the transistor relay driver to control the same tone generator circuit, remove the voltage regulator circuit and connect the base of the 2N3094 and electromechanical coil to any of the three output ports of the LEGO P-Brick. Figure 10-7 shows the circuit block diagram for wiring a transistor relay driver to the either of the three output ports.

The 1 KΩ resistor, in series with the base of the NPN transistor, limits the current flow to the switching semiconductor. As shown on the circuit block diagram, the relay's contacts are wired to the output port of the P-Brick, enabling it to switch the +8V DC source to the audio oscillator. Once the relay is energized, the contacts close, creating a closed circuit for the current to flow to the tone generator circuit. Figure 10-8 shows the two-transistor circuit responsible for managing the tone of the audible tone. This circuit block can be used as substitute for the tone generator brick discussed earlier in this chapter.

Testing the Transistor Relay Driver Circuit-controlled Audio Tone Generator A test procedure for validating the function of the transistor

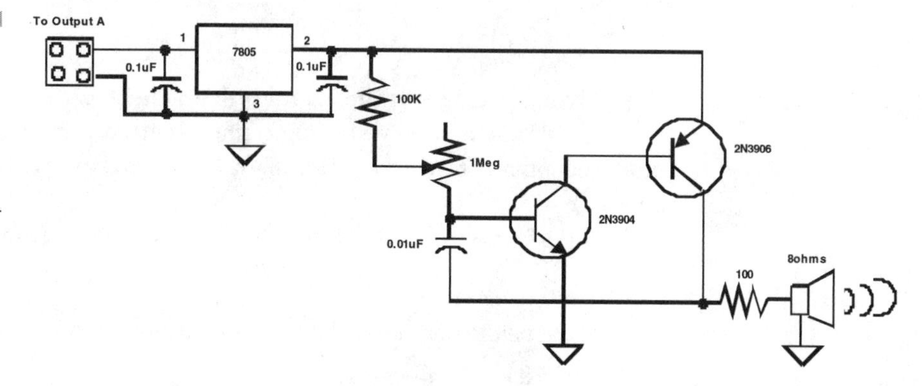

Figure 10-6 Tone generator circuit for C-Bot application, from LEGO Mindstorms Interfacing

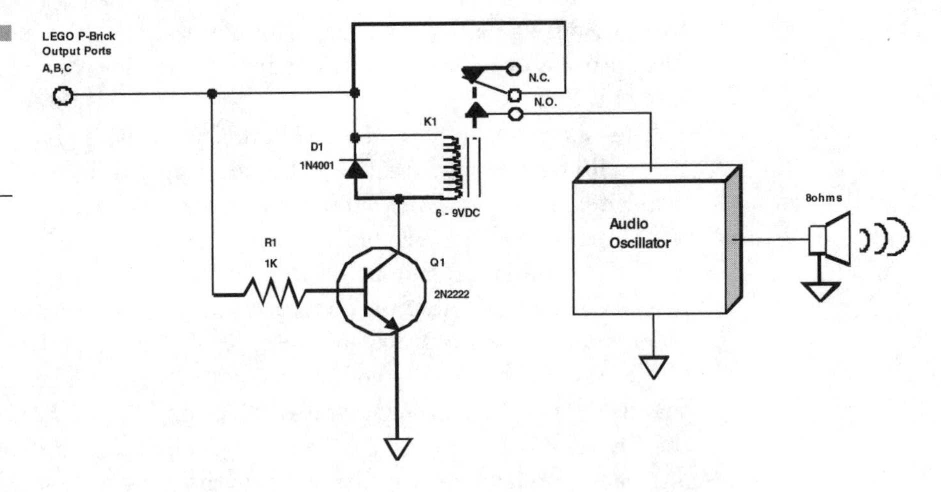

Figure 10-7 Transistor relay driver-controlled tone generator circuit

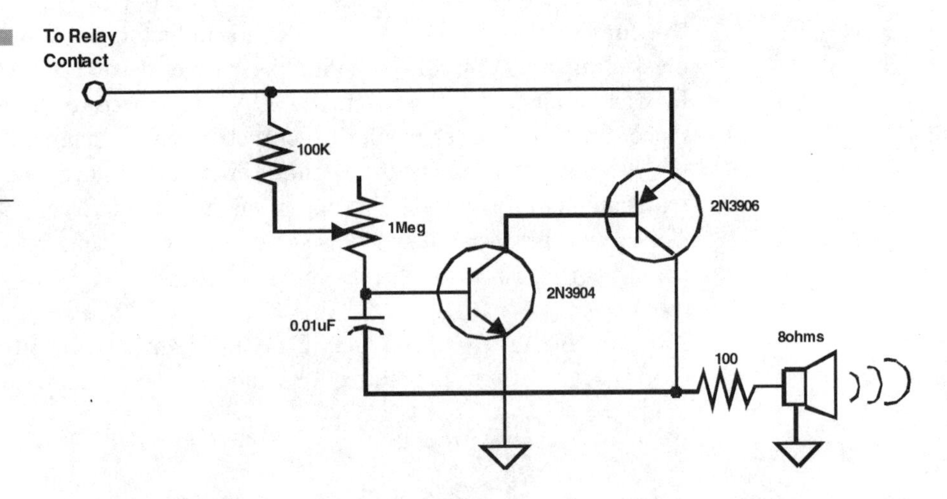

Figure 10-8 Two-transistor audio oscillator for the tone generator circuit

relay driver circuit-controlled audio tone generator is outlined. This test procedure should enable you to experiment with a variety tones and understand the operation of the transistor relay driver with an audio tone generator.

The *bill of materials* (BOM) for the procedure is as follows:

One 1 MΩ potentiometer

One 8 Ω speaker or equivalent audio output component

One 100 KΩ, 1/4-W resistor

One 1 KΩ, 1/4-W resistor

One 6 to -9V DC electromechanical relay

One 1N4001 silicon diode or equivalent

One 0.01 µf capacitor

One 2N3906 PNP transistor or equivalent

Two 2N3904 NPN transistors or equivalent

One touch sensor

One standard LEGO electric wire

One modified LEGO electric wire

One Radio Shack Electronics Learning Lab or equivalent circuit prototyping station

Assembly and Test procedure Use the following steps to complete this procedure:

1. Wire the audio tone generator, shown in Figures 10-7 and 10-8, using the Radio Shack Electronics Learning Lab or equivalent circuit prototyping station.
2. Attach a modified LEGO electric wire to output A of the P-Brick. Make sure the wire is pointing upward to ensure proper circuit operation.
3. Write a small switching program, using either RCX code, IC4, NQC, or Robolab software, so that output port A of the LEGO P-Brick turns on and stays on for 30 seconds after the touch sensor is released.
4. Connect the touch sensor to input port 1 of the LEGO P-Brick using a standard LEGO electric wire.
5. Activate the LEGO P-Brick and download the program written at step 3 into program slot 1.
6. Press *RUN* on the LEGO P-Brick. The audio tone generator circuit should sound immediately.
7. Press and release the touch sensor. The audio tone generator should sound. After 30 seconds, the audio tone generator should turn off.
8. Repeat step 7 and adjust the 1 MΩ potentiometer. Note the tone.

Testing an Audio Tone Generator Robot Using a Palm Handheld Computer The Palm handheld computer is a portable way to test the audio tone generator robot without using an input sensory device like a touch or light sensor. The assembly and test procedure outlined provides a wireless technique for validating the audio tone generator robot:

1. Attach the Palm handheld computer to the LEGO Mindstorms infrared tower using the cradle, as shown in Figure 3-13.
2. Place the infrared tower in front of LEGO P-Brick. Turn on the LEGO P-Brick.
3. Attach the audio tone generator circuit of Figure 10-7 to output port A.
4. Using the *LRemote* screen, touch your stylus to up arrow *A*. The audio tone generator should be activated.
5. With the stylus, touch the down arrow *A* on the application screen. The audio tone circuit should not be activated.
6. With the stylus, touch the P1 designator on the application screen. The audio tone generator should be activated. The tone will sound for 30 seconds and then turn off.
7. Modify the software-controlled program so that the operation will perform the function shown in Figure 9-16 of the RCX code.
8. Download the program into program slot 2.
9. With the stylus, touch the P2 designator on the application screen. The audio tone generator should be activated. The tone will sound for 10 seconds and then turn off. After 5 seconds elapse, the tone should be heard for another 10 seconds, then shutoff.
10. Repeat steps 6 and 7 to ensure proper operation of the code and the Palm handheld control function.

The Basic Stamp Sound Generator Function

As shown in Figure 10-1, the Basic Stamp drives the externally wired speakers with pulsating voltage for sound/tone generation. The two methods with which the Basic Stamp is able to produce sounds or tones are the instructions: *FREQOUT* and *PWM*. *FREQOUT* is a Basic Stamp instruction used to output a sine wave signal. The P-Basic programming style syntax is as follows:

```
FREQOUT, pin, duration, freq1, freq2
```

Pin is a variable or constant (0 through 15) that specifies the I/O pin to use. This P-Basic command puts the specific pin into output mode during the generation of tones, and remains in that state after the instruction finishes. The *duration* parameter is also a variable or constant that specifies how long in milliseconds (1 to 65,535) the output tones will remain on. The *freq1* and *freq2* are variables or constants that specify the tone's output frequency in hertz (0 to 32,767). The *FREQOUT* programmed for the audio concept in Figure 10-1 is as follows:

```
FREQOUT, 7, 1000, 800
```

Where 7 is pin P_7, 1,000 is the 1,000 milliseconds or (1 second), and the 800 value is 800 Hz. As shown in the Figure 10-9, a sine wave is produced using this P-Basic instruction.

The 800 Hz has enough output gain without connecting it to an audio amplifier. Driving an external hardwired speaker directly requires P_7, filtered by a *resistor-capacitor* (RC) circuit, to establish a clean sine wave signal. High impedance speakers may require a coupling capacitor and filter to help reduce high frequency noise.

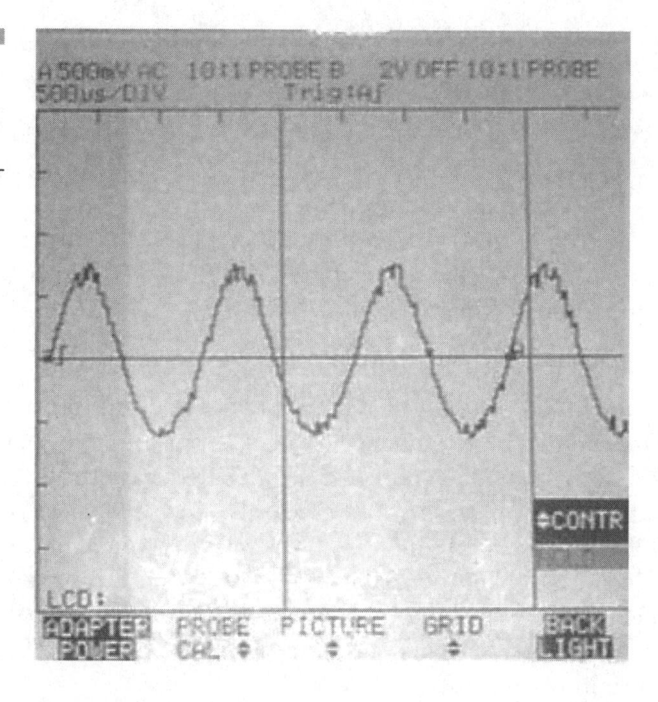

Figure 10-9 An 800-Hz signal produced by the Basic Stamp 2

To initiate this instruction with the Basic Stamp module, the following code listing is shown:

```
Loop:
   Button Swpin, ALow, DlyVal, SWrate, btnWK, SPKRpin, noPress
  FREQOUT SPKRpin, Durate, FREQ          'Emit 800Hz tone through pin
0 for 1sec
noPress: GOTO Loop                                    'Repeat endlessly
```

The first line of code enables the +5V signal from the voltage regulator to be used to activate the 800-Hz tone produced by the Basic Stamp. The input of the Basic Stamp requires a toggle output from the voltage regulator. Therefore, the output port of the LEGO P-Brick should be programmed to produce a one-shot pulse to input of Basic Stamp module. The complete Listing 10-1, for the 800 tone audio concept of Figure 101, is shown below:

Listing 10-1

```
SPKRpin        CON        0     ' Speaker connected to pin 0
Swpin          CON        7     ' Switch connected to Pin 7
DlyVal         CON      255     ' Delay Value
ALow           CON        0     ' Active Low switch rate
SWrate         CON      255     ' Switch cycles between autorepeats
Durate         CON     1000     ' Output On in milliseconds
FREQ           CON      800     ' Frequency in Hertz
btnWK          var     byte     ' Workspace for BUTTON instruction
btnWK =0                                  ' Clear the workspace variable

Loop:
   Button Swpin, Alow, DlyVal, Swrate, btnWK, SPKRpin, NoPress
  FREQOUT SPKRin, Durate, FREQ 'Emit 800Hz tone through pin 0 for
                               1sec
NoPress: GOTO Loop             'Repeat endlessly
```

The P-Basic *pulse width modulation* (PWM) instruction will provide a series of square waves to P_7 pin of the Basic Stamp module. The P-Basic programming style syntax is as follows:

```
PWM pin, duty, cycles
```

Pin will be, as discussed in the previous paragraph, an output that provides the modulated signal to the external hardwired speaker. The *duty* parameter is an 8-bit value (0-255) which specifies the duty cycle of the non-sinusodial waveform. The duty cycle of a square wave is the ratio of the time on T_{ON} over the total time T_{Total}. The cycles are the number of 256 pin update periods that take approximately 1 ms. When completed, the pin will be placed in input mode to enable the voltage to remain until another PWM is executed. To use this instruction replaces the ***FREQOUT*** with the PWM as shown in Listing 10-1.

Listing 10-2 an example of using the PWM P-Basic instruction for modifying the code in Listing 10-1.

Listing 10-2

```
SPKRpin     CON        0        ' Speaker connected to pin 0
Swpin       CON        7        ' Switch connected to Pin 7
DlyVal      CON      255        ' Delay Value
ALow        CON        0        ' Active Low switch rate
SWrate      CON      255        ' Switch cycles between autorepeats
PDuty       CON      256        ' Duty Cycle for PWM signal
Pcyle       CON      100        ' Cycles for PWM signal
btnWK       var     byte        ' Workspace for BUTTON instruction
btnWK =0                        ' Clear the workspace variable

Loop:
   Button Swpin, Alow, DlyVal, Swrate, btnWK, SPKRpin, NoPress
  PWM  SPKRin, PDuty, Pcyle     'Pulse width modulated output signal
NoPress: GOTO Loop              'Repeat endlessly
```

The pin, duty, cycles parameters will be experimentally determined in order to produce the appropriate PWM signal for driving an external hard-wired speaker.

Robotic Window/Door Alarm

Thomas Petruzzellis describes a window/door alarm as, "a compact circuit that can be used in many different alarm applications, including the protection of doors, windows or pieces of equipment or antiques."[2] If an interruption between the optical transmitter and receiver exists, the piezo-electric buzzer attached to the driver circuit will be activated. The audio tone generator circuit shown in Figure 10-7 can be used, along with a light sensor connected to any three input ports, to function as robotic window or door alarm.

A strip of dark construction paper would be placed on the window or door for monitoring an ajar (slightly open) condition. With the light sensor facing the dark strip, the alarm attached to any three output ports would not be activated. The audio tone generator circuit activates when light shines on the LEGO optical sensor. Therefore, setting the LEGO light sensor to detect light would enable the LEGO P-Brick to drive any three of the output ports to be activated. The robot could be built where it would advance forward, blaring the audible warning produced by the audio tone generator circuit.

A DCU-Controlled Audio Robot

This laboratory project illustrates how the DCU discussed in chapter 9 can operate an audio robot. The concepts discussed can be applied to the

[2]*The Alarm, Sensor and Security Circuit Cookbook*, Thomas Petruzzellis, Tab Books, 1994.

audible sound/tone generator while controlling the LEGO P-Brick with the DCU.

Assembly and Test Procedure Use the following steps to complete this procedure:

1. Wire the DCU output electromechanical switch on a Radio Shack Electronics Learning Lab breadboard or equivalent circuit prototyping station, using Figure 9-25.
2. Attach the modified LEGO electric wire to input port 1 of the LEGO P-Brick.
3. Install the *DCUINIT* and *DCUTOGGL* programs into the TI graphics calculator using the following procedure:
   a. Open the TI Graph Link software by double-clicking the icon on the desktop of the notebook or PC computer. The TI-82 keypad with the *Untitled1* window will appear on the screen. See Figure 9-18.
   b. Turn on the TI-82 calculator. Press the *2nd*-then *LINK* key on the calculator.
   c. With the right arrow key, move the cursor on the calculator to *RECEIVE*.
   d. Push the enter key. The calculator should be displaying *Waiting . . .* on the screen.
   e. On the notebook, select *LINK* from the main menu. From the pull-down screen, select *Send*.
   f. Select the *DCUINIT* from the accompanying CD-ROM.
   g. The CBL program will be displayed on the TI graphics calculator. Repeat the steps, installing the *DCUTOGGL* program onto the graphics calculator.
4. Run both the *DCUINIT* and *DCUTOGGL* programs in the TI graphics calculator.
5. Run the walking robot program in slot location 1.
6. Press *1* on the calculator. The status should be *ON*. The DCU LED should be illuminated. The software or hardware-based tone generator should be activated.
7. Press *1* on the calculator. The status should be *OFF*. The tone emitted from either the software-based or hardwired circuit should stop sounding.
8. Repeat steps 6 and 7 several times. Note the audio response to the input actuations.

9. Switch program 1 to program 2.
10. Run the program in slot location 2.
11. Repeat steps 6-8 for the new embedded code.
12. After validating the function of the tone generation-based software or hardwired circuit, stop the RCX code program on the LEGO P-Brick.

The audio bot concepts presented thus far were of a hybrid methodology to creating sound/tone-based robots. The laboratory projects and POCs presented in this chapter hopefully enabled you to see the many possibilities of how the LEGO P-Brick can generate an audible tone using either a software or a hardwired design concept.

Some of the testing procedures were based on previous laboratory projects in the book or from the author's first *Mindstorms* book. The intent of reviewing past applications is to illustrate how existing design concepts can enhance or introduce future robotics development projects.

The final pages of this chapter will discuss how special *integrated circuits* (ICs) can be used in the development of record-and-playback bots. A Radio Shack Science Fair® Digital Voice Recording and Sound Lab, an ISD -1000A Voice Record/Playback IC, and the ISD-VM1110A voice module will be discussed. These topics will illustrate how the LEGO P-Brick can be used to control the sound synthesis lab kit and IC components, respectively.

The Digital Voice Recording and Sound Lab

The Digital Voice Recording and Sound Lab kit was introduced by Radio Shack Science Fair in 1996. The kit is an all-inclusive way for the novice to learn about electronics through experimental play with voice record-and-playback technology. Figure 10-10 shows the Digital Voice Recording and Sound Lab kit.

The kit comes with an assortment of premounted electronic components electrically connected to spring terminals. The spring terminals enable the novice or experienced electronics circuit builder to wire all kinds of gadgets using the precut wires packaged with the kit. The laboratory manual that accompanies the voice recording and sound lab has information on the electronic parts, as well as wiring sequences. The wiring sequences help to wire the electronics projects by following a series of numbers. The numbers were associated with electronic component leads. Therefore, the circuit could be wired based on following the sequence of the numbering of the electronic component leads. This method of circuit building was the trademark of all

Figure 10-10 The Digital Voice Recording and Sound Lab kit

Radio Shack Science Fair electronic kits from the 1970s until 1998, when the solderless breadboard became the preferred method of building prototype circuits.

The essential component of the Digital Voice Recording and Sound Lab kit is the UM5583 IC. This 48-pin *dual-in-line* (DIL) semiconductor has all the necessary input/output and signal processor functions embedded within the IC. The important functions of the UM5583 are the following:

- **(Pin 22) Play** It enables the IC to output the recorded signal.
- **(Pin 24) Record** It enables MICIN to record the unit until the memory is full.
- **(Pin 33) MICIN** The built-in input terminal of the pre-amplifier.
- **(Pin 37) VDD** Power supply for the IC.
- **(Pin 32) *Speaker Positive Terminal* (SPKR)** The built-in output terminal of the operational amplifier, which is used to drive the speaker.
- **(Pin 34) *Speaker Negative Terminal* (SPKN)** Connects to the speaker's negative terminal.
- **(Pin 20) MODE** Selects either automatic or manual recording.
- **(Pin 9) VSS** The ground for the IC.

The manual that comes with the kit provides 50 projects, each increasing in complexity. In looking at the some of the projects, the key features to interfacing the LEGO P-Brick are the *Play* or *Record* inputs to the IC. Using a transistor relay driver, with its electromechanical relay contacts wired in series to either one of the input pins, would enable the P-Brick to have control over the UM5583 component.

Here are two examples of the how the LEGO P-Brick could be used to enable either record or playback functions on the Digital Voice Recording and Sound Lab kit.

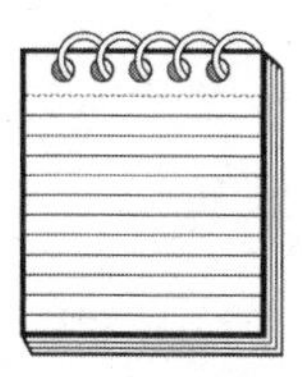

TECH NOTE: *The Digital Voice Recording and Sound lab kit is no longer available at Radio Shack. An Australia company, called Vidcam Pty Ltd., has limited number of these interesting kits in stock. The price is about $81 ($59.95 plus shipping). If you are interested in exploring this unique electronics lab kit, check out their web site at* `www.vidcam.com.au/sales/KJ7028.html`.

Robot Sound Effect

When making a science fiction movie or television show, people's voices sometimes are altered to sound as if they come from a robot. If a voice is recorded digitally and not restored to its original analog state before playing it back, the final audio is a machine-like sound. To create a robotics sound using the UM5583, pins 21 and 23 are connected to VDD. If the electromechanical contacts of the LEGO P-Brick's transistor relay driver are wired in series with the playback input of the IC, when the output is activated by the P-Brick, the recorded voice will have a robot effect. Figure 10-11 shows the circuit diagram of the robot sound effects generator.

Basic Record Circuit The circuit shown in Figure 10-12 is for recording sounds using the REC input of the IC. When the LEGO P-Brick's output energizes the transistor relay driver's electromechanical contacts, the background sound is recorded by the MICIN function. The voice IC then converts the analog signal from the microphone to a digital signal. The recorded signal is stored in the UM5583's memory. When the playback key is pressed, the IC plays back the through hardwired speaker terminals SPA and SPB.

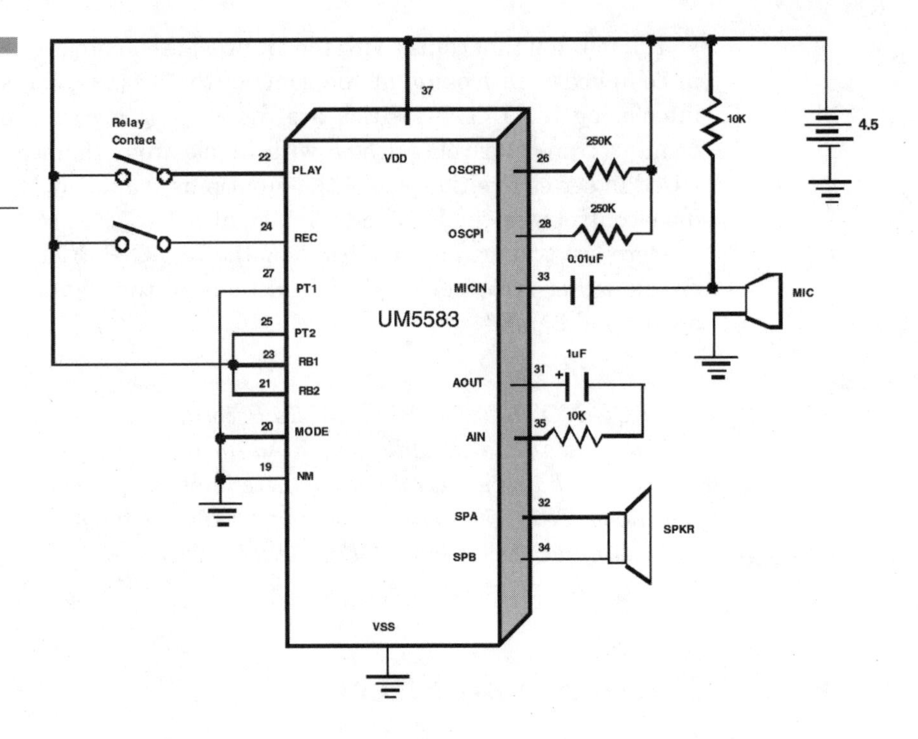

Figure 10-11 A LEGO P-Brick-controlled robot sound effect generator

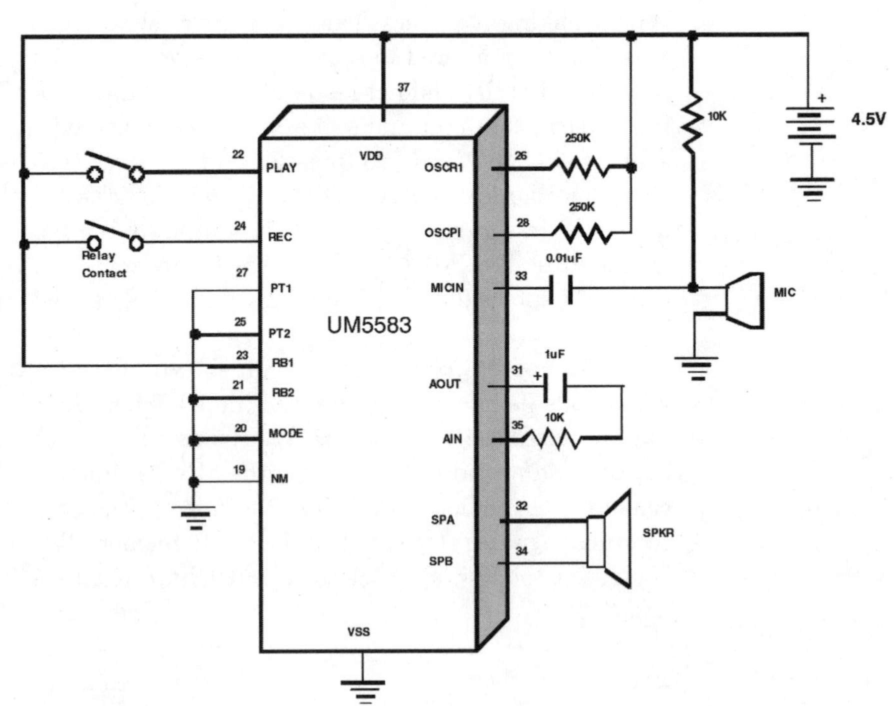

Figure 10-12 A recording LEGO Bot circuit schematic

ISD-1000A Voice Chip

The ISD-1000A voice chip, developed by Information Storage Devices, is a 28-pin IC that uses *Electrically Erasable Programmable Read Only Memory* (E^{2}PROM) technology to write analog data directly into a single cell. An *analog-to-digital* (A/D) or *digital-to-analog* (D/A) conversion is not required. This conversion process helps to increase density of analog data over equivalent methods of digital and non-volatile storage.

To use the voice chip, only a few external components are required: a microphone, loudspeaker, switches, resistors, capacitors, and a DC power supply or a battery. With this small quantity of electronic components available, a complete voice record-and-playback device can be built. All other electronic devices such as the preamplifier, filters, automatic gain control, power amplifier, control logic, and analog storage are embedded inside of the chip. Figure 10-13 shows the 48-pin IC.

How the ISD-1000A Voice Chip Works The ISD-1000A voice chip is an analog-sampled data system. It consists of the following components: an integral on-chip microphone, preamplifier circuit, *automatic gain control* (AGC) feature, anti-aliasing and smoothing filters, a storage array, speaker driver, an input control circuit interface, and an internal precision reference clock. The ISDA-1000A voice chip differs from the UM5583 IC in that no A/D or D/A converters are required. The ISD-1000A voice chip takes each sound sample and temporarily stores it in a sample-and-hold circuit that eventually records the data into a single E^{2}PROM cell. The recording of sound data has 8 bits of accuracy, requiring 1 Mb of digital memory storage space. Including the A/D and D/A converters, this gives the ISD voice chip 128,000 cells of storage arrays for capturing information. The ISD-1000A IC stores 20 seconds of sound using a sampling rate of 6.4 kHz.

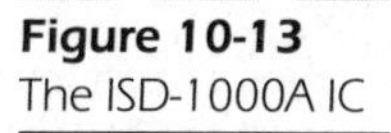

Figure 10-13
The ISD-1000A IC

Additional Information on the ISD-1000A Voice Chip Here is some additional information on the ISD-1000A voice chip and the important circuit blocks used in managing the recorded data. This information was compiled from the specification sheet that comes packaged with the IC:

- **Antialiasing and smoothing filters** The on-chip filters perform the anti-aliasing during record function and smoothing during playback. These features are required for a good signal sampling system. The filters are set at 2700 Hz, which defines the high frequency cutoff for the chip.
- **Storage array** The ISD-1000A E^2PROM storage array contains 128,000 memory locations. One analog voltage sample is stored at each memory location. This type of memory (non-volatile) does not require power to maintain memory.
- **Addressing** When a record or play cycle is initiated by the *Chip Enable(not)* pin, the ISDA reads the states of the address pins and the operation starts at the indicated address. When address pins 9 and 10 are both connected to VCC, the ISD-1000A voice chip goes into a special operating mode. This mode of operation lets the designer to use several special features, including automatic looping and sequenced messages.

The ISD-1000A Voice Chip and the LEGO P-Brick Using the transistor relay driver circuit electromechanical coil contacts as hardwired connecting points to the IC is the link to interfacing the LEGO P-Brick to the ISD-1000A voice chip. Like the UM5583, the record and playback pins are the key interfacing points for remotely controlling the IC using the LEGO P-Brick. The addressing pins on the ISD-1000A voice chip provide a programmable and interfacing opportunity, in terms of the LEGO P-Brick's effect on the function of the IC.

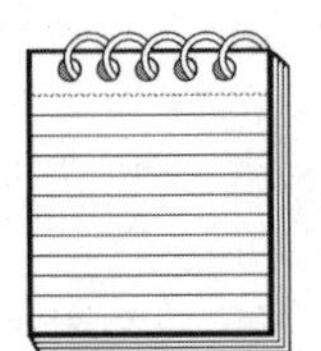

TECH NOTE: *Complete schematic diagrams for wiring the ISD-1000A can be found on Radio Shack's web support page,* `http://support.radioshack.com/support_supplies/doc16/16704.htm`.

Several application circuit schematics are included, as well as a parts list for building the basic record-and-play function.

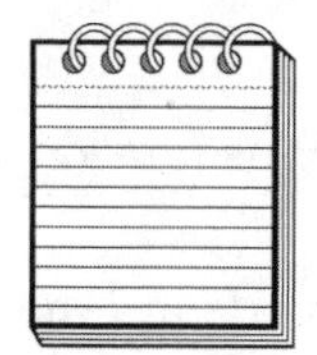

TECH NOTE: *Both the ISD-1000A voice record-and-play IC and the ISD-VM1110A voice module can be purchased from Radio Shack. Supporting documents on application circuits, and required parts can be found on the web page as well. Go to* `www.radioshack.com/` *for further info.*

The ISD-VM1110A Chip On-Board Module

This record-and-play voice module is controlled by a single signal for REC, and has either of two push-button control playback pins: PLAYE (edge-activated playback) and PLAYL (level-activated playback). The speech quality patented ChipCorder® technology provides natural sound quality for record and playback of recorded voice or sound. The input voice signals are stored directly in nonvolatile E^2PROM cells. They are reproduced without the synthetic effect often heard with solid-state speech solutions. A complete sample is stored in a single cell, minimizing the memory to store a recording of a given duration. The key pins for the ISD-VM1110A module are as follows:

- **(Pin 10) *Playback, Edge-Activated* (PLAYE)** When a low going transition is detected on this input signal, a playback cycle begins. Playback continues until an end-of-message marker has been reached or the end of memory space is reached. This pin has an internal pull-up device.
- **(Pin 9) *Playback, Level Activated* (PLAYL)** When this input signal transitions from high to low, a playback cycle is initiated. Playback continues until PLAYL is pulled high, an end-of-message marker is detected or the end of memory space is reached.
- **(Pin 12) *Record* (REC)** The REC input is an active-low record signal. The device records whenever REC is low. The signal must remain low for the duration of the recording. The REC function takes precedence over either playback (PLAYE or PLAYL) signal. A record cycle is completed when the REC is pulled high or the memory space is full. The end-of-marker is internally recorded, enabling a subsequent playback cycle to terminate appropriately.
- **(Pin 7) *Microphone Input* (MIC+)** An electric microphone (positive terminal) is connected to this pin.

- **(Pin 6)** ***Microphone Reference*** **(MIC−)** This pin is the negative connection for the two-terminal electric microphone.
- **(Pins 4, 5)** ***Speaker Outputs*** **(SPKR+**, SPKR−) Pins provide direct drive for loudspeakers with impedance as low as 16 Ω.
- **(Pins 13,18)** ***VCC*** **(BATT+**) Analog and digital circuits internal to the ISD1110 chip use separate power buses on the module to minimize noise on the chip.
- **(Pins 3,28)** ***VSS/GND*** **(BATT−)** Similar to VCC, the analog and digital circuits internal to the chip use separate ground buses on the module to minimize noise.

The ISD-VM1110A and the LEGO P-Brick The LEGO P-Brick interfacing technique as discussed for the ISD-1000A voice chip is applicable for the voice module. The ISD-VM1110A is a convenient way of experimenting with voice technology because all of the components are mounted on the board. The transistor relay driver's contact can be added in series to the REC, PLAYE, and PLAYL inputs of the module. By applying different sensors to the input of the P-Brick and writing the appropriate software, unique and interesting effects for controlling the voice module can be obtain with the P-Brick.

Additional Thoughts on Sound-Tone Generation and LEGO Robots

This chapter's intent was to provide concepts on how the LEGO P-Brick can be used to create unique applications and experiments for sound and tone generation. The discussions are ideas for further exploration into the audio bot concept. The purpose of the voice chip and module advanced technologies discussion is to briefly provide information about the design and development of technical concepts into fully functional robots and mechatronics-based devices. My philosophy is to use the section's material to further explore, develop and experiment with sound and tone applications using mechatronics techniques. Enjoy!

APPENDIX A

Glossary

Here is a compilation of some of the technical terms used throughout the book, with their definitions:

antialiasing The method of removing high frequencies from a lower frequency.

applet A small application that can be programmed "procedurally," whereby each line of code is sequenced from the previous command or instruction statement.

beta The ratio of the induced changes in collector current to the applied change-of-base current in a transistor.

bias The operating point of an electronic circuit.

circuit block A key component that provides a hardwired application-specific function to an electronic product.

decomposition The process of breaking something into smaller levels.

derivative A mathematical function used to find the difference between two values.

digitizing The process of converting a decimal-based signal to its equivalent binary value using a weighted place-holder method.

DMM The acronym for digital multi-meter.

duty cycle The mathematical relationship between a square wave or pulse train *time on* (t_{ON}) or *pulse width* (p_w) to its *total time* (t_{Total}).

E^2PROM The acronym for Electrically Erasable Programmable Read Only Memory.

event The request of the switching contacts of a detection device, triggered by either a single or a double-acting force initiated by the user (operator) of the machine.

gain The mathematical ratio between the output voltage and the input voltage.

integration A mathematical function used to find the derivative.

LCD The acronym for liquid crystal display.

LED The acronym for light emitting diode.

multimedia A computer technology that combines text with images.

mutual inductance The ability of a transformer to take the voltage or current on its primary winding and send it to its secondary winding using magnetic coupling.

partitioning The act of separating.

procedural programming A structured top-down method of writing code for embedded microcontrollers and processors.

PWM The acronym for pulse width modulation.

RDA The acronym for robot digital assistant.

supply bus rail A common electrical node used to provide a negative or positive source of DC voltage for an electronic circuit.

TRD The acronym for transistor relay driver.

transients Electrical noise.

UUT The acronym for unit under test.

Virtual instruments Also known by the acronym VI. Computer-based test equipment created using visual modeling techniques that behave like physical measuring instrumentation.

APPENDIX B

Tech Notes

Here is a list of resources presented in the book. It includes additional development plans, tools, and inspirations into building LEGO Mindstorms mechatronics robots and machines; compiled and centralized for easy accessibility.

Chapter 1, "Software Tools for LEGO Mindstorms Mechatronics Embedded Systems Development"

- Bill of Materials, or BOM, is an engineering term for a parts list. In order to manage the millions of parts used in electronic modules and mechanical assemblies of consumer products, computerized systems and special software are used to maintain the massive inventory of parts for these components.
- In *LEGO Mindstorms Interfacing*, the electronic circuits were built on the Radio Shack Electronic Learning Lab breadboarding system (Cat. No. 28-280) for ease of RCX interfacing and rapid prototyping. This breadboarding unit not only makes it convenient for building circuits, but has an ample supply of electronic components as well.
- A supply bus rail is quite easy to implement on a solderless breadboard. Take a positive supply voltage (in the case of the RCX P-Brick, the +7V connector stud), and wire it to a central point or strip of holes on the white board. All circuits that need the voltage can then be electrically connected to the DC source using a jumper wire.
- Creative Creation is a design philosophy that Alexander Slocum, Professor of Mechanical Engineer at MIT, uses to describe engineering problems as opportunities. Passion, Thought Processes, and the Golden Rectangle are just a few of the methods used in developing a Creative Creation design mentality. A PDF presentation by Professor Slocum on

Creative Creation can be found at `http://pergatory.mit.edu/perg/presentations.htm`. The information discussed in the presentation can provide additional Mindstorms robotic projects and lab experiments for the adventurous amateur roboticist.

- In the C programming language, one of the most common library functions is called *printf(string)*. This is a general output function used to display *strings,* or characters, on the screen. The *printf()*function for IC4 is used to output strings on the RCX P-Bricks' LCD screen.
- On pages 52-53 of the *Basic Electronics Workbook* that accompanies the Radio Shack Learning Lab, Forrest Mims shows the numbers and alpha-characters that can be created using a single seven-segment LED display. There is also an unfilled seven-segment display page that can be photocopied and used in the design and display of other non-standard characters.
- Keeping track of parentheses while writing code can be a software developer's worst nightmare. Fortunately, the IC4 IDE tool has a unique management feature for keeping track of parentheses. By clicking the Tools menu, a *Check parens* feature is available for keeping track of the parentheses used in your code.
- An excellent reference guide for using the *Inventor and Investigator* project area of Robolab is available online. If you purchased the Robolab software from Pitsco/LEGO Dacta, this online guide can be printed and used to unlock the data acquisition and controls features of this powerful graphical programming language. The web site address for the Robolab Inventor and Investigator guide is `www.lego.com/dacta/Robolab/investigatorprogam.htm`.
- Program Levels 4 and 5 allow the amateur roboticist to explore the LabView programming environment. Highlights of the these two levels include the following:
    - Advanced programming levels for the *Invent* feature of Robolab, using LabView's data-logging and motors control functions.
    - Uses the LabView VI programming language exclusively.
    - Systems engineering methods for *partitioning* (separating) and *decomposition* (breaking into small levels) can be easily implemented on mechatronic-based devices.
- To show labels on the VI block diagram, right click each motor control icon and move the cursor to *Visible Item* on the pull-down menu and select *Label*.

- The Direct Mode IDE can assist in the design and development of robotic or intelligent machine movements by turning the appropriate outputs on or off in a sequence, using the controls of the tool. By switching the RCX outputs in a preplanned manner, the actual autonomous motion of the Mindstorms built mechatronic device can be designed.
- John Hansen has a new version of BricxCC known as 3.3.6.5. This new version of the software supports programming the RCX (all versions), Scout, Cybermaster, and Spybot P-Bricks (all in one convenient package. It also supports programming the Scout, RCX2, and Spybot, using The LEGO Company's MindScript™ and *LEGO Assembly* (LASM™) languages via the RCX 2.0 SDK. Visit `http://members.aol.com/johnbinder/bricxcc.htm` to download this latest version of IDE software.

Chapter 2, "Electronic Hardware Add-Ons"

- The Radio Shack Sensor Lab kit (Catalog No. 28-278) is a great prototyping tool for mechatronics development because of the ample supply of electronic components, a small breadboarding area, and several sensor cards. Mindstorms robot sensory-detection designs can be implemented and tested easily, using the multitude of sensing devices (rotation, touch, magnetic, probe, induction coil, magnet, phototransistor, thermistor, and photoresistor) that come with the lab kit.
- A technical paper, written by three robotics engineers from Carnegie Mellon University, describes work on high quality rapid prototyping. The authors explain how to use servomotors, LEGO bricks, and electronics to achieve satisfying results. The paper can be downloaded from `www.cs.cmu.edu/~reshko/Publications/prototyping.pdf`.
- Typing *68HC11* within the web-search browser textbox will provide several links on programming, software, and how to build a 68HC11 EVB from scratch.
- The Scout can easily be used as a surrogate P-Brick for the lab projects. By using SAC or its own assembly language, the Scout P-Brick can be use in advance robotics projects quite easily.

- The Basic Stamp Windows Editor version 1.3 can be attained from Parallax, Inc.'s web site for free. Their web site address is www.parallaxinc.com/html_pages/downloads/software/software_basic_stamp.asp>.
- In addition to this development tool, other free software downloads are available as well.
- To make the Palm diagnostics controller an integral part of a robot or mechatronics-based machine, the handheld computer can be mounted on the LEGO P-Brick with standard Legos. See the following web site, www.beanos.com/~tsoutij/legopalm.php, for construction details.

Chapter 3, "Instrumentation and Robots"

- A computer-aided tool to assist in creating state machines can be downloaded free of charge from the CARMS web site www.tc.umn.edu/~puk/carms.htm. CARMS, which stands for *computer-aided rate modeling and simulation*, is an integrated tool for modeling and simulating time dependent prediction-oriented problems.
- The amateur roboticist can produce different walking motions for E-Beetle Bot by raising or lowering the legs of the robot. With one leg up and the other one down (offset position), the bot will emulate the movement of a beetle or insect.
- To learn more about the ICL7135 IC, go to Digikey's Web site, www.digikey.com. With the home page open, type in the part number inside the search text box and click *GO*. When the parametric table page is open, click on the Maxim data sheet to download the PDF document.
- The Multimedia Control Lab, discussed in Chapter 1, is an excellent way to create a unique testing tool for electronic circuits. By using the BricxCC or the Java RCXTools IDE applications, the LEGO P-Brick becomes a versatile test instrument for validating audio circuits used as audible warning devices for Mindstorms mechatronics-based devices.
- On the CD-ROM that accompanies the book, L-Remote is another Palm handheld computer software application written by Mike Kory. This software application emulates the handheld Infrared Remote Control

unit packaged with LEGO Mindstorms Ultimate Accessory kit. Features like turning on and off output ports A through C, beep control, forward/reverse motor direction, and selection of embedded programs are possible with the L-Remote software application.

Chapter 4, "Electronic Circuit Simulators and MLCAD"

- Electronic circuit interfaces can be prevalidated using the simulation techniques discussed in the basic integrator circuit laboratory project. Use the steps outlined in the software assembly and test procedures to develop unique circuit blocks for I/O interfacing to the LEGO or Scout P-Brick.
- The monostable circuit on page 262 in *LEGO Mindstorms Interfacing* shows little dots at every component and wiring connection. These dots are known as nodes. A node is an electrical junction that has two or more electrical elements connected together. During circuit analysis, these nodes assist in collecting data about the circuit when the test probe is used.
- In addition validating LEGO Mindstorms mechatronics-based electrical/electronics I/O interfaces, both CircuitMaker 6 and PSpice-OrCAD software packages make great tools for documenting circuits and sharing them with others. By copying the circuits to the Windows clipboard and pasting them to either an MS Word document or PowerPoint presentation, the information can be shared through Internet downloads or email.

Chapter 5, "Telerobotics"

- Carnegie-Mellon University teaches a class on *Rapid Design Through Virtual and Physical Prototyping*. A six-page document produced by this course is entitled "Introduction to Mechanisms." This document explains simple machines and is a must-have document for the amateur roboticist involved in designing mechanical drives for LEGO Mindstorms mechatronics-based projects. The web site is `www.2.cs.cmu.edu/People/rapidproto/mechanisms/chpt2.html`.

- The *Vision Command Constructopedia* serves as an excellent mechanical design aid in developing motorized drives for telerobots. The various motorized drives for *pan* (to turn) and *tilt* (to raise or lower) can be developed using the designs shown in the *Constructopedia*.
- Want to know how to improve the strength of your mechatronics-based robots? How about reducing or increasing the speed of a LEGO electric motor using gears, or changing its axis of rotation? There is an excellent mechanical design guide titled *The Art of LEGO Design*. Fred G. Martin, codeveloper of the LEGO RCX P-Brick, wrote the 19-page PDF document. Fred, a graduate of the MIT Media Lab, teaches Computer Architecture and Robotics at the University of Massachusetts, Lowell campus. *The Art of LEGO Design* is a treasure trove of mechanical construction techniques, from building stronger structures to reducing noise with pulley wheels. Download the guide from `http://nike.wellesley.edu/~rds/rds00/handouts/ArtOfLEGODesign.pdf`.
- In the instructions on adding reference and controls for Active X components, the author assumes the reader has some knowledge or familiarity in doing this software enhancement within the Excel spreadsheet environment. For those readers who are not familiar with this software addition, step-by-step instructions can be found in *LEGO Mindstorms Interfacing* on pages 30–45 of Chapter 2, "Developing GUIs: Software Control Basics." Use the *uportal 21.0 type library* for the reference and the *videoportal class* as the control.
- For Windows XP users, a software package named *Indeo 5.2 Codec* must be installed on the PC or notebook hard-drive in order to use the Logitech SDK. Download the software from `www.ligos.com/indeo/downloads`. The software costs $14.95.

Chapter 6, "Smart Hand Controllers"

- The Scout P-Brick can be programmed in a high-level language call *LEGO Assembler* (LASM). LASM enables the complex functions like logical *ANDing* two touch sensors. The Scout *integrated development environment* (IDE) tool enables you to write programs in LASM using its embedded text editor. The Scout IDE tool can be found on the CD-ROM packaged with the book.

- The Scout IDE is a wonderful tool for checking the mechatronics-based input devices. By using the advanced monitoring feature, the mechatronics-based input device can be validated easily and quickly.
- Turning a moving object requires one brake being applied to a rotating wheel while the other one spins freely. The *Forward-Brake* program used in the previous lab project can perform this turn function quite easily. With the one-switch hand controller latched on, output port A turns off while output B stays on. Therefore, an attached mobile platform will turn, based on the one wheel spinning and the other one being stopped.
- A wireless one-switch hand controller can be built quite easily, by following the build instructions in Chapter 1 of *LEGO Mindstorms Interfacing*. The "Wireless Basics" section of the book explains how to modify the transmitter-receiver controls of an RC car. Detail block diagrams, circuit schematics, and photos provide the technical information needed to make this wireless control modification.
- *High Tech Services* (HTS) has a wealth of technical information on their Web site. Information on inputs/outputs, motion control, .NET, robots, and companies specializing in HMI and operator interfaces can be found on this well-organized site. To learn more about operator interfaces and HMI systems, go to `www.htservices.com/Tools/OperatorInterfaces/index.htm`.
- The latching feature of the contact arm is accomplished by internal spring of the touch sensor. When the rotating arm makes contact with the yellow button on the touch sensor, the internal spring pushes up on the mini-electrical switch. The small yellow button makes contact with the rotating arm, providing a frictional force that minimizes movement from the rotating contact arm.

Chapter 7, "Hybrid Robots"

- Capsela is a wonderful building set that also teaches about science. The motors and working gears are encapsulated, hence the name Capsela. You can work with crown gears, an electrical switch box, the transformation of energy, and other mechanical principles. The

Capsela motorized building system provides the amateur roboticist with hands-on experiences. You can learn about energy and forces by actually handling batteries, creating motors, and seeing for yourself how energy changes with the help of visible gears and wheels. Smaller kits explore energy changes, electric circuits, motion energy, speed, and torque.

- Larger kits also provide introductory lessons on the following concepts: friction and traction, chain drive, inclined plane, crown gear, electrical switch box, buoyancy, electric circuitry, force, transformation of energy, vacuum, wheel-and-axle, propellers, and Newton's Third Law (to every action, or force, there is equal and opposite force).
- Reuse of software, mechanical subsystems, and electrical electronic circuits is an efficient way of expediting the mechatronics development process of robots and smart machines. This technique is illustrated in this project by making a minor change to high side driver circuit used in Chapter 6 laboratory projects. By building a bookshelf of proven hardware and software components, you eliminate a lot of debug time.
- If directional control is desired using a LEGO electric motor, the *bk(n);* function will reverse the shaft of the electromechanical component. By removing the DC-to-DC converter high side driver circuit and attaching the LEGO electric motor directly to the output port, the direction of the motor can be changed using the two-switch hand controller and *back* IC4 function.
- The Capsela 1000 Discovery set (Max Out) comes with 108 interlocking parts to construct over 100 land and water projects such as: a tug boat, water pump, crane, cable lift, generator, steam roller, tricycle, vacuum cleaner and as many simple machines as your imagination can conceive! It includes a full-color *Science Discovery Design Manual* with easy-to-follow assembly instructions, as well as an illustrated basic *Science Booklet* to explore 18 physical science principles. This motorized construction set can be purchased from Brain Builders. Their web site `www.brain-builders.com/12001.html`. See Figure 7-22.
- *Simple Capsela Autonomous Robot* (SCAR) was an online robotics project documented by a robot hobbyist in January 2002. His journal details the why and how of SCAR development. This online journal captures some design concepts and lessons learned. The web site of SCAR `www.intp.us/richardbthursby/scar.html`.
- The SAE AWIM Challenge 2 kit comes with a comprehensive lab manual. Although the material in the lab manual was written for a junior high

school classroom, the experiments and projects will provide plenty of useful information needed to build a hybrid mobile platform robot.

- On the Web, you can find an interesting article describing how the BRIO Corporation has brought back the Erector Set. The article provides a brief historical discussion on the Erector Set and the many corporations that helped evolve the 1950s construction kit for budding young engineers. To view the article, go to `www.trains.com/Content/Dynamic/Articles/000/000/001/109exais.asp`.

Chapter 8, "Mechatronic Bots"

- By drilling small holes into LEGO bricks, you can build up an ample supply of mechanical interfacing blocks. Drilling four small holes on each side of the brick enables you to mount additional blocks onto the core mechanical interface in any orientation.
- A web page called *ErectorWorld* has wealth of information on the Meccano Erector Set. It has information on the history of the metal construction kit, as well as several assembly instruction sheets that can be downloaded. The mechanical interface technique demonstrated in the previous laboratory project can be easily applied to these Erector models. You can find this interesting web site at `www.erectorworld.com/efiles.html`.
- To improve the rotational speed of the Erector motor-gear box, apply one drop of four-in-one household oil or equivalent small appliance lubricant to each gear. The oil helps reduce friction while increasing the speed of the electromechanical drive.
- The first walking-beam engine modeled using the Erector Set consisted of a wheel attached to a small electric motor. The reciprocating arms were attached to the wheel, providing direct transmission control of the rotating device. Go to `www.erectorworld.com/efiles.html` to see a PDF of this incredible machine.
- On the Web, a site has 233 posted messages regarding the LEGO Vision Command software and other image processing applications. Subjects range from seeing in the dark using infrared diodes, to a robot that sorts parts, to LEGO-supports XP. The wonderful resource web site is the *LEGO Users Group Network* (LUGNET), located at `www.lugnet.com/`.

- With each infrared control request made using the RCX Tools Direct Mode GUI, information relating to packets of data being received and sent by the LEGO P-Brick is available. The information is located at the bottom of the virtual control panel. This active toolbar can also be used as a real-time diagnostics tool when debugging circuit block add-ons to the P-Brick.
- The Build-It-Yourself web site, `www.built-it-yourself.com/`, has a good examples of developing hybrid mechatronics-based robots using purchased or ordinary products found around the home. Creator John Galinato has developed an online site where kids in third through twelfth grade can learn about math, science, and art using the LEGO Mindstorms *Robotic Invention System* (RIS) and common household products. You can find a wealth of hybrid-based ideas on this site free of charge.
- The LEGO Cam can be moved to help reduce the load of Robo Pump's reciprocating arm. By placing the digital camera to the center of the reciprocating arm, as shown in Figure 8-17, the load can be evenly distributed through the mechatronics bot mechanical arm.

Chapter 9, "Walking Robots"

- The modular design of the walking robot's platform mechanism is capable of being mounted to the LEGO RCX. All of the gears and leg-foot assemblies and supports enable the P-Brick to mount on top of the mechanical package that will thrust it into forward or reverse modes of direction. The Scout P-Brick can also be used with this design as well. To mount it on top of the platform mechanism, move the brick slowly until it stops. Lightly press the Scout onto the mechanism, to ensure a secure fit to the electromechanical package.
- The RCX code is a great low-fidelity prototype programming language for the quick development of robotics concepts. Initially, writing the robotics application in RCX code enables the mechatronics concept to be validated in a convenient and quick manner. After the concept has be validated, the RCX code can be translated into a robust programming language like *Not Quite C* (NQC), Java, or Robolab software for further enhancements and function expandability.
- If you're looking for some cool and sophisticated walking robots, check out Colins' LEGO Robot page. The web page has three prototype

walking robots built from LEGO Technic and Mindstorms RIS kits. Each prototype robot has clear colored pictures of these fascinating walking bots. Video clips showing the robots movements are also included. Colins' web page is at `http://filebox.vt.edu/users/cocampb2/LEGO/`.

- Vernier Software & Technology is a company driven by science education and computer-aided laboratory experimental methods. They have an extensive online resource of curriculum and laboratory experiments in the scientific disciplines of physics, physical science, chemistry, biology, and water quality. The physics measurement experiment outlined in the laboratory project can be implemented using the legacy CBL or CBL 2. Check out their web site at `www.vernier.com/` for additional sensors and robotics project ideas.

Chapter 10, "Audio Bot Concepts"

- Fred G. Martin teaches a robotics course at the University of Massachusetts-Lowell campus, where graduate students learn about design and construction of robotic systems. Course notes on various lab experiments convey his design philosophy on robotics. To find more info on Fred and his class, go to `www.cs.uml.edu/~fredm/`.
- The BricxCC is an *Integrated Development Environment* (IDE) for the *Not Quite C* (NQC) programming language. Besides enabling you to write, edit, and compile NQC code, BricxCC also has a piano keyboard that can play and record tones and sounds. The sound synthesize tool is interactive and enables all the musical notes to play on the LEGO P-Brick. The piano keyboard can be found under the *Tools* drop-down menu of the BricxCC GUI.
- The Digital Voice Recording and Sound lab kit is no longer available at Radio Shack. An Australian company, called Vidcam Pty Ltd., has limited number of these interesting kits in stock. The price is about $81.00 ($59.95 plus shipping). If you are interested in exploring this unique electronics lab kit, check out their web site at `www.vidcam.com.au/sales/KJ7028.html`.
- Complete schematic diagrams for wiring the ISD-1000A can be found on Radio Shack's web support page, `http://support.radioshack.com/support_supplies/doc16/16704.htm`. Several application

circuit schematics are included, as well as a parts list for building the basic record-and-play function.

- Both the ISD-1000A voice record-and-play IC and the ISD-VM1110A voice module can be purchased from Radio Shack. Supporting documents on application circuits and required parts can be found on the web page as well. Go to `www.radioshack.com/` for further info.

APPENDIX C

Web Sites

Here is a list of Web site resources for open source software, programming languages, and design and development tools for LEGO Mindstorms robots:

A Computer-aided Tool to Assist in Creating State Machines (CARMS)
`www.tc.umn.edu/~puk/carms.htm`

Basic Stamp Windows Editor Version 1.3
`www.parallaxinc.com/html_pages/downloads/software/software_basic_stamp.asp`

BricxCC Version 3.3.6.5 for NQC Programming Language
`http://members.aol.com/johnbinder/bricxcc.htm`

Capsela Mix Racer
`www.constructiontoys.com`

Capsela Motorized Construction Set Can Be Purchased from Brain Builders
`www.brain-builders.com/12001.html`

Capsela MX Trooper
`www.constructiontoys.com`

Carnegie Mellon University Class on Rapid Design Through Virtual and Physical Prototyping
`www.2.cs.cmu.edu/People/rapidproto/mechanisms/chpt2.html`

Carnegie Mellon University Robotics Rapid Prototyping Method
`www.cs.cmu.edu/~reshko/Publications/prototyping.pdf`

Colins' LEGO Robot Page
`filebox.vt.edu/users/cocampb2/LEGO/`

Complete Schematic Diagrams for Wiring the ISD-1000A Can Be Found on Radio Shack's Web Support Page
`http://support.radioshack.com/support_supplies/doc16/16704.htm`

Creating Creative Creation PDF Paper from MIT's Alexander Slocum
pergatory.mit.edu/perg/presentations.htm

Digital Voice and Sound lab kit Can Be Purchased from Vidcam Pty Ltd.
www.vidcam.com.au/sales/KJ7028.html

ErectorWorld Erector Set Web Site
www.erectorworld.com/efiles.html

Evolution of CircuitMaker
www.microcode.com/

First Walking-beam Engine Modeled Using the Erector Set
www.erectorworld.com/efiles.html

Fred G. Martin's Robotics Course at the University of Massachusetts -Lowell
www.cs.uml.edu/~fredm/

HTS (High Tech Services) has a Wealth of Technical Information on Inputs/outputs, Motion Control, .NET, and Robots on their Web Site
www.htservices.com/Tools/OperatorInterfaces/index.htm

How the BRIO Corporation has Brought Back the Erector Construction Set, An Interesting Article Describing
www.trains.com/Content/Dynamic/Articles/000/000/001/109exais.asp

IC4 Programming Language for the RCX Programmable Brick
www.kipr.org/ic/download/

Java (RCXTools Version 1.4) for the RCX Programmable Brick
rcxtools.sourceforge.net/

J2SE Version 1.3.1_01 Software
http://java.sun.com/j2se

LDraw Software for Creating LEGO CAD Models
www.ldraw.org/download/start/win/step1

LEGO Design, The Art of
nike.wellesley.edu/~rds/rds00/handouts/ArtOfLEGODesign.pdf

LEGO Mindstorms Ideas Using Household Items—The Build-It-Yourself Web Site
www.built-it-yourself.com/

LEGO Palm Computer
www.beanos.com/~tsoutij/legopalm.php

LEGO Users Group Network
www.lugnet.com/

leJOS Version 2.0 Software for the RCX Programmable Brick
http://lejos.sourceforge.net/download.html

MLCAD—The Front-end Software for LDraw
www.lm-software.com/mlcad/

Motorized Construction Set Can Be Purchased at the SAE AWIM's Challenge 2 Web Site
www.sae.org/students/awim2.htm

Orgler Electronics Web Site for Building a 68HC11 EVB
http://space.tin.it/computer/lorgler/sw-e.htm

Palm Programmable Brick Library
www.harbaum.org/till/palm/pbrick/

Pro/E (Engineer) Software, Information on
www.ptc.com/products/proe

Robolab Inventor and Investigator Guide
www.lego.com/dacta/Robolab/investigatorprogam.htm

SCAR-Simple Capsela Autonomous Robot
www.intp.us/richardbthursby/scar.html

SolidWorks Software
www.solidworks.com/

Student Version of PSpice-Orcad
www.cadencepcb.com/products/downloads/PSpicestudent/default.asp

To Learn More about the ICL7135 IC Go to Digikey's Web Site
www.digikey.com

Windows XP Users, in Order to Use the Logitech SDK. A Software Package Named *Indeo 5.2 Codec*
www.ligos.com/indeo/downloads

Vernier Software and Technology for CBL or CBL Tools
www.vernier.com/

APPENDIX D

Software Tools Installation Procedures

Some of the unique programming and control tools for LEGO Mindstorms mechatronics-based robots and machines are presented here for reference. These procedures can also be found in the appropriate chapters of the book.

Opening the Robolab Vision Center

The vision center is an advanced tool within the Robolab programming environment. Therefore, an Inventor 4 programming session must be opened to gain access to the vision software. Before running the following vision center procedure, connect the LEGO Cam to the USB of the PC or notebook computer. Place the Power Stand telerobot in a good viewing location and mechanical movement area.

The following steps will guide you to obtaining the vision center software:

1. Run the Robolab software by double-clicking its icon located on the Windows desktop.
2. Click the *Programmer* button.
3. Double-click *Inventor 4* under *Inventor*.
4. Within the Robolab *Untitled 1.vi diagram* main tool bar, click *Tools*. A drop-down menu box will be displayed on the screen. See Figure 5-18.
5. Scroll down to the *Select Camera* menu choice and click it. A *Select Camera* box will appear on the screen. See Figure 5-19.
6. Click the scroll down arrow once. The Microsoft WDM Image Capture (Win 32) will appear within the *Select Camera* box.
7. Click the check mark with the mouse. The *Select Camera* box will disappear from the screen.
8. Click *Tools*.
9. Scroll down to select *Vision Center* and click it. The *Vision Center* window will appear on the screen. See Figure 5-20. Note the image within the *Vision Center* window.

Loading the CBL.82 Software onto a TI Graphics Calculator

Before proceeding with the lab project, the CBL.82 software needs to be installed on a TI graphics calculator. The following procedure provides the installation steps for the CBL.82 software onto the graphics calculator:

1. Open the TI Graph Link software by double-clicking the icon on the desktop of the notebook or PC computer. The TI-82 keypad with the *Untitled1* window will appear on the screen. See Figure 9-18.
2. Turn on the TI-82 calculator and press the *2nd-then LINK* key on the calculator.
3. With the right arrow key, move the cursor on the calculator to *RECEIVE*.
4. Push the enter key. The calculator should be displaying *Waiting . . .* on the screen.
5. On the notebook select *LINK* from the main menu. From the pull-down screen, select *Send*.
6. Select the *CBL.82P* from the *Exploring Physics and Math with CBL System* floppy. Note: CBL.82P program can be found on the CD-ROM that accompanies this book.
7. The CBL program will be displayed on the TI graphics calculator.

Vernier Software & Technology's Digital Control Unit

The following procedure outlines the installation of the control software required to operate the DCU with a TI graphics calculator:

1. Open the TI Graph Link software by double-clicking the icon on the desktop of the notebook or PC computer. The TI-82 keypad with the

Untitled1 window will appear on the screen. See Figure 9-18.

2. Turn on the TI-82 calculator and press the *2nd-then LINK* key on the calculator.
3. Move cursor on the calculator to *RECEIVE*, using the right arrow key.
4. Push the enter key. The calculator should be displaying *Waiting . . .* on the screen.
5. On the notebook select *LINK* from the main menu. From the pull-down screen, select *Send*.
6. Select the *DCUINIT* from the accompanying CD-ROM.

   The CBL program will be displayed on the TI graphics calculator.
7. Repeat the steps, installing the *DCUTOGGL* program onto the graphics calculator.

Java-Based RCXTools

The following installation procedure outlines how to install the RCXTool and leJOS operating system for Java based LEGO Mindstorms software:

1. Go to Sun Microsystems web site (`http://java.sun.com/j2se`) to download and install J2SE version 1.3.1_01 software onto your notebook or PC hard-drive. Chapter 1's Figure 1-22 shows the home page for J2SE software.
2. Go to Jose Soloranzo's web site (`http://lejos.sourceforge.net/download.html`) to download and install leJOS version 2.0 software onto your notebook or PC hard-drive. Chapter 1's Figure 1-23 shows the download page for leJOS software.
3. Go to Tim Rinkens' web site (`http://rcxtools.sourceforge.net/`) to download and install the RCX Tools *integrated development environment* (IDE) software onto your notebook or PC hard-drive. Chapter 1's Figure 1-24 shows the download page for RCXTools software.

With the main software components installed on your notebook or PC hard-drive, the leJOS firmware is ready to be installed to the RCX P-Brick.

The leJOS Software Install Procedure

Use the following steps to install the leJOS software:

1. Go to the *RCXTools_1_4* directory created in the step 3 download/ install. Open the folder and double-click the *RCXDownload.bat* file. After a few seconds, the IDE tool should be on your notebook or PC video screen.
2. Click *Preferences* and check the pathway and COM port for your system. Make the appropriate changes and click *Accept*.
3. Turn on the LEGO P-Brick and have the infrared tower in front of it.
4. To send the leJOS firmware to the LEGO P-Brick, click *Download Firmware*. A dialog box will appear on the IDE tool, asking if you would like to download the firmware. Click *Yes*. See Chapter 1's Figure 1-25 for details on the dialog box and the firmware download confirmation.
5. After downloading the firmware to the LEGO P-Brick, close the Download RCXTool.

INDEX

Note: Boldface numbers indicate illustrations and tables.

B

C

D

E

F

G

H

I

J–K

N

O

S

T

W

SOFTWARE AND INFORMATION LICENSE

The software and information on this diskette (collectively referred to as the "Product") are the property of The McGraw-Hill Companies, Inc. ("McGraw-Hill") and are protected by both United States copyright law and international copyright treaty provision. You must treat this Product just like a book, except that you may copy it into a computer to be used and you may make archival copies of the Products for the sole purpose of backing up our software and protecting your investment from loss.

By saying "just like a book," McGraw-Hill means, for example, that the Product may be used by any number of people and may be freely moved from one computer location to another, so long as there is no possibility of the Product (or any part of the Product) being used at one location or on one computer while it is being used at another. Just as a book cannot be read by two different people in two different places at the same time, neither can the Product be used by two different people in two different places at the same time (unless, of course, McGraw-Hill's rights are being violated).

McGraw-Hill reserves the right to alter or modify the contents of the Product at any time.

This agreement is effective until terminated. The Agreement will terminate automatically without notice if you fail to comply with any provisions of this Agreement. In the event of termination by reason of your breach, you will destroy or erase all copies of the Product installed on any computer system or made for backup purposes and shall expunge the Product from your data storage facilities.

LIMITED WARRANTY

McGraw-Hill warrants the physical diskette(s) enclosed herein to be free of defects in materials and workmanship for a period of sixty days from the purchase date. If McGraw-Hill receives written notification within the warranty period of defects in materials or workmanship, and such notification is determined by McGraw-Hill to be correct, McGraw-Hill will replace the defective diskette(s). Send request to:

Customer Service
McGraw-Hill
Gahanna Industrial Park
860 Taylor Station Road
Blacklick, OH 43004-9615

The entire and exclusive liability and remedy for breach of this Limited Warranty shall be limited to replacement of defective diskette(s) and shall not include or extend any claim for or right to cover any other damages, including but not limited to, loss of profit, data, or use of the software, or special, incidental, or consequential damages or other similar claims, even if McGraw-Hill has been specifically advised as to the possibility of such damages. In no event will McGraw-Hill's liability for any damages to you or any other person ever exceed the lower of suggested list price or actual price paid for the license to use the Product, regardless of any form of the claim.

THE McGRAW-HILL COMPANIES, INC. SPECIFICALLY DISCLAIMS ALL OTHER WARRANTIES, EXPRESS OR IMPLIED, INCLUDING BUT NOT LIMITED TO, ANY IMPLIED WARRANTY OF MERCHANTABILITY OR FITNESS FOR A PARTICULAR PURPOSE. Specifically, McGraw-Hill makes no representation or warranty that the Product is fit for any particular purpose and any implied warranty of merchantability is limited to the sixty day duration of the Limited Warranty covering the physical diskette(s) only (and not the software or information) and is otherwise expressly and specifically disclaimed.

This Limited Warranty gives you specific legal rights; you may have others which may vary from state to state. Some states do not allow the exclusion of incidental or consequential damages, or the limitation on how long an implied warranty lasts, so some of the above may not apply to you.

This Agreement constitutes the entire agreement between the parties relating to use of the Product. The terms of any purchase order shall have no effect on the terms of this Agreement. Failure of McGraw-Hill to insist at any time on strict compliance with this Agreement shall not constitute a waiver of any rights under this Agreement. This Agreement shall be construed and governed in accordance with the laws of New York. If any provision of this Agreement is held to be contrary to law, that provision will be enforced to the maximum extent permissible and the remaining provisions will remain in force and effect.